Singularities in Fluids, Plasmas and Optics

AF894440

NATO ASI Series

Advanced Science Institutes Series

A Series presenting the results of activities sponsored by the NATO Science Committee, which aims at the dissemination of advanced scientific and technological knowledge, with a view to strengthening links between scientific communities.

The Series is published by an international board of publishers in conjunction with the NATO Scientific Affairs Division

A Life Sciences **B Physics**	Plenum Publishing Corporation London and New York
C Mathematical and Physical Sciences **D Behavioural and Social Sciences** **E Applied Sciences**	Kluwer Academic Publishers Dordrecht, Boston and London
F Computer and Systems Sciences **G Ecological Sciences** **H Cell Biology** **I Global Environmental Change**	Springer-Verlag Berlin, Heidelberg, New York, London, Paris and Tokyo

NATO-PCO-DATA BASE

The electronic index to the NATO ASI Series provides full bibliographical references (with keywords and/or abstracts) to more than 30000 contributions from international scientists published in all sections of the NATO ASI Series.
Access to the NATO-PCO-DATA BASE is possible in two ways:

– via online FILE 128 (NATO-PCO-DATA BASE) hosted by ESRIN, Via Galileo Galilei, I-00044 Frascati, Italy.

– via CD-ROM "NATO-PCO-DATA BASE" with user-friendly retrieval software in English, French and German (© WTV GmbH and DATAWARE Technologies Inc. 1989).

The CD-ROM can be ordered through any member of the Board of Publishers or through NATO-PCO, Overijse, Belgium.

Series C: Mathematical and Physical Sciences - Vol. 404

Singularities in Fluids, Plasmas and Optics

edited by

Russel E. Caflisch
Department of Mathematics,
University of California at Los Angeles,
Los Angeles, California, U.S.A.

and

George C. Papanicolaou
Courant Institute of Mathematical Sciences,
New York University,
New York, New York, U.S.A.

Springer Science+Business Media, B.V.

Proceedings of the NATO Advanced Research Workshop on
Singularities in Fluids, Plasmas and Optics
Heraklion, Greece
July 6-10, 1992

Library of Congress Cataloging-in-Publication Data

Singularities in fluids, plasmas, and optics : proceedings of a NATO advanced research workshop July 6-10, 1992, Heraklion, Greece / edited by Russel E. Caflisch and George C. Papanicolaou.
p. cm. -- (NATO ASI series. Series C, Mathematical and physical sciences ; vol. 404)
ISBN 978-94-010-4894-1 ISBN 978-94-011-2022-7 (eBook)
DOI 10.1007/978-94-011-2022-7
1. Singularities (Mathematics)--Congresses. 2. Fluid mechanics--Mathematics--Congresses. 3. Plasma (Ionized gases)--Mathematics--Congresses. 4. Optics--Mathematics--Congresses. 5. Mathematical physics--Congresses. I. Caflisch, Russel E. II. Papanicolaou, George. III. Series: NATO ASI series. Series C, Mathematical and physical sciences ; no. 403.
QC20.7.S54S56 1993
530.1'5--dc20 93-1726

ISBN 978-94-010-4894-1

Printed on acid-free paper

All Rights Reserved
© 1993 Springer Science+Business Media Dordrecht
Originally published by Kluwer Academic Publishers in 1993
Softcover reprint of the hardcover 1st edition 1993
No part of the material protected by this copyright notice may be reproduced or utilized in any form or by any means, electronic or mechanical, including photocopying, recording or by any information storage and retrieval system, without written permission from the copyright owner.

This book contains the proceedings of a NATO Advanced Research Workshop held within the programme of activities of the NATO Special Programme on Chaos, Order and Patterns as part of the activities of the NATO Science Committee.

Other books previously published as a result of the activities of the Special Programme are:

ABRAHAM, N.B., ALBANO, A.M., PASSAMANTE, A. and RAPP, P.E. (Eds.) - *Measures of Complexity and Chaos* (B208) 1990 Plenum Publishing Corporation. ISBN 0-306-43387-7.

BUSSE, F.H. and KRAMER, L. (Eds.) - *Nonlinear Evolution of Spatio-Temporal Structures in Dissipative Continuous Systems* (B225) 1990 Plenum Publishing Corporation. ISBN 0-306-43603-5.

CHARMET, J.C., ROUX, S. and GUYON, E. (Eds.) - *Disorder and Fracture* (B235) 1991 Plenum Publishing Corporation. ISBN 0-306-43688-4.

MARESCHAL, M. (Ed.) - *Microscopic Simulations of Complex Flows* (B236) 1991 Plenum Publishing Corporation. ISBN 0-306-43687-6.

MacDONALD, G.J. and SERTORIO, L. (Eds.) - *Global Climate and Ecosystems Change* (B240) 1991 Plenum Publishing Corporation. ISBN 0-306-43715-5.

CHRISTIANSEN, P.L. and SCOTT, A.C. (Eds.) - *Davydov's Soliton Revisted: Self-Trapping of Vibrational Energy in Protein (B243) 1991 Plenum Publishing Corporation. ISBN 0-306-43734-1*

HOLDEN, A.V., MARKUS, M. and OTHMER, H.G. (Eds.) - *Nonlinear Wave Processes in Excitable Media* (B244) 1991 Plenum Publishing Corporation. ISBN 0-306-43800-3.

LING-LIE CHAU, NAHM, W. (Eds.) - *Differential Geometric Methods in Theoretical Physics: Physics and Geometry* (B245) 1991 Plenum Publishing Corporation. ISBN 0-306-43807-0.

ATMANSPACHER, H. and SCHEINGRABER, H. (Eds.) - *Information Dynamics* (B256) 1991 Plenum Publishing Corporation. ISBN 0-306-43912-3.

BABLOYANTZ, A. (Ed.) - *Self-Organisation, Emerging Properties, and Learning* (B260) 1991 Plenum Publishing Corporation. ISBN 0-306-43930-1.

PELITI, L. (Ed.) - *Biologically Inspired Physics* (B263) 1991 Plenum Publishing Corporation. ISBN 0-306-44000-8.

BISHOP, A.R., POKROVSKY, V.L. and TOGNETTI, V. (Eds.) - *Microscopic Aspect of Nonlinearity in Condensed Matter* (B264) 1992 Plenum Publishing Corporation. ISBN 0-306-44001-6.

JIMENEZ, J. (Ed.) - *The Global Geometry of Turbulence: Impact of Nonlinear Dynamics* (B268) 1992 Plenum Publishing Corporation. ISBN 0-306-44014-8.

MOSEKILDE, E. and MOSEKILDE, L. (Eds.) - *Complexity, Chaos and Biological Evolution* (B270) 1992 Plenum Publishing Corporation. ISBN 0-306-44026-1.

ROY, A.E. (Ed.) - *Predictability, Stability and Chaos in N-Body Dynamical Systems* (B272) 1992 Plenum Publishing Corporation. ISBN 0-306-44034-2.

BEN AMAR, M., PELCE, P. and TABELING, P. (Eds.) - *Growth and Form: Nonlinear Aspects* (B276) 1992 Plenum Publishing Corporation. ISBN 0-306-44046-6.

JENA, P., KHANNA, S.N. and RAO, B.K. (Eds.) - *Physics and Chemistry of Finite Systems: From Clusters to Crystals* (C374) 1992 Kluwer Academic Publishers. ISBN 0-7923-1818-8

SEGUR, H. (Ed.) - *Asymptotics Beyond All Orders* (B284) 1992 Plenum Publishing Corporation. ISBN 0-306-44112-8

DAVIS, S.H., HUPPERT, H.E., MÜLLER, U. and WORSTER, M.G. (Eds.) *Interactive Dynamics of Convection and Solidification* (E219) 1992 Kluwer Academic Publishers. ISBN 0-7923-1910-9

BOUNTIS, T. (Ed.) *Proton Transfer in Hydrogen-Bonded Systems* (B291) 1992 Plenum Publishing Corporation. ISBN 0-306-44216-7

MARESCHAL, M. and HOLIAN, B.L. (Eds.) *Microscopic Simulations of Complex Hydrodynamic Phenomena* (B292) 1992 Plenum Publishing Corporation. ISBN 0-306-44226-4

BOUNTIS, T. (Ed.) *Chaotic Dynamics: Theory and Practice* (B298) 1992 Plenum Publishing Corporation. ISBN 0-306-44247-7

TABLE OF CONTENTS

II. Singularities in Optics and Plasmas

PREFACE

The articles in this volume represent presentations at the NATO Advanced Research Workshop (ARW) on Singularities in Fluids, Plasmas and Optics, which was held at the Institute of Applied and Computational Mathematics in Heraklion, Greece, on July 6-10, 1992. The purpose of this Workshop was to survey current knowledge on singularities in fluids, plasmas and optics and to stimulate further research on the subject. The scope of the talks included mathematical, numerical and physical aspects of singularities and their applications.

Singularities or nearly singular behavior occur in a wide range of physical and mathematical problems, such as nonlinear focusing of laser beams, curvature singularities on vortex sheets and magnetic reconnection. As a coherent and distinct phenomenon, a singularity can provide a relatively simple solution that describes a complex physical state.

Topics discussed at this meeting included the following:

- Singularities for the nonlinear Schrödinger equation, as a model for the propagation of a laser beam in a nonlinear optical material.
- Near blowup for solutions of the nonlinear Schrödinger equation with saturation, or the Zakharov equations.
- Singularity formation on fluid interfaces and vortex sheets.
- Possible singularities in solutions of the Euler equations for 3D incompressible, inviscid flow.
- Computation of the formation and propagation of singularities in the complex plane.
- Computational methods for singularities using adaptive rescaling.

The Workshop was attended by an international group of mathematicians, physicists and engineers. Two specific objectives of the meeting were to stimulate the exchange of ideas between researchers in fluid mechanics and those in optics and plasma physics, and to promote interactions between scientists from Russia and those from western countries.

Scientific and financial support for this Workshop was provided by the NATO Science Committee as a part of their program on Chaos, Order and

Patterns. We are sincerely grateful to the Science Committee and, in particular, to David Crighton, who initially suggested an ARW on this topic. It is a pleasure to also thank the Institute of Applied and Computational Mathematics for its warm hospitality and stimulating environment. John Papadakis, the Director of the Institute, and Renia Papanastasiou deserve special thanks for their efforts in planning and running the Workshop.

Russel E. Caflisch
UCLA, Los Angeles
March 8, 1993

Well-posed Numerical Calculations for Free-Surface Flows

Greg Baker, and Saleh Tanveer
Department of Mathematics
The Ohio State University

ABSTRACT. We introduce a new numerical procedure for tracking interfaces between immiscible, incompressible fluids. The motion of such interfaces is typically ill-posed; small deviations in initial conditions lead to dramatic changes in the subsequent motion. Our method is based on the analytic continuation of the equations for the location of the interface into the complex unphysical plane. Specifically in two-dimensions, we let $(x(\theta,t), y(\theta,t))$ be the parametric representation for the location of the interface at time t, and we consider θ a complex-valued variable. Calculations are performed far from the real axis of θ. In particular, we keep track of the nearest singularities of $z(\theta,t)$ to the real axis of θ. The reconstruction of the location of the interface is a well-posed numercial procedure, involving the anayltic interpolation of $z(\theta,t)$ to the real axis of θ. As an example, we consider two-dimensional Hele-Shaw flow in a channel in the absence of surface tension effects. The results from our numerical method compare extremely well with a known analytic solution for the location of the interface.

1. Introduction

Free-surface flows of incompressible, inviscid liquids have received considerable attention in recent years for several reasons. Their practical importance has always kept them high on the list of challenging problems. However, it is the advent of high-speed computers that have allowed numerical solution for the motion, typically by using boundary integral techniques or generalized vortex methods. In particular, numerical studies have examined the evolution of the classical Kelvin-Helmholtz instability (Moore 1985; Krasny 1986a; Shelley 1992), the Rayleigh-Taylor instability (Baker, Meiron & Orszag 1980, 1984; Pullin 1982; Kerr 1986; Tryggvason 1988), and the Saffman-Taylor instability in Hele-Shaw flow (Tryggvason & Aref 1983; DeGregoria & Schwartz 1986; Meiburg & Homsy 1988). This list is not comprehensive, but serves simply to indicate the broad interest in numerical simulations of free-surface flows.

The general feature of these instabilities is that in the absence of some physical regularization, infinitessimal perturbations with smallest wavelength

R. E. Caflisch and G. C. Papanicolaou (eds.), Singularities in Fluids, Plasmas and Optics, 1–10.
© 1993 *Kluwer Academic Publishers.*

grow the fastest. If this pattern of growth persists into the nonlinear regime, then the motion is ill-posed and the free-surface will form singularities in finite time. Numerical calculations do exhibit difficulties, typically in the rapid development of a saw-tooth pattern. The trend has been to introduce various regularizations in the numerical procedure in order to continue the calculations for longer times (for example, Tryggvason & Aref 1983, Kerr 1986, Krasny 1986b).

However, several asymptotic (Moore 1985; Caflisch, Orellana & Siegel 1990) and numerical (Meiron, Baker & Orszag 1982; Krasny 1986a; Pugh 1989; Baker, Caflisch, & Siegel 1992, Cowley & Pugh 1992; Shelley 1992) studies have provided convincing evidence that typical free-surface calculations are truly ill-posed. Curvature singularities occur in finite time. A particularly pleasing mathematical interpretation of the occurrence of curvature singularities is achieved by regarding the parametrization variable for the free-surface location as a complex-valued variable. In particular, let $(x(\theta,t), y(\theta,t))$ be the location of the free-surface at time t. The analytic extension of x and y for complex θ will be valid in a strip containg the real axis of θ. The boundaries of the strip are determined by the nearest singularity in x or y to the real axis. Usual studies of the classical free-surface instabilities correspond to isolated singularities, often branch points or logarithmic singularities, that move in towards the real axis of θ and reach it in finite time. These singularites then account for the spontaneous appearance of tangent and curvature singularities in the physical domain.

Obviously, it is the neglect of physically important effects that cause the mathematical difficulties. For example, a finite, but thin layer of vorticity may replace a vortex sheet undergoing Kelvin-Helmholtz instability (Baker & Shelley 1990). The thin layer does develop regions of high curvature, but the curvature remains finite for finite time. The analytic extension of the boundaries of the thin layer into the complex plane of their parametrization variable does contain branch point singularities, but they slow down as they approach the real axis, and never reach it in finite time. Nevertheless, their close proximity to the real axis induces regions of very high curvature which require high resolution in the numerical method. In addition, the thinner the layer, the more modes that are susciptible to growth: round-off errors can promote the growth of high wavenumbers and the numerical calculations soon break down. Consequently, there are severe limitations on thinness of the layer that can be studied reliably by boundary integral techniques.

These same numerical difficulties plague all methods that attempt to follow the motion of the free-surface in the presence of weak physical regularizations. Examples include studies of thin viscous shear layers (Tryggvason, Dahm & Sbeih 1991), surface tension effects in Rayleigh-Taylor instability (Pullin 1982) or in Hele-Shaw flow (Degregoria & Schwartz 1986). In all these cases, the close approach of singularities in the complex unphysical plane to the real axis induce high sensitivity in the calculations and require high resolution for accurate results.

In this paper, we describe a different approach from standard boundary integral techniques. The equations are continued analytically into the complex unphysical plane and solved off the real axis. Singularities, if present may be handled explicitly, so that the calculations may be performed in a wider strip of analyticity. Calculations are well-posed off the real axis, and the analytic interpolation to recover data, for example, the location of the free-surface, is well-posed. Consequently, calculations can proceed for as long as the singularities do not reach the real axis. The extension of this method may be suitable for the long time evolution of free-surfaces in the presence of physical regularizations. However, this method does require certain conditions to be met: the equations of motion for the interface must be analytically continued into the complex unphysical plane, a step that may be hard to do in some cases; the generic nature of all singularities should be known for explicit representation in the numerical procedure; and the initial data must be known in the full complex unphysical plane.

In essence, it is the last requirement that removes the ill-posedness in the formulation of the problem. Normally, small perturbation to data on the real axis will produce new initial singularities in the complex unphysical plane. These singularities may be nearer the real axis and cause failure in the solution for earlier times. When we prescribe the initial behavior in all of the complex unphysical plane, we preclude the numerical method from introducing those perturbations which would cause difficulties.

We select a particular problem, fingering instability in Hele-Shaw flow in a channel, to illustrate our ideas for several reasons. Much is known mathematically about the motion of free-surfaces in Hele-Shaw flows (some recent reviews are Hohlov 1990; Tanveer 1991; Howison 1992): the equations have been extended to the complex unphysical plane, the singularities retain their form as they move in the complex unphysical plane; special solutions are known which may be used as an aid to test the numerical method; and, in a sense, the equations are the simplest for free-surface motion that exhibits fingering instabilities.

In the next section, we present the equations. Then we describe our numerical method and present the results for a test case using a well-known solution due to Saffman (1959). Our results confirm our expectations that we can calculate the long time evolution of the finger for far longer times than previously possible with a high degree of accuracy.

2. The Analytically Extended Equations

Free-surface motion in Hele-Shaw cells may be described in different mathematical forms. We follow the approach in Tanveer (1992). We introduce a conformal map $z(\zeta, t)$ that takes the interior of a unit semi-circle into the region of fluid in the channel as shown in Figure 1. e may write

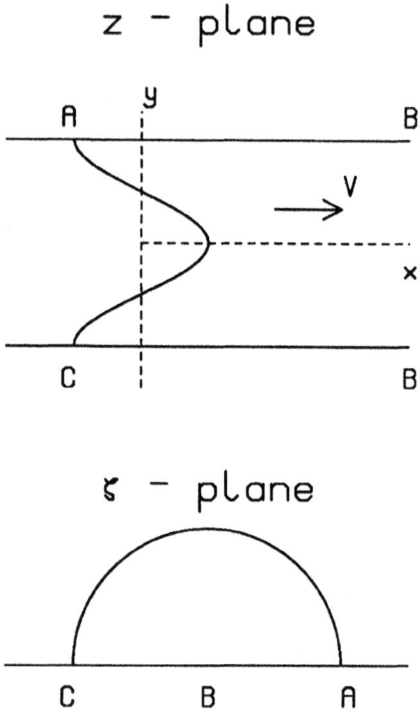

Fig. 1. An Illustration of the Conformal Map, $z(\zeta)$

$$z(\zeta, t) = -\frac{2}{\pi} \ln \zeta + i + f(\zeta, t) \tag{1}$$

where f is analytic in the unit semi-circle. We require

$$\Im\{f\} = 0 \tag{2}$$

in order for the channel walls to map to the diameter of the unit semi-circle. The free-surface is located along the arc of the semi-circle;

$$\begin{aligned} x(\theta, t) &= \Re\{f(e^{i\theta}, t)\} \\ y(\theta, t) &= 1 - \frac{2\theta}{\pi} + \Im\{f(\theta, t)\} \end{aligned} \tag{3}$$

We introduce the complex velocity potential, $W(\zeta, t)$. Then the kinematic requirement that the free-surface moves with the fluid may be stated as

$$\Re\left\{\frac{z_t}{\zeta z_\zeta} - \frac{\zeta W_\zeta}{\mid z_\zeta \mid^2}\right\} = 0 \tag{4}$$

on $\mid \zeta \mid = 1$. Dynamical considerations determine W. In particular, in the absence of surface tension effects and with the viscosity of the less viscous fluid neglected, continuity of pressure at the free-surface implies

$$W(\zeta, t) = -\frac{2}{\pi} \ln \zeta + i \tag{5}$$

where the velocity V far downstream of the channel is taken to be unity, and there is no fluid velocity through the channel walls.

Equation (4) may be analytically continued into the ζ plane outside the semi-circle, $\mid \zeta \mid > 1$ (Tanveer 1992); [1]

$$z_t + q_1 z_\zeta = q_2 \tag{6}$$

where q_1, q_2 are analytic functions of ζ determined by the conditions,

$$q_2(\zeta) = -\frac{4\zeta}{\pi z_\zeta(1/\zeta)} \tag{7}$$

and on $\mid \zeta \mid = 1$,

$$\Re\left\{\frac{q_1}{\zeta}\right\} = -\frac{2}{\pi \mid z_\zeta \mid^2} \tag{8}$$

Equation (6) has similar properties to a first-order hyperbolic equation in the complex plane with coefficients that depend globally on z_ζ. Detailed analysis of this equation establishes the following properties for the singularities in the conformal map $z(\zeta)$. Isolated singularities and zeros track inwards towards the unit circle, and may reach it in finite time depending on their nature (Tanveer 1992). Specifically, isolated singularities of the form,

$$z_\zeta = A(t)(\zeta - \zeta_s(t))^{-\beta} \tag{9}$$

move with speed $q_1(\zeta_s(t))$ and will reach the unit circle in finite time provided $\beta < 1/2$. Otherwise, these singularities approach the unit circle very closely in time but don't reach it. In either case, the close proximity of these singularities to the unit circle during some period of time induce strong variations in the actual physical profile. The contorted geometry places strong constraints on the ability of standard numerical methods to follow the evolution reliably. Attempts to increase the resolution face the underlying ill-posedness of the motion; either smoothing, filtering or increased precision

[1] Analytic continuation to $\mid \zeta \mid > 1$ corresponds to complexifying θ to $\Im\{\theta\} < 0$.

must be used (Dai & Shelley 1992). Clearly, a numerical procedure that captures the presence of these singularities explicitly has the potential to far excced the capabilities of other numercial approaches. In the next Section, we describe one such numerical procedure.

3. Numerical Method

At first sight, (6) needs to be solved in all the complex plane outside the unit circle. However, this is not the case for the following reason. Any function $f(z)$, analytic inside a closed curve, is determined by its value on the closed curve. In particular, consider the Taylor series for $f(z)$ evalued on a circle of radius R.

$$\tilde{f}(\theta) \equiv f(Re^{i\theta}) = \sum_{k=0}^{\infty} a_k R^k e^{ik\theta} \tag{10}$$

The coefficients for the Taylor series are determined by

$$a_k R^k = \frac{1}{2\pi}\int_0^{2\pi} \tilde{f}(\theta) e^{-ik\theta} d\theta \tag{11}$$

Now if $f(z)$ is required at a point $r\exp(i\theta)$, $r < R$, then we may use

$$f(re^{i\theta}) = \sum_{k=o}^{\infty} a_k R^k \frac{r^k}{R^k} e^{ik\theta} \tag{12}$$

The Fourier transforms in (10,11,12) may be replaced by their discrete analogues. The first step is the calculation of $a_k R^k$ from (11). These coefficients are then multiplied by the factor r^k/R^k before (12) is used. Clearly, for $r < R$, the tail of the Fourier spectrum is kept at round-off levels if the number of terms in the discrete transform is big enough. Consequently, this numerical procedure is well-posed. Conversely, for $r > R$, the tail of the spectrum is multiplied by a big factor so that any round-off errors in $a_k R^k$ will be exaggerated. Extrapolation of analytic functions is ill-posed. Our aim then is to update $z(\zeta, t)$ on a circle of large radius through (6). The actual physical profile on $r = 1$ is then easily obtained by analytic interpolation as described above.

Unfortunately, the circle on which we would like to update $z(\zeta, t)$ must lie inside any singularities of $z(\zeta, t)$. Since these singularities will move towards the unit circle, our circle will have to contract and reach the unit circle before any truly interesting behavior in the motion of the interface occurs. Consequently, we decompose our representation for $f(\zeta, t)$ into two parts, $f_s(\zeta, t) + g(\zeta, t)$, where f_s gives the nearest singularities to the unit circle and g is analytic up to a much larger radius, determined by the nearest neglected singularity.

We keep our description of the method brief and specific by choosing an example with only two singularities.

$$f(\zeta,t) = \frac{1}{\pi}\ln(1-a(t)\zeta^2) + \frac{2}{\pi}g(\zeta,t) \tag{13}$$

This form is an exact solution (Saffman 1959) with a prescribed behavior for $a(t)$ and $g(\zeta,t) = d(t)$, allowing us to determine the precise error in our numerical results. The singularities are located at $\pm 1/\sqrt{a}$ and are of the form (9) with $\beta = 1$. These singularities do not reach the unit circle in finite time, so the calculation of the motion of the interface for long times is a severe test of the method. Upon substitution of (13) into (6) we are led to equations for the evolution of $a(t)$ and $g(\zeta,t)$.

$$\frac{da}{dt} \equiv \dot{a} = -2a\sqrt{a}\, q_1(1/\sqrt{a}) \tag{14}$$

and

$$g_t + q_1 g_\zeta = Q \equiv \frac{\pi}{2}\, q_2 + \frac{1}{2}\frac{\dot{a}\zeta^2}{1-a\zeta^2} + \frac{q_1}{\zeta}\left(1 + \frac{a\zeta^2}{1-a\zeta^2}\right). \tag{15}$$

where subscripts with ζ or t imply differentiation.

We solve (14,15) by the method of backwards characteristics. We assume that we know $g_{j,n} = g(r_n \exp(ij\Delta\theta), n\Delta t)$. By integrating along characteristics with the modified Euler's method we obtain the following equations; the first step is a simple Euler step given by

$$\begin{aligned} r_{n+1}e^{ij\Delta\theta} &= \bar{\zeta}_j + \Delta t\, q_1(\bar{\zeta}_j, n\Delta t) \\ \tilde{g}_j &= g(\bar{\zeta}_j, n\Delta t) + \Delta t\, Q(\bar{\zeta}_j, n\Delta t) \\ \tilde{a} &= a_n - \Delta t\, 2a_n\sqrt{a_n}\, q_1(1/\sqrt{a_n}) \end{aligned} \tag{16}$$

The starting point of the characteristic, $\bar{\zeta}_j$, is determined through the first equation by the requirement that the charcteristic reaches $r_{n+1}\exp(ij\Delta\theta)$ at time $t + \Delta t$. We solve the first equation by Newton's method to get $\bar{\zeta}_j$. The second equation gives an estimate $\tilde{g}_j$ for the new value of g. However, this first step gives only $O(\Delta t)$ accuracy. We add the correction step,

$$\begin{aligned} r_{n+1}e^{ij\Delta\theta} &= \hat{\zeta}_j + \frac{\Delta t}{2}\left[q_1(\hat{\zeta}_j, n\Delta t) + \tilde{q}_1(r_{n+1}e^{ij\Delta\theta}, (n+1)\Delta t)\right] \\ g_{j,n+1} &= g(\hat{\zeta}_j, n\Delta t) + \frac{\Delta t}{2}\left[Q(\hat{\zeta}_j, n\Delta t) + \tilde{Q}(r_{n+1}e^{ij\Delta\theta}, (n+1)\Delta t)\right] \\ a_{n+1} &= a_n - \Delta t\left[a_n\sqrt{a_n}\, q_1(1/\sqrt{a_n}) + \tilde{a}\sqrt{\tilde{a}}\, \tilde{q}_1(1/\sqrt{\tilde{a}})\right] \end{aligned} \tag{17}$$

Once again we solve the first equation in (17) by Newton's method to obtain the improved estimate $\hat{\zeta}$ for the starting point of the characteristic. The remaining two equations are then explicit. We use the notation $\tilde{q}_1, \tilde{Q}$ to indicate that these functions are evaluated by using the estimates $\tilde{\zeta}_j, \tilde{a}$. The accuracy of the combined steps (16,17) is $O(\Delta t^2)$.

The condition for stability requires that $\bar{\zeta}, \hat{\zeta}$ lie inside the circle with radius r_n. So r_{n+1} must be chosen subject to $r_{n+1} \leq r_n + \max(S)\Delta t$, where $S = \Re\{q_1 \exp(-i\theta)\}$ is the radial speed of the characteristics. This bound will ensure that analytic interpolation of quantities at $\bar{\zeta}_j$ and $\hat{\zeta}_j$ will be stable. We choose r_{n+1} to be close to the bound in order to keep the rate down at which the circle contracts ($S < 0$) towards the unit circle.

We now describe how q_1, Q may be evaluated with spectral accuracy. Substitution of (13) into (1) leads to

$$\frac{1}{z_\zeta(\zeta)} = -\frac{\pi}{2} \frac{1 - a\zeta^2}{1 - \zeta(1 - a\zeta^2) g_\zeta(\zeta)} \tag{18}$$

Consequently,

$$q_2(\zeta) = \frac{2\zeta(\zeta^2 - a)}{\zeta^3 - (\zeta^2 - a) g_\zeta(1/\zeta)} \tag{19}$$

Given the values $g_{j,n}$, we may use a discrete Fourier representation correpsonding to (10) to determine the derivative g_ζ and to find its interpolated value at $1/\zeta$ by using (12). We may use this same approach to evaluate $1/z_\zeta$ on the unit circle. Then we may use the fourier series,

$$\frac{1}{| z_\zeta |^2} = c_0 + \sum_{k=1}^{\infty} c_k \cos(k\theta) + \sum_{k=1}^{\infty} d_k \sin(k\theta) \tag{20}$$

to find

$$q_1(\zeta) = -\frac{2\zeta}{\pi} \left[c_0 + \sum_{k=1}^{\infty} (c_k + i d_k) \frac{1}{\zeta^k} \right] \tag{21}$$

All these expressions may be approximated by using discrete Fourier series.

e make the specific choice, $a(1) = 0.25, g(\zeta, 0) = 1.0$, as a test case for the numerical method. The exact solution is given by $a^2 = 1/(1 + 15\exp(-2))$, $\pi g = 2\pi t + \pi \ln(4) - \ln(a)$. We use $N = 16$ modes in the discrete Fourier representation for g, even though g remains a constant in ζ. In this way we can check whether round-off errors accumulate in the high modes in time. We use a time step, $\Delta t = 0.01$. Our results are shown in Figure 2 at time $t = 2.0$. We take 512 points on the unit circle in the ζ plane to generate the profile; the data at these points are determine by analytic interpolation (12). The error in the profile is less than 7.0×10^{-3}. We find no growth of round-off errors in our results. The initial circle has radius $R = 2000$. At $t = 2.0$ the circle has a radius of 6.862 and the singularities are at $\zeta = \pm 1.0000017$. The calculations take about half a minute on a SUN SPARC 1. By taking a larger initial radius, we can easily run our code for as long as we care to.

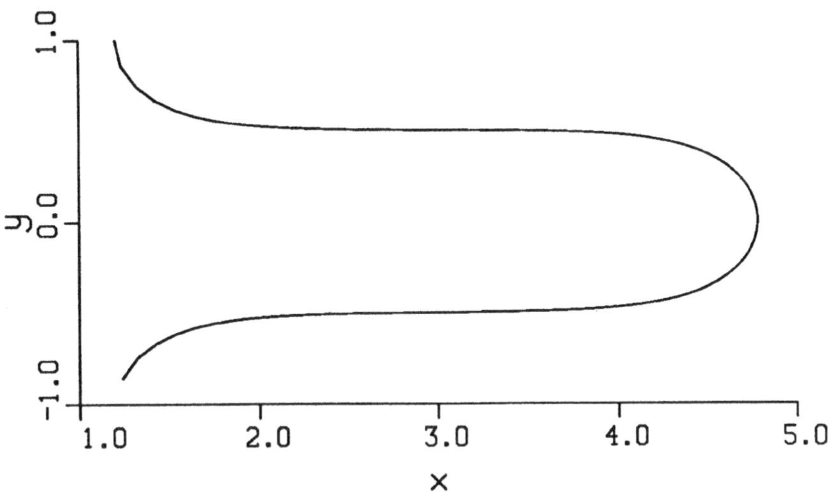

Fig. 2. The shape of the interface at $t = 2$

4. Conclusions

We have demonstrated the feasibility of computing the motion of a free-surface that is ill-posed in the usual sense by extending the equations into the complex unphysical plane. The method is spectrally accurate and well-posed. It can be easily extended to handle several singularities of the form considered in (13).

Acknowledgements

This work has been partially supported by The National Science Foundation through grants DMS-9005932 and DMS-9107608, and by the Department of Energy through grant DE-FG02-92ER14270.

References

G.R. Baker, R.E. Caflisch & M. Siegel, Singularity formation during Rayleigh-Taylor instability, to appear in *J. Fluid Mech.*, (1992).

G.R. Baker, D.I. Meiron, & S.A. Orszag, Vortex simulations of the Rayleigh-Taylor instability, *Phys. Fluids* 23, p.1485, (1980).

G.R. Baker, D.I. Meiron, & S.A. Orszag, Boundary integral methods for axisymmetric and three-dimensional Rayleigh-Taylor instability problems, *Physica D*, 12, p.19, 1984.

G.R. Baker & M.J. Shelley, On the connection between thin vortex layers and vortex sheets, *J. Fluid Mech.*, 215, p.161, (1990).

R. Caflisch, O. Orellana, & M. Siegel, A localized approximation method for vortical flows, *SIAM J. Appl. Math.*, 50, p. 1517, (1990).

S. Cowley & D. Pugh, On the formation of an interface singularity in the rising 2-d Boussinesq bubble, to appear in *J. Fluid Mech.*, (1992).
W. Dai & M. Shelley, Numerical study of the effect of surface tension and noise on an expanding Hele-Shaw bubble, submitted to *Phys. of Fluids*, (1992).
A.J. Degregoria & L.W. Schwartz, A boundary-integral method for two-phase displacement in Hele-Shaw cells, *J. Fluid Mech.* 164, p.383, (1986).
E. Hohlov, Time dependent free boundary problems: explicit solutions. I, *M.I.A.N. Preprint*, 14, Steklov Institute, Moscow, (1990).
S.D. Howison, Complex variable methods in Hele-Shaw moving boundary problems, to appear in *European J. Appl. Math.*, (1992).
R.M. Kerr, Analysis of Rayleigh-Taylor flows using vortex blobs", *Lawrence Livermore National Laboratory Report*, UCID-20915, (1986).
R. Krasny, On singularity formation in a vortex sheet and the point vortex approximation, *J. Fluid Mech.* 167, p.65, (1986a).
R. Krasny, Desingularization of periodic vortex sheet roll-up, *J. Comput. Phys.*, 65, p.292, (1986b).
E. Meiburg & G.M. Homsy, Nonlinear unstable viscous fingers in Hele-Shaw flows. II Numerical simulation, *Phys. Fluids*, 31, p. 429, (1988).
D.I. Meiron, G.R. Baker, & S.A. Orszag, Analytic structure of vortex sheet dynamics. Part 1. Kelvin-Helmholtz instability, *J. Fluid Mech.*, 114, p. 283, (1982).
D.W. Moore, Numerical and analytical aspects of Helmholtz instabilty, *Theoretical and Applied Mechanics* (Ed. F.I. Niordson, & N. Olhoff, Elsevier, New York) (1985).
D.A. Pugh, Development of vortex sheets in Boussinesq flows - formation of singularities, Ph. D. Thesis, Imperial College, London, (1989).
D. Pullin, Numerical studies of surface-tension effects in nonlinear Kelvin-Helmholtz and Rayleigh-Taylor instability, *J. Fluid Mech.*, 119, p.507, (1982).
P.G. Saffman, *Q. J. Mech. Appl. Math.*, 12, p.146, (1959).
M.J. Shelley, A study of singularity formation in vortex sheet motion by a spectrally accurate method, to appear in *J. Fluid Mech.*, (1992).
S. Tanveer, Viscous displacement in a Hele-Shaw cell, *Asymptotics beyond all orders*, NATO ASI Series B284, (ed. H. Segur, S. Tanveer, & H. Levine, Plenum Publishing Co.), (1991)
S. Tanveer, Evolution of Hele-Shaw interface for small surface tension, to appear in *Phil Trans. R. Soc. London*, (1992).
G. Tryggvason, Numerical simulations of the Rayleigh-Taylor instability, *J. Comp. Phys.*, 75, p.253, 1988.
G. Tryggvason & H. Aref, Numerical experiments on Hele-Shaw cell with a sharp interface, *J. Fluid Mech.*, 136, p.1, (1983).
G. Tryggvason, W.J.A. Dahm, & K. Sbeih, Fine structure of vortex sheet roll-up by viscous and inviscid simulation, *ASME J. Fluids Engin.*, 113, p.31, (1991).

On the Well-Posedness of Two Fluid Interfacial Flows with Surface Tension

J.Thomas Beale
Duke University
Thomas Y. Hou
Courant Institute
and
John S. Lowengrub
University of Minnesota

ABSTRACT.
We apply a technique to linearize the equations of motion about any smooth solution, even far from equilibrium. This technique was developed in our recent study of water waves (Beale, Hou & Lowengrub 1992). We show that these new linearized equations have a beautiful form when projected into the tangential and normal coordinates of the underlying smooth interface. We prove that these equations are well-posed in the appropriate Sobolev spaces. We investigate the full nonlinear behavior numerically. We describe several methods to compute these flows. We prove that of these, only suitably modified spectral methods are nonlinearly stable and converge. The other methods are proved to be linearly unstable, even far from equilibrium. (Baker & Nachbin 1992) independently demonstrated these instabilities near equilibrium for vortex sheets with surface tension. Finally, we present numerical computations. Our results include a sequence of computations of vortex sheets with surface tension using successively smaller surface tensions. The interfaces remain smooth and there seems to be no finite time physical singularity. Their structure also differs from the zero surface tension case.

1. Introduction

In this paper, we discuss the well-posedness of interfacial fluid flows with surface tension. We assume the flow to be inviscid, incompressible and irrotational so that the entire motion is determined by the position and strength of the interface, which is a vortex sheet.

It is well known that surface tension is a dispersive regularization of the Kelvin-Helmholtz and Rayleigh-Taylor instabilities near equilibrium (Drazin & Reid 1981). The equilibrium solution corresponds to a flat sheet with constant strength. However, in the case of two fluids, the interface will typically

R. E. Caflisch and G. C. Papanicolaou (eds.), Singularities in Fluids, Plasmas and Optics, 11–38.
© 1993 *Kluwer Academic Publishers.*

deform significantly before strong nonlinear effects such as singularity formation become important. Therefore, the near flat (equilibrium) approximation may be poor even for relatively short times. Further, the linear stability of solutions far from equilibrium is intimately related to the stability of numerical methods for computing such flows.

We apply a general technique to linearize the equations of motion, in boundary integral form, about any smooth solution. The solution may be far from equilibrium. This technique was developed in our recent study of water waves (Beale, Hou & Lowengrub 1992), hereafter referred to as (BHL1). We show that these new linearized equations have a beautiful form when projected into the normal and tangential coordinates of the underlying smooth interface. Viewed in these coordinates, the linearized equations have a qualitative structure surprisingly like that for the case near equilibrium. An interesting consequence is that we are able to prove that in the *absence* of surface tension, these linearized equations are almost always ill-posed, even in the case of stable stratification.

We prove that these linearized equations, far from equilibrium and with *positive* surface tension, are well-posed in the appropriate Sobolev spaces. A delicate balance of terms is required to do this. In the process, we also show how the time dependence of the smooth, underlying solution affects the dynamic regularizing behavior of surface tension. This cannot be seen near equilibrium. The linear well-posedness does not preclude the growth of linear disturbances, and thus departure from the original motion, but it does imply that the rate of growth is bounded. While, this well-posedness may be taken as an indication of the predictive behavior of the full nonlinear equation for short times, it is possible that strong nonlinear interactions could dominate the linearly regularizing effect of surface tension and may yet lead to singularity formation.

Currently, we only study the full nonlinear behavior of these flows numerically. It is well known that the numerical computation of interfacial flows with surface tension is a difficult and delicate problem. In fact, there has been much discussion as to whether the instabilities that previous investigators encountered are numerical or physical (Pullin 1982; Rangel & Sirignano 1988). We show that many of the instabilities are strictly numerical. We describe several methods to compute the full nonlinear equations. We prove that of these, only suitably modified spectral methods are nonlinearly stable and converge. Specifically, the convergent methods involve the modified point vortex approximation and the singularity subtraction approximation combined with the spectral derivative. The other methods involve lower order accurate discrete derivative operators as well as the usual point vortex approximation. These other methods and the unmodified spectral methods are proved to be linearly unstable, even far from equilibrium. (Baker & Nachbin

1992) independently demonstrated these instabilities near equilibrium for vortex sheets with surface tension (one fluid flow, no density difference). We were actually led to noticing these instabilities because of the special balance of terms required to prove the well-posedness in the continuous case. Thus, it is natural to ask whether this balance is maintained for the discrete equations. Clearly, special properties are required of the numerical scheme in order for this to be so.

We also present computational results. We demonstrate the numerical instability of some of the unstable methods. We also demonstrate the physical instability of a stably stratified flow in the absence of surface tension.

We are particularly interested in answering the following physical questions. First, do vortex sheets with surface tension form singularities in finite time? Second, what is the limiting behavior of vortex sheets with surface tension in the limit of small surface tension? We present a sequence of computations of vortex sheets using successively smaller surface tensions. For our initial data, the answers to our questions are: first, it does not appear that vortex sheets form physical singularities in finite time. And second, the structure of the interfaces differs from the zero surface tension solution. Two well-separated peaks form in the curvature of the interface for non-zero surface tensions. In the case of zero surface tension, there are also two peaks, but with no distance separating them (see section 4). Thus, surface tension seems to split the zero surface tension peaks into two new separated ones. This can be interpreted via singularities in the analytic continuation of the interface (see section 4). These new peaks approach each other in the limit of small surface tension. Although it seems likely that these peaks will merge in the limit of vanishing surface tension, it is difficult to identify their limiting behavior. More refined computations are necessary. The numerical solutions are difficult to obtain due to the stiffness arising from the surface tension term. This is currently under study (Lowengrub, Shelley & Hou).

The organization of the paper is as follows. The formulation and the classical theory near equilibrium is given in section 2. The derivation of the linearized equations, far from equilibrium, and the well-posedness is given in section 3. The numerical results are presented in section 4. Some conclusions are given section 5.

2. The Equations of Motion

In this section, we give the equations of motion of an interface separating two inviscid, incompressible and irrotational fluids in the presence of gravity and surface tension in two space dimensions. In the following, the subscript 2 denotes the fluid above the interface and 1 denotes the fluid below the interface.

In each fluid, we have Euler's equation

$$\rho_i\left[\partial_t u_i + (u_i \cdot \nabla) u_i\right] = -\nabla p_i - \rho_i g \hat{y} \tag{1}$$

where $\hat{y} = (0,1)^t$, $g > 0$ and $i = 1,2$. The incompressibility constraint is $\nabla \cdot u_i = 0$ and $\nabla \times u_i = 0$. At the interface, we impose Laplace's boundary condition which relates the pressure jump to the curvature of the interface, κ, by

$$\pi_1 - \pi_2 = -\tau\kappa \tag{2}$$

where π_1, π_2 are the limiting pressures of p_1, p_2 from below and above the interface respectively. τ is the surface tension coefficient. We also require the interface to move with the fluid. This gives the additional boundary condition

$$(v_1 - v_2) \cdot N = 0 \tag{3}$$

where N is the upward normal vector to the interface and v_1, v_2 are the limiting velocities of u_1, u_2. It is well known that these flows have vortex sheet representations. Here we only give the equations, the details of their derivation can be found in (Birkhoff 1962; Baker, Meiron & Orszag 1982; Shelley & Vinson 1992). Moreover, we present the formulation in complex notation.

Assume that the velocity of the sheet, U, is given by the average of the limiting velocities from above and below, i.e. in real coordinates, $U = \frac{1}{2}(v_1 + v_2)$. Let the position of the sheet be $z(\alpha,t) = x(\alpha,t) + iy(\alpha,t)$ where α is a Lagrangian parameter along the sheet. Then, the sheet evolves according to

$$\frac{d\bar{z}(\alpha,t)}{dt} = \bar{U} = \frac{1}{2\pi i} P.V. \int \frac{\gamma(\alpha',t)}{z(\alpha,t) - z(\alpha',t)} d\alpha' \tag{4}$$

where $P.V.$ denotes the principal value integral and γ is the unnormalized vortex sheet strength. The sheet strength γ evolves by

$$\frac{d\gamma}{dt} = -2A\left[Re\left\{\frac{d^2\bar{z}}{dt^2} z_\alpha\right\} + \frac{1}{8}\partial_\alpha\left(\frac{\gamma^2}{|z_\alpha|^2}\right) + g y_\alpha\right] + \tau\kappa_\alpha \tag{5}$$

where $A = (\rho_1 - \rho_2)/(\rho_1 + \rho_2)$ is the Atwood number. The curvature, κ, is evaluated by

$$\kappa = \frac{x_\alpha y_{\alpha\alpha} - x_{\alpha\alpha} y_\alpha}{(x_\alpha^2 + y_\alpha^2)^{3/2}} \tag{6}$$

Notice that equation (5) is actually a Fredholm integral equation of the second kind for $d\gamma/dt$. It has been shown to be invertible and it is globally contractive in the periodic case (Baker, Meiron & Orszag 1982; BHL1). Finally, (4) and (5) completely determine the motion of the entire flow.

It is instructive now to consider the linear stability of equilibrium solutions to (4),(5). Classically, it is known that surface tension is a dispersive regularization near equilibrium (Drazin & Reid 1981). This is seen as follows. The equilibrium solution is the flat sheet

$$z^e(\alpha,t) = \alpha \quad and \quad \gamma^e(\alpha,t) = \gamma_o \tag{7}$$

where γ_o is a constant. Looking for solutions of (4), (5) of the form $z = z^e + \epsilon\dot{z}$ and $\gamma = \gamma^e + \epsilon\dot{\gamma}$ and keeping only the $O(\epsilon)$ terms gives the linearized equations

$$\begin{aligned}
\frac{d\dot{x}}{dt} &= -\frac{\gamma_o}{2}H(\dot{y}_\alpha) \\
\frac{d\dot{y}}{dt} &= \frac{1}{2}H(\dot{\gamma}) - \frac{\gamma_o}{2}H(\dot{x}_\alpha) \\
\frac{d\dot{\gamma}}{dt} &= -A\gamma_o\dot{\gamma}_\alpha + A\gamma_o^2\dot{x}_{\alpha\alpha} - 2Ag\dot{y}_\alpha + \tau\dot{y}_{\alpha\alpha\alpha}
\end{aligned} \tag{8}$$

where $\dot{z} = \dot{x} + i\dot{y}$. Notice that $d\dot{\gamma}/dt$ is determined explicitly in the third equation. Linearly, the integral equation contribution has dropped out. The operator H is the Hilbert transform

$$H(f) = H[f](\alpha) = \frac{1}{\pi}P.V.\int \frac{f(\alpha')}{\alpha - \alpha'}d\alpha' \tag{9}$$

The system is easily diagonalized in Fourier transform space. Let

$$\hat{f}(k) = \int f(\alpha)e^{-ik\alpha}d\alpha \quad \text{so that} \quad f(\alpha) = \frac{1}{2\pi}\int \hat{f}(k)e^{ik\alpha}d\alpha \tag{10}$$

Then, it is easy to see $\hat{H} = -isgn(k)$. The eigenvalues (growth rates) of the transformed system are

$$\omega(k) = 0, -\frac{A\gamma_o}{2}ik \pm \sqrt{\frac{\gamma_o^2k^2}{4}(1-A^2) - Ag|k| - \tau|k|^3} \tag{11}$$

The $\omega = 0$ growth rate simply reflects that translating Lagrangian points along the interface does not change its shape. For large k and $\tau > 0$, the other growth rates become pure imaginary. Since solutions of (8) are basically superpositions of $e^{\omega(k)t} \cdot e^{ik\alpha}$, an imaginary growth rate produces oscillations rather than growth. Therefore, the low modes can grow, but the high modes just oscillate. This results in an overall bounded growth (in k) of solutions. On the other hand, if $\tau = 0$ then the growth may be unbounded in k due to the the Kelvin-Helmholtz and Rayleigh-Taylor instabilities. Thus, surface tension is a dispersive regularization of those instabilities.

It is natural to ask how generic is this behavior. That is, if the interface is far from flat (equilibrium), then is surface tension still regularizing? Of course, we are interested in the full nonlinear behavior, but we begin by investigating the linear well-posedness of more general interfaces. We address this question in the next section.

3. Linear Well-Posedness Far from Equilibrium

In this section, we show the linear well-posedness of the equations (4),(5) far from equilibrium. We begin by showing that the linearized equations have a beautiful form when viewed in the appropriate coordinates. The appropriate coordinates are tangential and normal to the smooth underlying sheet.

In the following, we use both real and complex notation interchangably. The meaning should be clear from the context. Let the smooth, time dependent solution to (4),(5) be given by z and γ. We will be more precise about the conditions z and γ must satisfy later. As before, let $\dot{z}, \dot{\gamma}$ be the perturbations in z, γ. Replace z, γ in (4),(5) by $z + \epsilon\dot{z}, \gamma + \epsilon\dot{\gamma}$. Define tangent and normal vectors by

$$T = \sigma(x_\alpha, y_\alpha)^t \quad and \quad N = \sigma(-y_\alpha, x_\alpha)^t \tag{12}$$

where $\sigma = |z_\alpha|^{-1} = (x_\alpha^2 + y_\alpha^2)^{-\frac{1}{2}}$ is the arclength metric. The tangential and normal components of the perturbation $\dot{z}$ are

$$\begin{aligned} \dot{x}^T &= (\dot{x}, \dot{y})^t \cdot T = \sigma(\dot{x}x_\alpha + \dot{y}y_\alpha) \\ \dot{x}^N &= (\dot{x}, \dot{y})^t \cdot N = \sigma(\dot{y}x_\alpha - \dot{x}y_\alpha) \end{aligned} \tag{13}$$

We obtain the following

Lemma: Linearized Equations Far from Equilibrium

Let $\dot{\Gamma} = \sigma\dot{\gamma}$ be a rescaled vortex sheet strength and suppose that $|z_\alpha| \geq c_o > 0$. Then, to $O(\epsilon)$, the equations for $\dot{x}^N$, $\dot{x}^T$ and $\dot{\Gamma}$ are

$$\begin{aligned} \frac{d\dot{x}^T}{dt} &= -\frac{\gamma\sigma^2}{2}H(\dot{x}_\alpha^N) + E_o(\dot{x}^N + \dot{x}^T) + E_{-\infty}(\dot{\Gamma}) \\ \frac{d\dot{x}^N}{dt} &= \frac{1}{2}H(\dot{\Gamma}) - \frac{\gamma\sigma^2}{2}H(\dot{x}_\alpha^T) + E_o(\dot{x}^N + \dot{x}^T) + E_{-\infty}(\dot{\Gamma}) \\ \frac{d\dot{\Gamma}}{dt} &= -A\gamma\sigma^2\dot{\Gamma}_\alpha + A\gamma^2\sigma^4\dot{x}_{\alpha\alpha}^T + \tau\left[\sigma\partial_\alpha\sigma\partial_\alpha\sigma\right]\dot{x}_\alpha^N + E_o(\dot{x}_\alpha^T + \dot{x}_\alpha^N + \dot{\Gamma}) \end{aligned} \tag{14}$$

where the H is the Hilbert transform, the $E_o(\cdot)$ are bounded operators of their argument and $E_{-\infty}(\cdot)$ is a smoothing operator. The properties of E_o and *$E_{-\infty}$ are understood to be with respect to the Sobolev space norm.*

Notice the remarkable similarity between the equations (14) and the near linear equations (8). The integral equation for $d\dot{\Gamma}/dt$ has again dropped out, although now this is a consequence smoothing properities of the double layer potential, which is the kernel of the integral equation (Kellogg 1929; BHL1). This argument will be presented elsewhere (BHL4). In addition, except for the lower order operators E_o and $E_{-\infty}$, the equations (8) are essentially *frozen* coefficient versions of (14). This is because the smooth vortex sheet, z, when viewed in its tangential and normal coordinates looks locally flat.

The time dependence of the coefficients of (14) is extremely important. We will now be able to see how the smoothing effect of the surface tension will vary in time. This is, of course, not possible from the near equilibrium equation. Clearly, the time and spatial dependence of the underlying vortex sheet strength γ and the metric σ play a critical role. We will quantify this later.

Further, notice that the gravity term is missing from the $\dot{\Gamma}$ equation in (14). This is because in that equation, the gravity term involves only one spatial derivative as compared to the three from the surface tension and the two from the generation of vorticity due to density stratification (the second term in the $\dot{\Gamma}$ equation). We therefore consider it to be lower order compared to these other terms. In general, we call terms lower order than others if their corresponding Fourier multipliers are lower order in wavenumber k, for large k.

Finally, equation (14) has the following immediate and somewhat surprising consequence. It shows that if $\tau = 0$, then the linear equations for the perturbations are essentially always ill-posed, even in the stably stratified case where the Atwood number $A > 0$! This is made precise in the following corollary. We first introduce some notation. Let the space H^s, be the standard Sobolev space equipped with the usual norm. If s is an integer

$$||f||_{H^s}^2 = \sum_{j=0}^{s} ||\partial^j f||_{L^2}^2 \tag{15}$$

If s is not an integer, then the definition is extended in the usual way through Fourier transforms (Folland 1976).

Corollary: Ill-posedness of Linearized Equations if $\tau = 0$

Let the smooth solution z, γ be such that there exists α^*, t^* such that $\gamma(\alpha^*, t^*) \neq 0$. Let S be the solution operator of the system (14). *Then, for any Atwood ratio, A, with $-1 \leq A < 1$,* and for any $t > t^*$, *there does not exist a constant M for all data $\dot{z}, \dot{\gamma} \in H^{s'}$ such that*

$$||S[\dot{z}, \dot{\gamma}](\cdot, t, t')||_{H^s} \leq M||(\dot{z}, \dot{\gamma})(\cdot, t')||_{H^{s'}} \tag{16}$$

for $0 \leq t' \leq t$ and $s \leq s'$. This is ill-posedness in the sense of Hadamard.

Proof of Corollary

The proof of the corollary is based on comparing the variable coefficient problem with a reduced frozen coefficient system. The main tool is the following lemma. It is based on generalizing an argument of (Strang 1966).

Lemma: Frozen Coefficient Implications

Freeze the the coefficients of the principal part of (14) and ignore the lower order terms to get the reduced, constant coefficient system

$$\begin{aligned}
\frac{d\dot{x}^T}{dt} &= -\frac{\gamma\sigma^2}{2}H(\dot{x}^N_\alpha)\\
\frac{d\dot{x}^N}{dt} &= \frac{1}{2}H(\dot{\tilde{\gamma}}) - \frac{\gamma\sigma^2}{2}H(\dot{x}^T_\alpha)\\
\frac{d\dot{\tilde{\gamma}}}{dt} &= -A\gamma\sigma^2\dot{\tilde{\gamma}}_\alpha + A\gamma^2\sigma^4\dot{x}^T_{\alpha\alpha} + \tau\sigma^3\dot{x}^N_{\alpha\alpha\alpha}
\end{aligned} \tag{17}$$

If the Fourier modes of the solution to this problem ever exhibit unbounded growth in wavenumber, then the variable coefficient problem (14) is necessarily ill-posed.

Of course this lemma holds much more generally. The proof is now as follows. Consider, the system (17) where the coefficients are evaluated at α^*, t^*. If the Atwood number satisfies $-1 \leq A < 1$, then the classical formula (11) implies unbounded growth in wavenumber for this reduced system since $\gamma(\alpha^*, t^*) \neq 0$. Hence, by the lemma, the full variable coefficient problem is ill-posed. Notice that unless $z(\alpha, 0) \equiv \alpha$ and $\gamma(\alpha, 0) \equiv 0$ such an α^* and t^* will exist. This completes the discussion of the proof. The details will appear elsewhere (BHL4).

We now turn to the proof of the linearization lemma. The proof is rather long and computational. However, we illustrate how to identify the most important terms. We developed this approach in our recent study of water waves. This is a very general procedure and is useful in other contexts (BHL3; Lowengrub, Shelley & Hou). We begin with a well-known result about Hilbert transforms. We present it here as it captures the spirit of our computations.

Hilbert Transform Lemma

Let f be a smooth function. Then, for any function g, say in L^2, we get

$$H(fg) = fH(g) + E_{-\infty}(g) \tag{18}$$

That is, smooth functions can be passed in and out of Hilbert transforms at the minor cost of introducing smoothing operators.

Proof of Lemma:

Observe that

$$\begin{aligned} H(fg) &= \frac{1}{\pi} P.V. \int \frac{f(\alpha')g(\alpha')}{\alpha - \alpha'} d\alpha' \\ &= \frac{f(\alpha)}{\pi} P.V. \int \frac{g(\alpha')}{\alpha - \alpha'} d\alpha' + \frac{1}{\pi} P.V. \int g(\alpha') \left[\frac{f(\alpha') - f(\alpha)}{\alpha - \alpha'}\right] d\alpha' \end{aligned} \tag{19}$$

Since f is smooth, the second integral in (19) is a smoothing operation on g so that we are justified to write

$$\frac{1}{\pi} P.V. \int g(\alpha') \left[\frac{f(\alpha') - f(\alpha)}{\alpha - \alpha'}\right] d\alpha' = E_{-\infty}(g) \tag{20}$$

which completes the proof of the lemma.

For clarity, we present only the linearization of the velocity integral. This linearization, along with the Hilbert transform lemma, contains the crux of our method. The linearization of the velocity integral is technically simpler than the linearization for the vorticity equation, which contains more terms, and will be presented elsewhere (BHL4).

Proof of Velocity Linearization:

It is straightforward to see that

$$\frac{d\bar{\dot{z}}}{dt} = \frac{1}{2\pi i} P.V. \int \frac{\dot{\gamma}(\alpha')}{z(\alpha) - z(\alpha')} d\alpha' - \frac{1}{2\pi} P.V. \int \frac{\gamma(\alpha')(\dot{z}(\alpha) - \dot{z}(\alpha'))}{(z(\alpha) - z(\alpha'))^2} d\alpha' \tag{21}$$

For a justification, see (BHL1). Consider the first integral

$$\begin{aligned} I_1 &= \frac{1}{2\pi i} P.V. \int \frac{\dot{\gamma}(\alpha')}{z(\alpha) - z(\alpha')} d\alpha' \\ &= \frac{1}{2iz_\alpha} H(\dot{\gamma}) + \frac{1}{2\pi i} \int \dot{\gamma}(\alpha') \left[\frac{1}{z(\alpha) - z(\alpha')} - \frac{1}{z_\alpha(\alpha)(\alpha - \alpha')}\right] d\alpha' \\ &= \frac{1}{2iz_\alpha} H(\dot{\gamma}) + \frac{1}{2\pi i} \int \dot{\gamma}(\alpha') G_1(\alpha, \alpha') d\alpha' \end{aligned} \tag{22}$$

where G_1 is defined by

$$G_1(\alpha, \alpha') = \frac{1}{z(\alpha) - z(\alpha')} - \frac{1}{z_\alpha(\alpha)(\alpha - \alpha')} \tag{23}$$

Since z is smooth and $|z_\alpha| \geq c_o > 0$, $G_1(\alpha, \alpha')$ is smooth in α, α'. Thus, integration against G_1 is a smoothing operation and we are justified in calling

$$\frac{1}{2\pi i}\int \dot{\gamma}(\alpha')G_1(\alpha,\alpha')d\alpha' = E_{-\infty}(\dot{\gamma}) \tag{24}$$

The second integral is handled similarly.

$$\begin{aligned} I_2 &= \frac{1}{2\pi i}P.V.\int \frac{\gamma(\alpha')(\dot{z}(\alpha)-\dot{z}(\alpha'))}{(z(\alpha)-z(\alpha'))^2}d\alpha' \\ &= \frac{\gamma}{2iz_\alpha^2}H(\dot{z}_\alpha) + \frac{1}{2\pi i}P.V.\int(\dot{z}(\alpha)-\dot{z}(\alpha'))\left[\frac{\gamma(\alpha')}{(z(\alpha)-z(\alpha'))^2} - \frac{\gamma(\alpha)}{z_\alpha^2(\alpha)(\alpha-\alpha')^2}\right] \\ &= \frac{\gamma}{2iz_\alpha^2}H(\dot{z}_\alpha) + \frac{1}{2\pi i}P.V.\int(\dot{z}(\alpha)-\dot{z}(\alpha'))G_2(\alpha,\alpha')d\alpha' \end{aligned}$$

where G_2 is defined by

$$G_2(\alpha,\alpha') = \frac{\gamma(\alpha')}{(z(\alpha)-z(\alpha'))^2} - \frac{\gamma(\alpha)}{z_\alpha^2(\alpha)(\alpha-\alpha')^2} \tag{27}$$

and in (25) we have used the fact that

$$\partial_\alpha H(\dot{z}) = H(\dot{z}_\alpha) = \frac{1}{\pi}P.V.\int\frac{\dot{z}(\alpha)-\dot{z}(\alpha')}{(\alpha-\alpha')^2}d\alpha' \tag{28}$$

Since $G_2(\alpha,\alpha')$ has a singularity of $O(1/(\alpha-\alpha'))$ near $\alpha=\alpha'$, integration against G_2 has the same order of singularity as the Hilbert transform. This is a bounded operation (Stein & Weiss 1971) and so we are justified to write

$$\frac{1}{2\pi i}P.V.\int(\dot{z}(\alpha)-\dot{z}(\alpha'))G_2(\alpha,\alpha')d\alpha' = E_o(\dot{z}) \tag{29}$$

Putting (22)-(29), together with (21), we get

$$\frac{d\bar{\dot{z}}}{dt} = \frac{1}{2iz_\alpha}H(\dot{\gamma}) - \frac{\gamma}{2iz_\alpha^2}H(\dot{z}_\alpha) + E_o(\dot{z}) + E_{-\infty}(\dot{\gamma}) \tag{30}$$

Now, we put (30) in the tangential and normal coordinates of the underlying sheet z. Define tangent and normal vectors and the corresponding components of the perturbation as before. Notice that

$$\begin{aligned} \bar{\dot{z}}\cdot z_\alpha &= \frac{1}{\sigma}(\dot{x}^T - i\dot{x}^N) && (31) \\ \dot{z}_\alpha\cdot\bar{z}_\alpha &= \frac{1}{\sigma}\left[\dot{x}_\alpha^T + i\dot{x}_\alpha^N + N_\alpha\dot{x}^N + iT_\alpha\dot{x}^T\right] && (32) \\ &= \frac{1}{\sigma}\left[\dot{x}_\alpha^T + i\dot{x}_\alpha^N\right] + E_o(\dot{x}^T + \dot{x}^N) && (33) \end{aligned}$$

Similarly,

$$\frac{d\bar{\dot{z}}}{dt} \cdot z_\alpha = \frac{d(\bar{\dot{z}} \cdot z_\alpha)}{dt} - \dot{z}\frac{dz_\alpha}{dt} = \frac{1}{\sigma}\left(\frac{d\dot{x}^T}{dt} - i\frac{d\dot{x}^N}{dt}\right) + E_o(\dot{z}) \tag{34}$$

using (31). Now, multiply (30) by z_α, use (31) -(34) and split the result into real and imaginary parts to yield

$$\frac{d\dot{x}^T}{dt} = -\frac{\gamma\sigma^2}{2}H(\dot{x}_\alpha^N) + E_o(\dot{x}^T + \dot{x}^N) + E_{-\infty}(\dot{\gamma}) \tag{35}$$

$$\frac{d\dot{x}^N}{dt} = \frac{\sigma}{2}H(\dot{\gamma}) - \frac{\gamma\sigma^2}{2}H(\dot{x}_\alpha^T) + E_o(\dot{x}^T + \dot{x}^N) + E_{-\infty}(\dot{\gamma}) \tag{36}$$

This completes the velocity linearization.

We now consider the well-posedness of the linear equations (14). For presentational purposes, we consider the nonperiodic case. The periodic case can, in some sense, be treated as a special case of the nonperiodic problem. This will be presented elsewhere (BHL4; BHL5). We will need the following properties of the underlying smooth solution.

Necessary Properties of Smooth Solution:

(1). *The sheet can not self-intersect, i.e.* $z(\alpha, t) \neq z(\alpha', t)$ for $\alpha \neq \alpha'$

(2). $z(\alpha, t) - \alpha$ *is continuous in time with values in* H^m *with* $m \geq 4$.

(3). $z_\alpha(\alpha, t) \neq 0$ *for all* α, t.

These are rather mild assumptions about the underlying solutions and they have the following important consequences. Property (2) implies that

$$\partial_\alpha^j(z(\alpha, t) - \alpha) \to 0 \quad as \quad |\alpha| \to \infty, \quad j \leq 2 \tag{37}$$

Properties (2) and (3) together with (37) imply the existence of c_o and C_o such that

$$|z_\alpha(\alpha, t)| \geq c_o > 0 \quad and \quad |z(\alpha, t) - z(\alpha', t)| \geq C_o|\alpha - \alpha'| \tag{38}$$

for all α, α', t. This allows us to use the Linearization lemma for this data. We are now ready to present our main result of this section.

Theorem: Linear Well-Posedness Far from Equilibrium

Let the smooth underlying solution z, γ *satisfy properties* (1)-(3). *Then, we conclude that for* $0 \leq t \leq T$, *there exists a constant* M *such that the perturbation solutions of* (14) *satisfy*

$$\begin{aligned}&\|\dot{x}^T(\cdot, t)\|_{H^{3/2}} + \|\dot{x}^N(\cdot, t)\|_{H^{3/2}} + \|\dot{\gamma}(\cdot, t)\|_{L^2} \leq \\ &\qquad M\left(\|\dot{x}^T(\cdot, 0)\|_{H^{3/2}} + \|\dot{x}^N(\cdot, 0)\|_{H^{3/2}} + \|\dot{\gamma}(\cdot, 0)\|_{L^2}\right)\end{aligned} \tag{39}$$

Moreover, the wavenumbers at which the dispersive effect of surface tension is felt include those k *such that* $|k| > \max_\alpha \{\gamma^2\sigma/4\tau\}$.

Our result for the dispersive wavenumbers is a straightforward generalization of the near linear result corresponding to $A = 0$, where A is the Atwood number. See the formula (11). The $A = 0$ case is the most singular and consequently our result can be interpreted as a worst case scenario. Notice that the arclength metric plays an important role as well. We see that vortex sheet stretching may greatly enhance or diminish the effects of surface tension in time.

The proof of the theorem follows from several manipulations of the equations (14) and energy estimates. The estimates are obtained by correctly balancing the terms. The equations are split into high and low wavenumber components. In the high modes, surface tension dominates, but in the low modes, the Kelvin-Helmholtz and/or the Rayleigh-Taylor instabilities dominate. This is a necessary and physical approach as the solutions, themselves, behave differently in these regimes. We are able to partially localize the equations by introducing Hilbert transforms as variables (see also (Caflisch & Semmes 1990)). The details of the proof will be presented elsewhere (BHL4). However, we indicate the way in which the dispersive wavenumbers are found.

Discussion of Proof:

The equations can be partially localized as follows. Let new variables be

$$\begin{aligned} \dot{\phi} &= \dot{x}_\alpha^T \\ \dot{\psi} &= H(\dot{x}_\alpha^N) \\ \dot{\Gamma} &= \dot{\Gamma} \end{aligned} \tag{40}$$

Then, (14) becomes

$$\frac{d\dot{\phi}}{dt} = -\frac{\gamma\sigma^2}{2}\dot{\psi}_\alpha + E_o(\dot{\phi}+\dot{\psi}) + E_{-\infty}(\dot{\Gamma}) \tag{41}$$

$$\frac{d\dot{\psi}}{dt} = -\frac{1}{2}\dot{\Gamma}_\alpha + \frac{\gamma\sigma^2}{2}\dot{\phi}_\alpha + E_o(\dot{\phi}+\dot{\psi}) + E_{-\infty}(\dot{\Gamma}) \tag{42}$$

$$\frac{d\dot{\Gamma}}{dt} = -A\gamma\sigma^2\dot{\Gamma}_\alpha + A\gamma^2\sigma^4\dot{\phi}_\alpha - \tau\sigma\partial_\alpha(\sigma\partial_\alpha\sigma H(\dot{\psi})) + E_o(\dot{\phi}+\dot{\psi}+\dot{\Gamma}) \tag{43}$$

Notice that the Kelvin-Helmholtz instability shows up as the first term in (41) and the second term in (42). These terms, together with the time derivatives, form an elliptic system for $\dot{\phi}$ and $\dot{\psi}$ in α, t. This is the worst case as we are ignoring the role of the Atwood number. We know that for special Atwood numbers ($A = \pm 1$), the Kelvin-Helmholtz instability disappears even without surface tension (BHL1). However, our point of view is that if we control this case, then certainly the less singular cases are controlled as well. By changing variables again several times, we can move the instability to the third equation. Writing the final variables as $\tilde{\dot{\phi}}, \tilde{\dot{\psi}}$ and $\tilde{\ddot{\Gamma}}$, we get

$$\frac{d\tilde{\dot{\phi}}}{dt} = -\frac{\gamma\sigma^2}{2}\tilde{\dot{\psi}}_\alpha + E_o(\tilde{\dot{\phi}} + \tilde{\dot{\psi}}) + E_{-\infty}(\tilde{\ddot{\Gamma}}) \tag{44}$$

$$\frac{d\tilde{\dot{\psi}}}{dt} = -2\sigma\tilde{\ddot{\Gamma}}_\alpha - \frac{\gamma\sigma^2}{2}\tilde{\dot{\phi}}_\alpha - A\gamma\sigma^2\tilde{\dot{\psi}}_\alpha + E_o(\tilde{\dot{\phi}} + \tilde{\dot{\psi}}) + E_{-\infty}(\tilde{\ddot{\Gamma}}) \tag{45}$$

$$\frac{d\tilde{\ddot{\Gamma}}}{dt} = \tau\sigma\partial_\alpha\sigma\left[\frac{\gamma^2\sigma}{4\tau} - \Lambda\right]\tilde{\dot{\psi}} + E_o(\tilde{\dot{\phi}} + \tilde{\dot{\psi}} + \tilde{\ddot{\Gamma}}) \tag{46}$$

The operator $\Lambda = \partial_\alpha H$ and is positive definite, i.e. $\hat{\Lambda} = |k|$. Notice that the second term in (45) changed its sign from that in (42). This has the effect of changing the type of the system from elliptic to hyperbolic. Thus, the instability has been successfully put into the third equation and appears as the $\gamma^2\sigma/4\tau$ term that is added to the $-\Lambda$. The $-\Lambda$ term represents the dispersive effect of surface tension. Consequently, we see that the dispersive effect will dominate for those wavenumbers k satisfying $|k| > \max_\alpha \{\gamma^2\sigma/4\tau\}$. This completes our discussion of the proof.

4. Nonlinear Behavior and Numerical Results

In this section, we investigate the nonlinear behavior of interfacial flows with surface tension numerically. Our main interest is to understand the competition between nonlinear effects and the linearly regularizing effect of surface tension. It is possible that strong nonlinear interactions may yet lead to singularity formation.

The numerical computation of interfacial flows with surface tension is a difficult and challenging problem. As we saw on the continuous level, there is a delicate balance of terms that is required to prove the linear well-posedness. Part of the challenge is to design discrete schemes that maintain this balance. Not surprisingly, this requires special schemes. Previous investigators noticed numerical instabilities because their schemes did not preserve this balance (Pullin 1982; Rangel & Sirignano 1988).

We begin by presenting several numerical methods for computing two fluid interfacial flows. There are three main ingredients to the computation of these flows. First, there is the computation of the velocity integral. Second, there is a choice of which discrete derivative to use in the vorticity equation. Third, the integral equation for the time derivative of the vortex sheet strength must be solved.

(1). Discretizations of the Velocity Integral:

We now consider the case of periodic vortex sheets. We suppose that $z(\alpha + 2\pi, t) = 2\pi + z(\alpha, t)$. The velocity integral becomes

$$I[\gamma, z](\alpha) = \frac{1}{4\pi i} P.V. \int_0^{2\pi} \gamma(\alpha') \cot\left(\frac{z(\alpha) - z(\alpha')}{2}\right) d\alpha' \tag{47}$$

We consider the following discretizations of this integral:

(a). Point Vortex Approximation

The point vortex approximation is the oldest quadrature rule for this problem and dates to (Rosenhead 1932). It is just the trapeziodal rule applied to I, omitting the self-induction term. It is O(h) accurate.

$$I_j^{pva}[\gamma, z] = \frac{h}{4\pi i} \sum_{k \neq j} \gamma_k \cot\left(\frac{z_j - z_k}{2}\right) \tag{48}$$

where z_j approximates $z(jh, t)$.

(b). Singularity Subtraction Approximation

This method provides a $O(h)$ correction to the point vortex approximation which makes this quadrature rule spectrally accurate. It has been used in this and other forms by (Van de Vooren 1980; Moore 1981; Baker 1980; Pullin 1982).

$$I_j^{ssa}[\gamma, z] = \frac{h}{4\pi i} \sum_{k \neq j} \gamma_k \cot\left(\frac{z_j - z_k}{2}\right) + \frac{h}{4\pi i} \frac{\gamma_j z_{\alpha\alpha j} - 2\gamma_{\alpha j} z_{\alpha j}}{2 z_{\alpha j}^2} \tag{49}$$

(c). Modified Point Vortex Approximation

This method is a spectrally accurate discretization of the velocity integral. Unlike the singularity subtraction method, no derivatives of the sheet position or strength are required. It is essentially the midpoint rule on a grid size of $2h$ and arises from one Richardson extrapolation of the singularity subtraction approximation. It has been used in this context by (Roberts 1983; Baker 1983; Shelley 1992).

$$I_j^{mpva}[\gamma, z] = \frac{h}{4\pi i} \sum_{k+j \ odd} \gamma_k \cot\left(\frac{z_j - z_k}{2}\right) \tag{50}$$

We now turn to the type of derivatives that may be used.

(2). Types of Derivatives

(a). Second order differences: $D_2 z_j = \frac{z_{j+1} - z_{j-1}}{2h}$.

(b). Cubic spline derivatives: $D_3 z_j$, see (Stoer & Burlisch 1980).

(c). Fourth order differences: $D_4 z_j$ see (Stoer & Burlisch 1980).

(d). Spectral derivatives: $D_s z_j$ see (Brigham, 1974).

Finally, the integral equation is easily inverted by iteration, as it is contractive, and $d\gamma/dt$ is integrated in time to find γ. Now, for a given quadrature rule $I^{(\cdot)}$ and a discrete derivative $D_{(\cdot)}$, the general semi-discrete numerical discretization of the periodic form of (4) and (5) is given by

$$\frac{d\tilde{z}_j}{dt} = I_j^{(\cdot)}[\tilde{\gamma}, \tilde{z}] \tag{51}$$

$$\begin{aligned}\left(\frac{d\tilde{\gamma}_j}{dt}\right)^{n+1} &= -2A\left[Re\left\{I_j^{(\cdot)}[\left(\frac{d\tilde{\gamma}}{dt}\right)^n, \tilde{z}]D_{(\cdot)}z_j\right\}\right] \\ &\quad -2A\left[R_j^{(\cdot)}[\tilde{\gamma}, \tilde{z}] + \frac{1}{8}D_{(\cdot)}\left(\frac{\tilde{\gamma}_j^2}{|D_{(\cdot)}\tilde{z}_j|^2}\right) + gD_{(\cdot)}\tilde{y}_j\right] + \tau D_{(\cdot)}\tilde{\kappa}_j\end{aligned} \tag{52}$$

The curvature is evaluated by

$$\kappa_j = \frac{D_{(\cdot)}\tilde{x}_j D_{(\cdot)}^2\tilde{y}_j - D_{(\cdot)}^2\tilde{x}_j D_{(\cdot)}\tilde{y}_j}{|D_{(\cdot)}z_j|^3} \tag{53}$$

and D^2 means the derivative operator is applied twice. By R in equation (52) we mean the remaining terms from $Re\left\{(d^2\bar{\tilde{z}}/dt^2)D_{(\cdot)}\tilde{z}_j\right\}$ which do not depend on $d\tilde{\gamma}/dt$. These terms involve integrals which are evaluated with the same quadrature rule as the velocity. The terms $(d\tilde{\gamma}/dt)^{n+1}$ and $(d\tilde{\gamma}/dt)^n$ refer to the level of the iteration.

Finally, we define the modified versions of M.P.V.A., S.S.A., P.V.A. to be M.P.V.A.*, S.S.A.* and P.V.A.* respectively. In M.P.V.A.*, S.S.A.* and P.V.A.*, the discretizations M.P.V.A., S.S.A. and P.V.A. are still used, but the Nyquist frequency (the $N/2$, and highest, discrete Fourier mode) is set to zero thoughout the full approximation (51) and (52). It is is not unusual to remove the $N/2$ mode in spectral methods as it is dominated by aliasing errors anyway (Brigham, 1974). Of course, M.P.V.A.* and S.S.A.* with D_s are still spectrally accurate. Our main analytical result, in this section, is the following theorem.

Theorem: Numerical Stability and Convergence

Let the initial data for the numerical methods described in (51)-(53) *be* $z(jh,0)$ *and* $\gamma(jh,0)$ *such that their corresponding solutions to the spatially continuous problem* $z(\alpha,t)$ *and* $\gamma(\alpha,t)$ *satisfy properties* (1)-(3) *of the previous section. Then, our results are summarized by the following table.*

Numerical Stability and Convergence Properties of Different Algorithms			
	P.V.A. & P.V.A.*	M.P.V.A. & S.S.A.	M.P.V.A.* & S.S.A.*
D_2	lin. unstable	lin. unstable	lin. unstable
D_3	lin. unstable	lin. unstable	lin. unstable
D_4	lin. unstable	lin. unstable	lin. unstable
D_s	lin. unstable	lin. unstable	nonlin. stable and conv.

Thus, almost all methods are numerically unstable, at least linearly! Further, no additional numerical filtering (for example, see (Krasny 1986b; Shelley 1992)) is required to prove the convergence of the S.S.A.* and M.P.V.A.* with D_s. Only the $N/2$ mode needs to be removed.

We make some remarks here. First, although the removal of the $N/2$ mode is a very mild numerical filtering, it differs from other filtering, or regularizing, schemes in an important way. In a typical scheme, the maximum growth rate is bounded independently of physical parameters such as N, at least for fixed times. The growth rate is actually bounded by artificial parameters that are introduced by the scheme. However, simply removing the $N/2$ mode does **not** introduce such an artificial parameter. The effective growth rate depends entirely on the physical parameters, see (Lowengrub 1992) for a discussion of filtering. See also (Baker & Nachbin 1992) for alternative methods to M.P.V.A.* and S.S.A.*

Second, the convergence of the S.S.A.* and M.P.V.A.* quadrature rules with the spectral derivative D_s is obtained in the discrete analogues of the continuous Sobolev norms of the linear well-posedness theorem. The exact details will be presented elsewhere (BHL5).

Third, neither the initial data, nor the time dependent solution, is assumed to be near equilibrium. This is again a far from equilibrium result, in fact it is a nonlinear one for the S.S.A.* and M.P.V.A.* with D_s. It is nonlinear in the sense that we are able to show the nonlinear numerical stability of each of these methods.

Fourth, we note that (Baker & Nachbin 1992) independently demonstrated these numerical instabilities near equilibrium for the case of vortex sheets with surface tension ($A = 0$). It turns out that a near equilibrium analysis is actually sufficient to show the linear instability in the general case. This is seen as follows. As in the continuous case, the near equilibrium discrete linearization is essentially the frozen coefficient version of the principal part of the general discrete linearization. Thus, the frozen coefficient lemma of the previous section can be adapted to the discrete case to imply the general linear instability. Again, the details of this will be presented elsewhere (BHL5).

Discussion of Proof:

We give a heuristic discussion of the ideas of the proof. First, the numerical instability arises from the following facts. The surface tension is stabilizing at the high modes. And, surface tension arises through differentiation. The numerical derivatives $D_2, \ldots, D_4$ damp the high modes artificially. For some function g, we have

$$\widehat{D_{(\cdot)} z}_k = ik\hat{z}_k g(k) \tag{54}$$

where by $\widehat{(\cdot)}_k$ we mean the discrete Fourier transform. The function g satisfies $g(0) = 1$ and $g(k) \to 0$ as $k \to N/2$. Therefore, the effect of surface tension is actually lost at the highest modes using these derivatives. This upsets the balance of terms! In the case of D_s, the symbol $g(k) = 1$ if $k \neq N/2$, but due to aliasing errors $g(N/2) = 0$. Thus, with D_s, *only* at the highest mode, is the effect of surface tension actually lost. This is why the modified schemes are necessary. The band of instability for $D_2, \ldots, D_4$ is too much to overcome by simply removing the $N/2$ mode. Traditional filtering methods can by used to make these schemes $D_2, \ldots, D_4$ converge though (Hou, Lowengrub & Caflisch)! So, for the unmodified and lower order schemes with $D_2, \ldots, D_s$, there is a actually a high mode instability present that looks like a discrete form of the Kelvin-Helmholtz instability! It is **not** physical though. It is purely numerical. So, the only derivative that can be expected to work is the spectral derivative D_s where the $N/2$ mode is removed from the entire calculation. It turns out that the role of the quadrature rule is very important as well. In fact, the P.V.A. has the effect of artificially damping the highest modes also. That is why it is unstable even if it is modified and combined with the spectral derivative!

The linear stability of the S.S.A.* and M.P.V.A.* are obtained in an analogous fashion to that in the continuous case. It is crucial though, that the discrete symbol of the discretized Hilbert transform be $-isgn(k)$; exactly what it is in the continuous case. This is also necessary to maintain the correct balance of terms. This is satisfied by the S.S.A. and M.P.V.A. quadrature rules. It is also crucial as we have seen above, that the methods be modified and the spectral derivative D_s be used. The nonlinear stability of the scheme is obtained straightforwardly as these methods are spectrally accurate. We use the smallness of the error to control the nonlinear terms. This completes the discussion of the proof.

We now present some numerical computations. We remark that the presence of the surface tension term makes the time discretization quite stiff. There are constraints of the form $\tau\sigma^3\Delta t \le h^3$, $\tau\kappa_\alpha\Delta t \le h$ and $\tau\kappa^2\Delta t \le h$ when explicit time discretizations are used. Thus, for large κ and κ_α, the time step must be controlled as well. We use fourth order Runge-Kutta time integration with an adaptive time step. The adaptivity depends on the size of the interfacial curvature. In the following, the time step has been taken small enough so that time discretization errors are negligible. In some cases, as we will see, the time step becomes prohibitively small due to the adaptivity and stiffness. With one exception, all of the following computations will be for $A = 0$ (vortex sheet with surface tension). This is a good test problem as it contains the essential features of these flows, and, it is particularly convenient numerically as there is no integral equation to solve. $d\gamma/dt$ is determined explicitly in this case. Further, our computations are on $[0, 1]$

rather than $[0, 2\pi]$. This essentially rescales the effective surface tension by a factor of 2π. Finally, the initial data for our computations is

$$\begin{aligned} x(\alpha, 0) &= \alpha + .01 sin(2\pi\alpha) \\ y(\alpha, 0) &= -.01 sin(2\pi\alpha) \\ \gamma(\alpha, 0) &= 1 \end{aligned} \tag{55}$$

This corresponds to an unstable eigenfunction of the linear system near equilibrium with $\tau = 0$. It has been used by Krasny and others.

In figure 1, we demonstrate the numerical instability by presenting a computation of the M.P.V.A. with $D_2, \ldots, D_s$. The surface tension is $\tau = .1$. The logarithm of the power spectrum of the y component, $\log|\hat{y}_k|$, is shown at $t = .12$. A linear profile with negative slope in $\log|\hat{y}_k|$ vs. k corresponds to exponential decay of $|\hat{y}|$ in k. Notice the spurious growth in the high modes. The 10-mode to the 20-mode of D_2 and D_4 have begun to become polluted. D_3 and D_s still agree in this region. All the schemes agree for modes 1-10. This is clear evidence of a *numerical* instability as we are well within the linear regime. At this short time we know analytically that the sheet is stable. Figure 1 is contrasted with figure 2 which shows the time evolution of $\log|\hat{y}_k|$ the M.P.V.A.* and D_s for $N = 128$. The different curves correspond to $\log|\hat{y}_k|$ at different times. The time step of the calculation in figures 1 and 2 is 1×10^{-3} The last time shown in figure 2 is actually $t = .2$. It is clear that the latter method is stable.

We now turn to asking **physical** questions. First, are stably stratified flows with $\tau = 0$ physically unstable? Second, do vortex sheets with surface tension form singularities in finite time? And third, what is the limiting behavior of vortex sheets with surface tension in the limit of small τ?

Consider the stably stratified case. In figure 3, we show $\log|\hat{\gamma}_k|$ for the case $A = .8$ and $\tau = 0$ using the M.P.V.A. and D_s at $t = .05$ and $t = .1$. $N = 256$ and $g = 9.8$. Notice that since $\tau = 0$, there is no reason to use M.P.V.A.*. There is clear evidence that this flow is physically unstable as *all* the high modes appear to grow unboundedly with wavenumber. If just the numerical method was unstable, we would expect to see *only* the highest modes growing so fast since the time of the computation is so short (recall figure 1). This is contrasted with figure 4, which is the same calculation except with numerical filtering. The filtering scheme eliminates all Fourier modes that are less than a certain minimum level (Krasny 1986b; Shelley 1992). In this calculation, the filter level is 10^{-14}. The last plot is at $t = .15$. The numerical filtering has suppressed the physical instability. This is typical for ill-posed problems.

We now consider the long time behavior of vortex sheets with surface tension. We present a series of computations with successively smaller surface tensions. We begin by presenting a calculation, for reference, of a vortex sheet without surface tension ($\tau = 0$). Figures 5-8 show the time evolution of z, $\log|\hat{y}|$, κ and $(max_\alpha \kappa)^{-1}$ respectively. On each figure, the different curves represent the appropriate numerical quantities at different times. This computation uses M.P.V.A. with $N = 128$, $\Delta t = 8 \times 10^{-5}$ and $0 \leq t \leq .366$. Numerical filtering is used at a level of 10^{-14}. Two peaks form in the curvature, but the width between them is zero. This, combined with the spectral plot, indicates that a single singularity forms in its curvature.

Consider the case now of a relatively large surface tension $\tau = .1$. The remaining computations use M.P.V.A.* with D_s. Figures 9-11 show the time evolution of z, γ and κ respectively. N=64, $\Delta t = 1 \times 10^{-3}$, and $0 \leq t \leq 2$. This is a completely resolved computation. Notice that the sheet exhibits wave-like behavior. The curvature is uniformly bounded and it seems very unlikely that a singularity will form at some later time.

Consider now $\tau = .075$. Figures 12-14 show z, γ and κ at the final time $t = 2$. Notice that two well-spearated peaks have formed in γ and κ. This is different from the $\tau = 0$ case where the peaks seen in κ have merged! The time evolution of $\log|\hat{\gamma}_k|$ is shown in figure 15. The exponential decay (of $|\hat{\gamma}|$) is clearly being lost in time. Further, the oscillations in the spectrum indicate the presence of two singularities in the analytic extension of z into the upper half of the complex α-plane (Carrier, Krook & Pearson 1966). This is opposed to the single singularity in the case of $\tau = 0$. In figures 16-17 we show the time evolution of $(max_\alpha \kappa)^{-1}$ and $\log(max_\alpha \kappa)^{-1}$. Figure 16 indicates that the curvature is blowing up, but the straight line in figure 17 indicates that this blow up is only exponential in time. There seems to be no finite time singularity for this flow either.

Clearly there is transition in behavior between $\tau = .1$ and $\tau = .075$. This is explained as follows. For $\tau = .1$, from linear theory, all modes are stable. However, for $\tau = .075$, there are several unstable modes. These modes contribute to the growth of the curvature.

Lastly, consider the case $\tau = .01$. Figure 18 shows the time evolution of z. By the last time $t = .526$, the interface has turned over slightly, perhaps to form the beginning of a spiral. The point positions of z at this time are shown in figure 19. In this calculation, N=256 and the initial time step is $\Delta t = 2 \times 10^{-4}$. It shrinks to $\Delta t = 1 \times 10^{-8}$ by the end of the computation! Filtering is used at a level of 10^{-14} to ensure greater numerical stability as the surface tension is so small. Actually, spatial filtering also serves to increase the stability region for time discretizations (Hou, Lowengrub & Caflisch). Figures 20-21 show γ and κ at $t = .526$. Notice how peaked the solutions are and that the peaks have moved closer together. Figure 22 shows the time evolution of $\log|\hat{\gamma}_k|$. The 'o' denotes the results from a $N = 128$ calculation. There is spectral accuracy even at the last time, $t = .526$. Notice that the frequency of the oscillations has increased indicating further that the singularities are closer together. The exponential decay of $|\hat{\gamma}_k|$ is being lost rapidly in time. Figures 23 shows the time evolution of $(max_\alpha \kappa)^{-1}$. It appears as if κ will blow up in finite time, but at the last moment its rate of increase slows (decrease in $(max_\alpha \kappa)^{-1}$). Figure 24 shows the logarithm of the last part of figure 24. The circles indicate the results from a $N = 128$ computation. The drop at the end indicates that resolution is lost. It is a good sign that the $N = 256$ computation survives for a slightly longer time than $N = 128$. This is evidence that a singularity hasn't yet formed in the physical solution for real α. The fact that the plot in figure 24 is so close to linear, again seems to indicate that the growth in κ is again only exponential in time and there is no finite time singularity.

Consequently, in answer to our questions, it does not appear that vortex sheets with surface tension form singularities in a finite time. Of course, to be really confident of this, we will have to do even longer time computations. This involves finding ways to deal with the stiffness. This is currently under study (Lowengrub, Shelley & Hou).

The solutions with surface tension exhibit two separated peaks in γ and κ. This is different from the case with $\tau = 0$ where there is no distance between the peaks. This indicates that the one singularity in the upper half of the complex α-plane, when $\tau = 0$, has split into two for $\tau > 0$. Like the $\tau = 0$ case, they travel towards the real α axis, but they do seem not to hit in finite time. The existence of two singularities is further evidenced by the oscillations in the spectrum of y seen in figures 15 and 22. In the limit of small τ, these singularities move closer together. It is difficult to accurately predict what will happen for even smaller surface tension on the basis of our current computations. More refined computations are necessary to do this. It seems possible though, that the singularities may merge as in the zero surface tension problem. We remark that Cowley has recently obtained some asymptotic results on the structure of the singularities in the complex α-plane for vortex sheets with surface tension (Cowley 1992).

5. Conclusions

We have investigated several aspects of two phase flows with surface tension. Using a general technique (BHL1; BHL2) to obtain the linearization of the equations of motion far from equilibrium, we demonstrated that these equations are well-posed. A special balance of terms was required to do this. Thus, even far from equilibrium, surface tension linearly regularizes the Kelvin-Helmholtz and Rayleigh-Taylor instabilities in a dispersive manner.

Led by the balance of terms necessary in the continuous case, we proved the nonlinear stability and convergence of two, spectrally accurate, numerical schemes to smooth solutions of the fully nonlinear problem. We showed that for a class of other schemes, this balance was lost and that even far from equilibrium, these methods were numerically unstable. (Baker & Nachbin 1992) independently demonstrated these instabilities near equilibrium and for vortex sheets with surface tension ($A = 0$). We demonstrated the instability numerically.

We also obtained the following numerical results. We showed that vortex sheet solutions, with our initial data and in the presence of surface tension, seemed to remain smooth in time and did not develop finite time singularities. It was difficult to maintain accuracy of the scheme due to the stiffness introduced by the surface tension term and the growing size of the curvature. This is currently under study and is certainly a direction for future progress. It would be very interesting to prove either a global existence theorem or conclusively demonstrate the existence of singularities. While there is the possibility that singularities may form in very long times, much longer than we have been able to compute (Siegel 1992), we believe that the solutions will remain smooth.

We showed that the singularity structure of the complex extension (unphysical) of the solutions differed from that in the zero surface tension case. Again, the stiffness makes it difficult to resolve these calculations. Future work should concentrate on determining the type of these complex singularities as well as developing new methods to determine the true limiting behavior of vortex sheets with surface tension. This is currently under study. In particular, it will be very interesting to discover the relationship between the limiting behavior of the vortex blob methods for small blobs (Krasny 1986a), the layer method for thin layers (Baker & Shelley 1990), and the small surface tension limit. Of course, this question is only interesting after the time the vortex sheet would normally form a singularity. If these limits differ, this would give a dynamic way to generate non-uniqueness, possibly that found by Pullin in the case of self-similar vortex sheets (Pullin 1989).

6. Acknowledgements

It is a pleasure to thank Michael Shelley for many interesting and fruitful discussions. It is also a pleasure to thank the organizers of this conference. The research of the first author was supported by N.S.F. Grant No. DMS-9102782. The second author was supported by a Sloan Foundation Research Fellowship, N.S.F. Grant No. DMS-9003202, the Air Force Office of Scientific Research under A.R.O. A.F.O.S.R. AFOSR-90-0090. He was also supported by a N.S.F. grant No. DMS-9100383 while at the Institute for Advanced Study. The third author was supported by a N.S.F. Postdoctoral Fellowship. The second and third authors are pleased to thank the hospitality of the Institute for Advanced Study in Princeton.

7. Figure Captions

Figure 1. $log|\hat{y}_k|$, for M.P.V.A. with $D_2 - D_s$ is shown at $t = .12$. $A = 0$, $\tau = .1$ and $\Delta t = 1 \times 10^{-3}$.

Figure 2. The time evolution of $log|\hat{y}_k|$ is shown for $N = 128$. $A = 0, \tau = .1$ and $\Delta t = 1 \times 10^{-3}$. Here $0 \leq t \leq .2$. This is using the M.P.V.A.* with D_s.

Figure 3. The time evolution of $log|\hat{\gamma}_k|$ is shown for $N = 256$. $A = .8, \tau = 0$, $g = 9.8$ and $\Delta t = 1 \times 10^{-3}$. The times shown are $t = .05, .10$. This is using the M.P.V.A. with D_s.

Figure 4. The time evolution of $log|\hat{\gamma}_k|$ is shown for $N = 256$. Numerical filtering is used with a filter level of 10^{-14}. $A = .8$, $\tau = 0$, $g = 9.8$ and $\Delta t = 1 \times 10^{-3}$. The times shown are $t = .05, .10, .15$. This is using the M.P.V.A. with D_s.

Figures 5-8. The time evolution of z, $log|\hat{y}_k|$, κ and $(max_\alpha \kappa)^{-1}$ are shown, respectively. $A = 0$, $\tau = 0$, $N = 128$, $\Delta t = 8 \times 10^{-5}$ and $0 \leq t \leq .366$. Numerical filtering was used at a level of 10^{-14}. This is using the M.P.V.A., no derivative is needed since $\tau = 0$.

Figures 9-11. The time evolution of z, γ and κ are shown, respectively. $A = 0, \tau = .1, N = 64$ and $\Delta t = 1 \times 10^{-3}$. Here $0 \leq t \leq 2$. This is using the M.P.V.A.* with D_s.

Figures 12-14. z, γ, and κ are shown at $t = 2$, respectively. $A = 0, \tau = .075$, $N = 128$ and $\Delta t = 2.5 \times 10^{-4}$. The 'o' denote the point positions. This is using the M.P.V.A.* with D_s.

Figures 15-17. The time evolution of $log|\hat{\gamma}_k|$, $(max_\alpha \kappa)^{-1}$ and $log(max_\alpha \kappa)^{-1}$ are shown, respectively, for the calculation in figures 12-14.

Figure 18. The time evolution of z is given. $A = 0$, $N = 256$, $\Delta t = 2 \times 10^{-4}$ for small times. At $t = .526$, the final time, $\Delta t = 1 \times 10^{-8}$. This is the M.P.V.A.* with D_s. Filtering is done at a level of 10^{-14}.

Figures 19-21. z, γ and κ are shown at $t = .526$ for the calculation in figure 18. The 'o' indicate point positions.

Figure 22. The time evolution of $log|\hat{\gamma}_k|$ is shown for the calculation in figure 18. The 'o' are the results of a $N = 128$ computation.

Figure 23-24. The time evolution of $(max_\alpha \kappa)^{-1}$ and $log(max_\alpha \kappa)^{-1}$ for the calculation in figure 18. Note that the time in figure 24 lies in the interval $.45 \leq t \leq .526$ and the 'o' denote the results from $N = 128$.

References

C. Anderson, *A Vortex Method for Flows with Slight Density Variations*, J. Comp. Phys., v. 61, 1985, p. 417.

G. Baker, *A Test of the Method of Fink and Soh for Following Vortex Sheet Motion*, J. Fluid Mech., v. 100, 1980, p. 209.

G. Baker, *Generalized Vortex Methods for Free Surface Flows*, in **Waves on Interfaces**, ed. R. Meyer, Academic Press, 1983.

G. Baker, D. Meiron and S. Orszag, *Generalized Methods for Free Surface Flow Problems*, J. Fluid Mech., v. 123, 1982, p. 477.

G. Baker, D. Meiron and S. Orszag, *Vortex Simulations of the Rayleigh-Taylor Instability*, v. 23, 1980, p. 1485.

G. Baker and A. Nachbin, *Stable Methods for Vortex Sheet Motion with Surface Tension*, preprint, 1992.

G. Baker and M. Shelley, *On the Connection between Thin Vortex Layers and Vortex Sheets*, J. Fluid Mech., v. 215, 1990, p. 161.

BHL1: J.T. Beale, T. Hou and J. Lowengrub, *Growth Rates for the Linear Motion of Fluid Interfaces Away from Equilibrium*, submitted to C.P.A.M., 1992.

BHL2: J.T. Beale, T. Hou and J. Lowengrub, *On the Convergence of Suitably Modified Spectrally Accurate Vortex Methods for Water Waves*, in preparation.

BHL3: J.T. Beale, T. Hou and J. Lowengrub, *On the Stability and Convergence of Suitably Modified Vortex Methods for Hele-Shaw Flows*, in preparation.

BHL4: J.T. Beale, T. Hou, and J. Lowengrub, *The Stability of Two Phase Flows with Surface Tension, Part 1: Growth Rates for the Linear Motion far From Equilibrium*, in preparation.

BHL5: J.T. Beale, T. Hou and J. Lowengrub *The Stability of Two Phase Flows with Surface Tension, Part 2: Convergence of Suitably Modified Vortex Methods*, in preparation.

G. Birkhoff, *Helmholtz and Taylor Instability*, in **Proc. Symp. Appl. Math.**, v. 13, 1962, p. 55.

E. Brigham, **The Fast Fourier Transform**, Prentice Hall, 1974.

R. Caflisch and S. Semmes, *A Nonlinear Approximation for Vortex Sheet Evolution and Singularity Formation*, Physica D, v. 41, 1990, p. 197.

G. Carrier, M. Krook and C. Pearson, **Functions of a Complex Variable**, Mcgraw-Hill, 1966.

A. Chorin and P. Bernard, *Discretization of a Vortex Sheet, with an Example of Roll-up*, J. Comp. Phys., v. 13, 1973, p. 423.

S. Cowley, private communication, 1992.

P. Drazin and W. Reid, **Hydrodynamic Stability**, Cambridge University Press, 1981.

G. Folland, **Introduction to Partial Differential Equations**, Princeton University Press, 1976.

T. Hou, J. Lowengrub and R. Caflisch, *Interfacial Instabilities, Numerical Filtering and Convergence*, in preparation.

D. Kellogg, **Foundations of Potential Theory**, Dover, 1929.

(a). R. Krasny, *Desingularization of Periodic Vortex Sheet Rollup*, J. Comp. Phys., v. 65, 1986, p. 65.

(b). R. Krasny, *On the Singularity Formation in a Vortex Sheet and the Point Vortex Approximation*, J. Fluid Mech., v. 167, 1986,

L. Landau and E. Lifshitz, **Fluid Mechanics**, Pergamon Press, 1959.

J. Lowengrub, *A Note on the Role of Numerical Filtering*, preprint.

J. Lowengrub, M. Shelley and T. Hou, *Removing the Stiffness from Interfacial Flows with Surface Tension*, in preparation.

D. Moore, *On the Point Vortex Method*, SIAM J. Sci. Stat. Comp., v. 2, 1981, p. 65.

D. Pullin, *On Similarity Solutions for the Self-Induced Motion of Two-Branched Vortex Sheets*, in **Mathematical Aspects of Vortex Dynamics**, ed. R. Caflisch, SIAM, 1989, p. 97.

D. Pullin, *Numerical Studies of Surface Tension effects in Nonlinear Kelvin-Helmholtz and Rayleigh-Taylor Instabilities*, J. Fluid Mech., v. 119, 1982, p. 507.

R. Rangel and W. Sirignano, *Nonlinear Growth of the Kelvin-Helmholtz Instability: Effect of Surface Tension and Density Ratio*, Phys. Fluids, v. 31, no. 7, 1988, p. 1845.

A. Roberts, *A Stable and Accurate Numerical Method to Calculate the Motion of a Sharp Interface Between Two Fluids*, I.M.A. J. Appl. Math., v. 31, 1983, p. 13.

L. Rosenhead, *The Point Vortex Approximation of a Vortex Sheet*, Proc. Roy. Soc. London Ser. A, v. 134, 1932, p. 170.

M. Shelley, *A Study of Singularity Formation in Vortex Sheet Motion by a Spectrally Accurate Method*, J. Fluid Mech., v. 244, 1992, p. 493.

M. Shelley and M. Vinson, *Coherent Structures on a Boundary Layer in Rayleigh-Bernard Turbulence*, Nonlinearity, v. 5, 1992, p. 323.

M. Siegel, private communication, 1992.

J. Stoer and R. Burlisch, Introduction to Numerical Analysis, Springer-Verlag, 1980.

E. Stein and G. Weiss, Introduction to Fourier Analysis on Euclidean Spaces, Princeton University Press, 1971.

G. Strang, *Necessary and Insufficient Conditions for Well-Posed Cauchy Problems*, J. Diff. Eq., v. 2, 1966, p. 107.

A. Van de Vooren, *A Numerical Investigation of the Rolling up of Vortex Sheets*, Proc. R. Soc. London Ser. A, v. 373, 1980, p. 67.

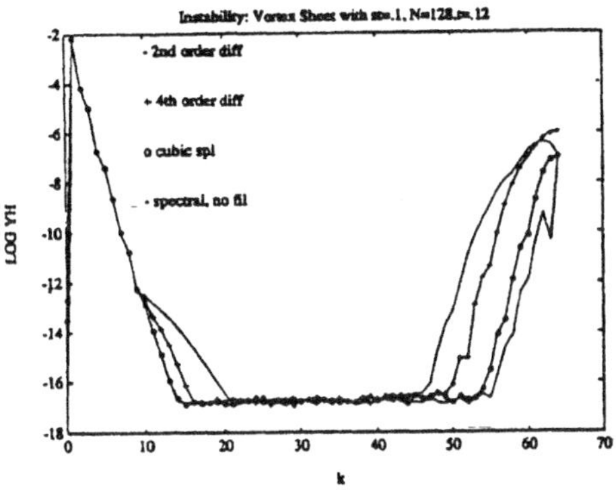

Figure 1

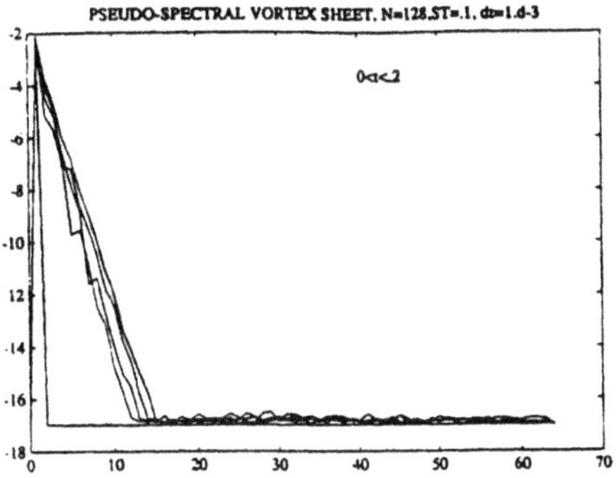

Figure 2

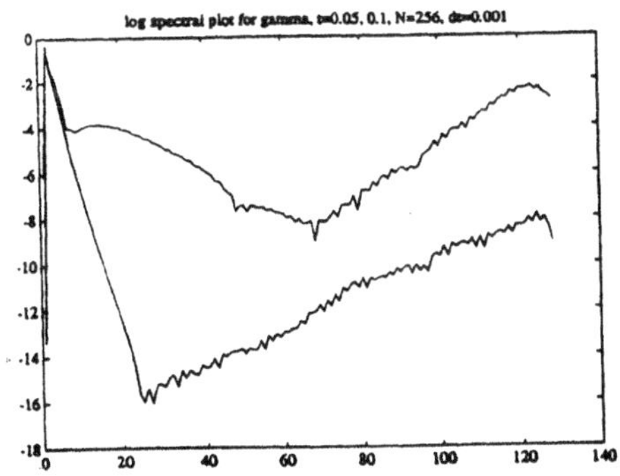

At=0.8, tau=0.0, spectral der., no filtering
Figure 3

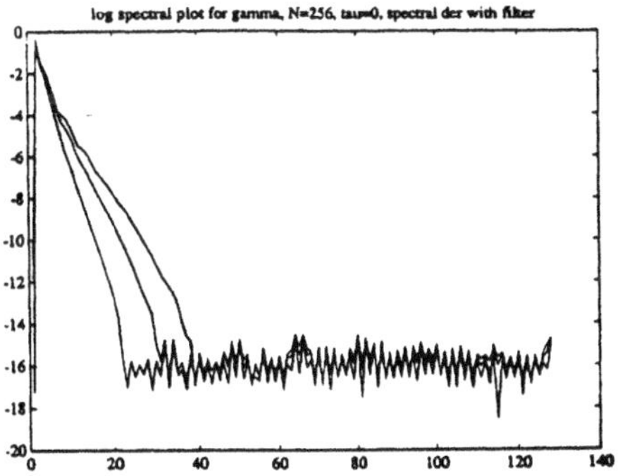

t=0.05, 0.1, 0.15, At=0.8, g=9.8
Figure 4

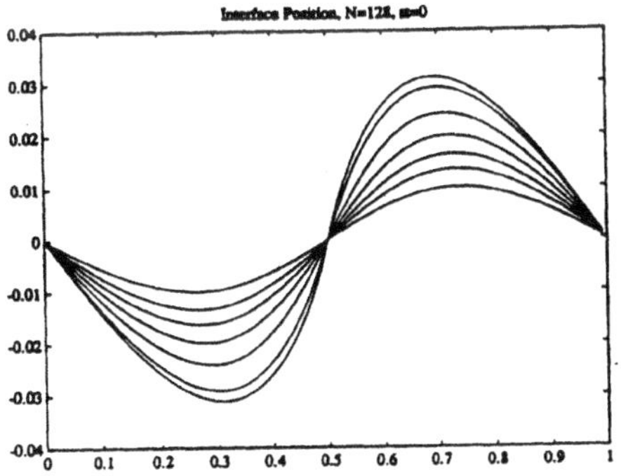

Figure 5

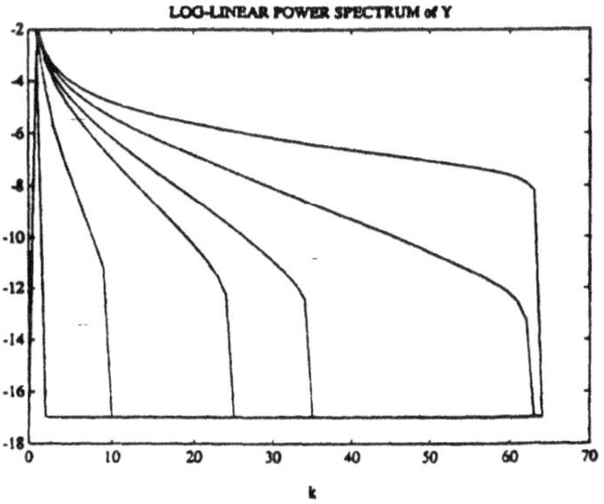

Figure 6

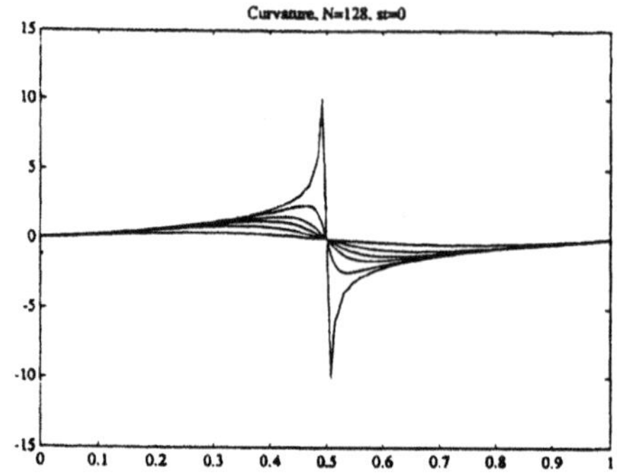

Figure 7

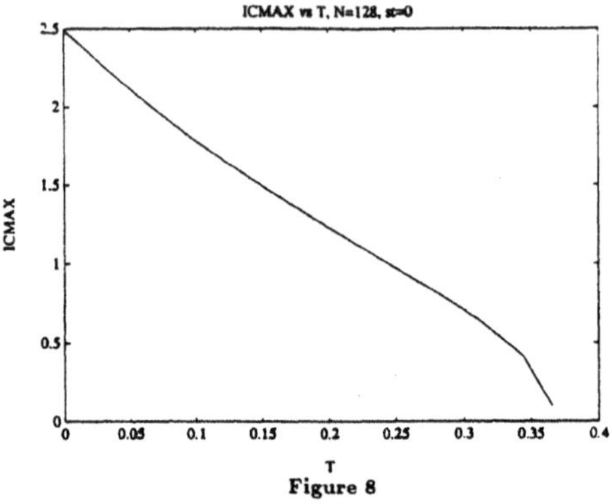

Figure 8

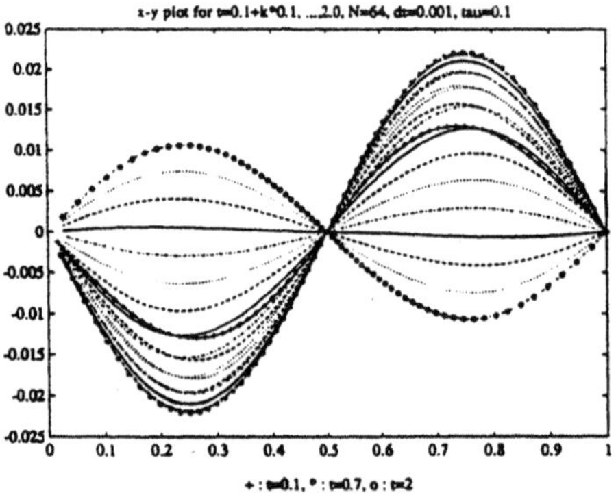

Figure 9

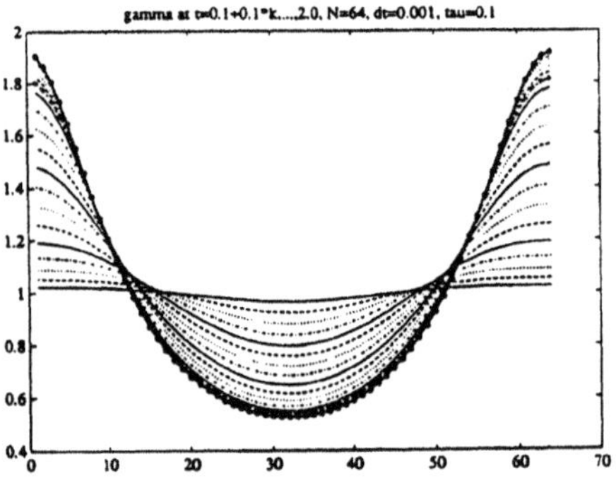

Figure 10

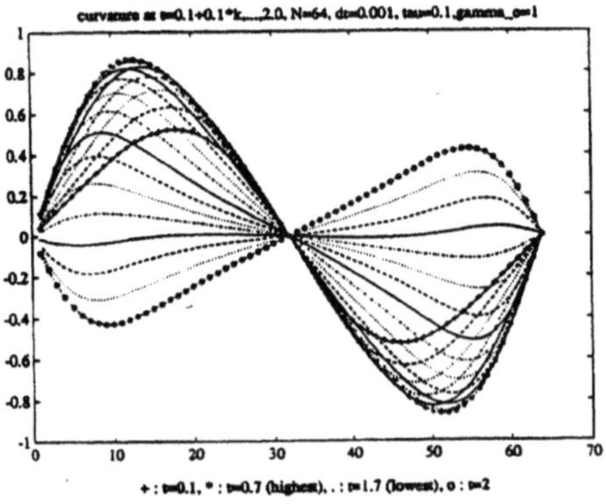

Figure 11

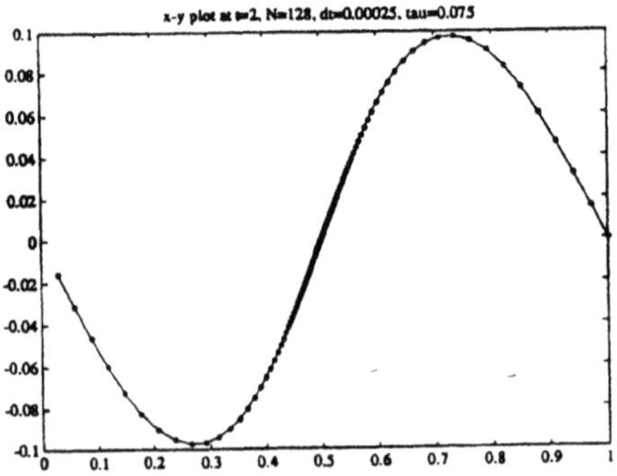

Figure 12

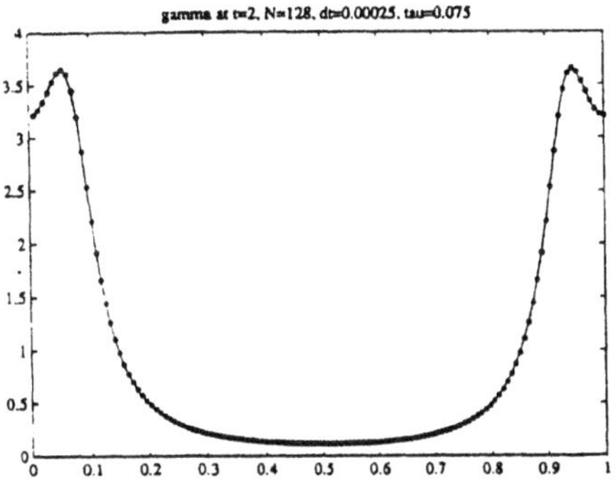

Figure 13

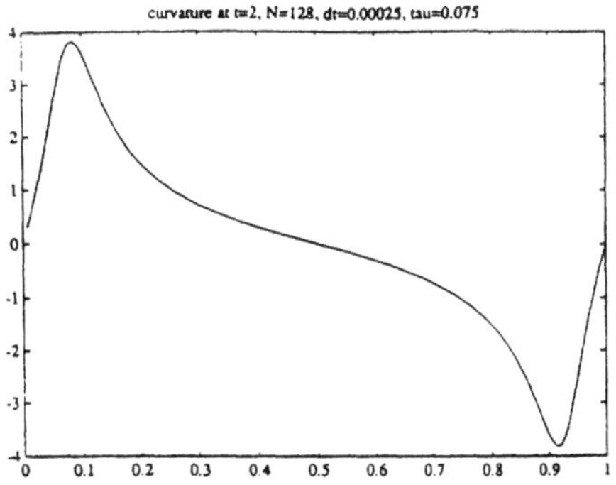

Figure 14

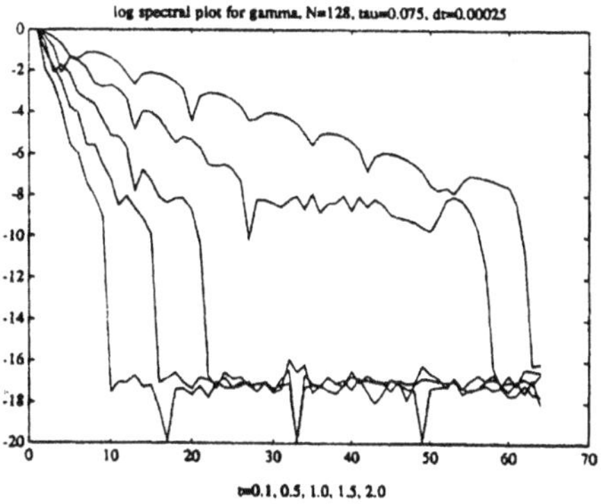

Figure 15

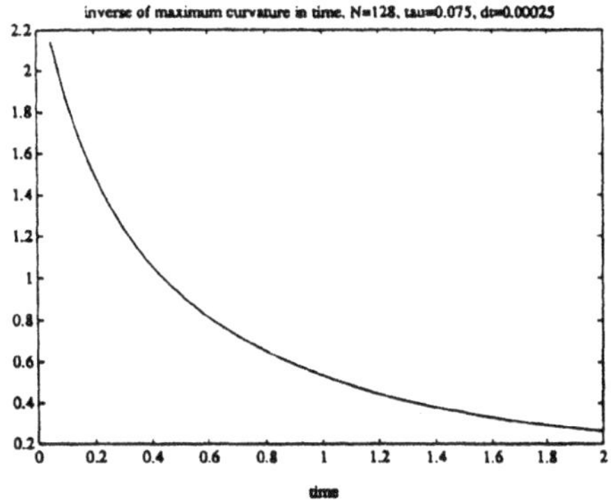

Figure 16

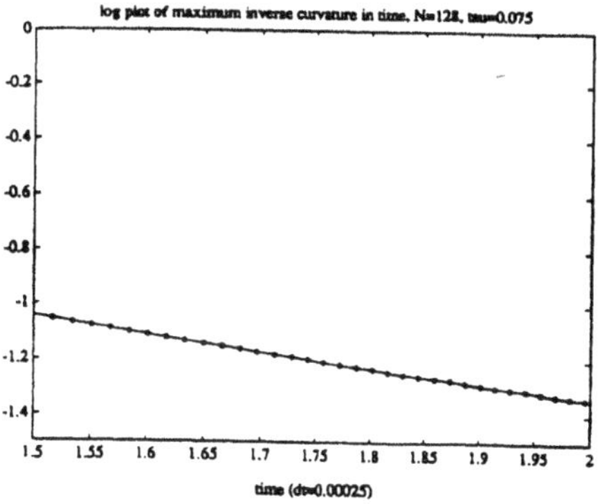

Figure 17

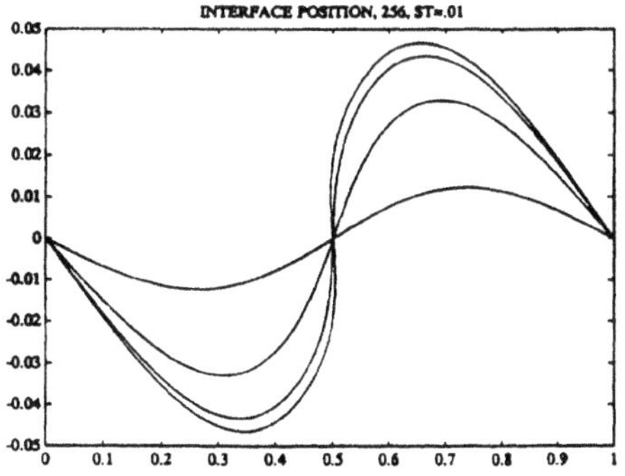

Figure 18

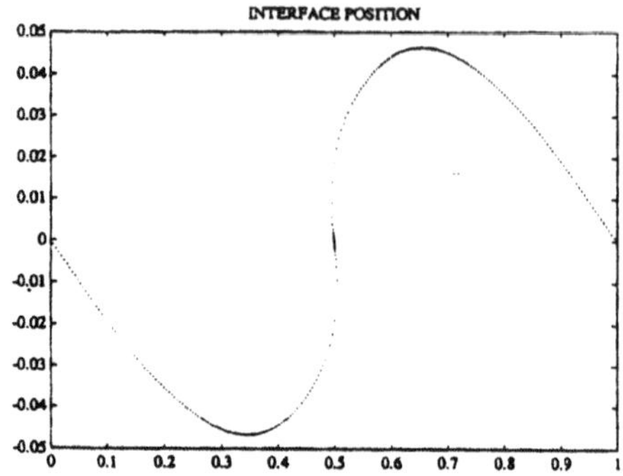

Figure 19

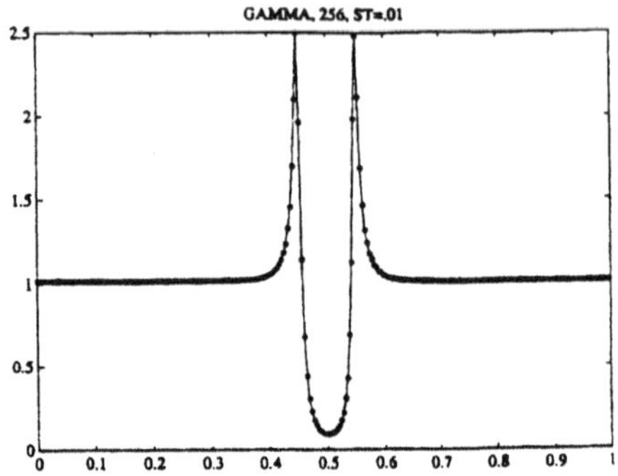

Figure 20

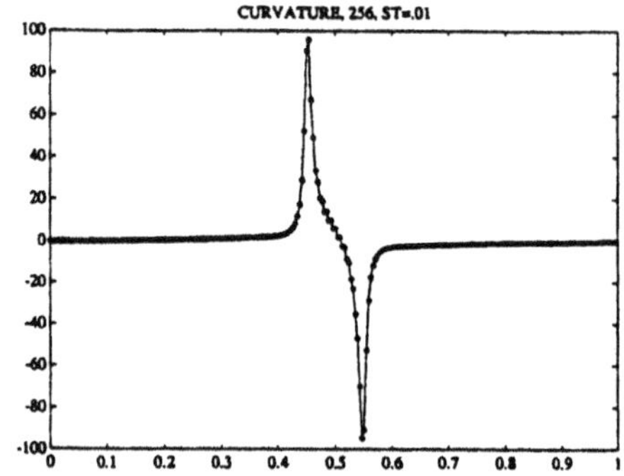

Figure 21

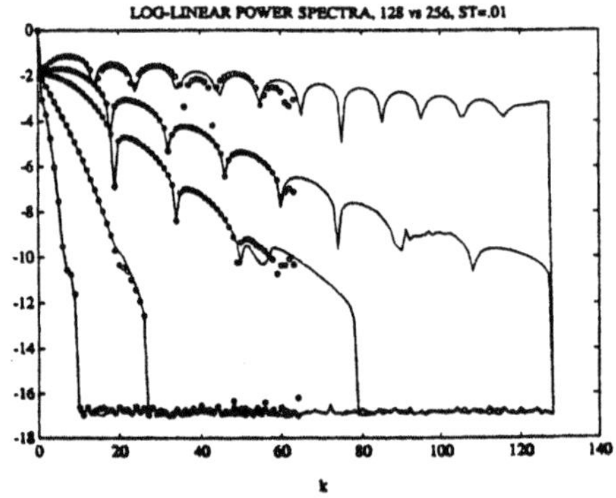

Figure 22

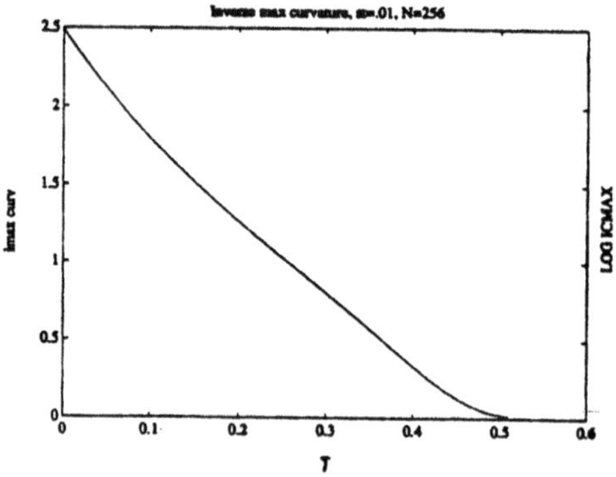

Figure 23

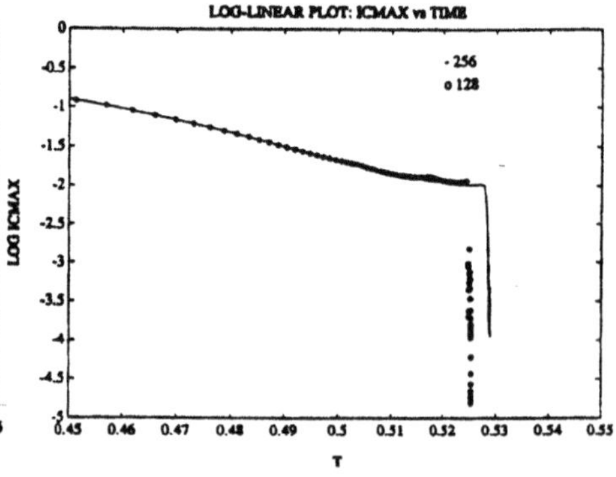

Figure 24

CAN THREE-DIMENSIONAL IDEAL FLOWS BECOME SINGULAR IN A FINITE TIME?

M.E. Brachet
LPS-ENS
24 rue Lhomond
75231 Paris Cedex 05
France
M. Meneguzzi, A. Vincent
CERFACS
42 Av. Coriolis
31 000 Toulouse
France
H. Politano and P.L. Sulem
Observatoire de la Côte d'Azur
BP 229
06304 Nice Cedex 4
France

ABSTRACT. The dynamics of three-dimensional ideal flows is investigated by direct numerical simulations of the Euler equations at resolutions up to 256^3 for general periodic flows and 864^3 for the symmetric Taylor-Green vortex. The spontaneous emergence of flat pancake-like structures that shrink exponentially in time is observed. A simple self-similar model that fits these observations is presented. Focusing instabilities similar to those leading to streamwise vortices in the context of free shear layers ([15]), are expected to subsequently concentrate the vorticity and produce isolated vortex filaments. A finite time singularity for the Euler equations is not excluded as the result of interactions among these filaments.

1. Numerical approach to singular problems

The Euler equations

$$\partial_t \mathbf{v} + (\mathbf{v} \cdot \nabla)\mathbf{v} = -\nabla p \tag{1}$$

$$\nabla \cdot \mathbf{v} = 0 \, , \tag{2}$$

R. E. Caflisch and G. C. Papanicolaou (eds.), Singularities in Fluids, Plasmas and Optics, 39–51.
© 1993 *Kluwer Academic Publishers.*

that describe the dynamics of three dimensional ideal flows, is a nonlocal PDE with quadratic nonlinearities. When integrated numerically, this PDE is replaced by a large number of ODE's. Generically, quadratic ODE's blow up in a finite time. Therefore, to numerically investigate possible singularities of the PDE, special care is requested to insure convergence of the solutions of the ODE's towards the solution of the original PDE.

In this context, spectral methods for the space variables are of great efficiency, especially in the case of analytic solutions, for which the convergence is exponential. In the simple case of a periodic function expanded in Fourier series, the convergence rate is controlled by the width δ of the analyticity strip. The Fourier coefficients indeed decay like $e^{-\delta k}$ (up to a possible prefactor). This relation gives a practical method to estimate δ. It is easily established by noting that, when a function $f(x)$ is analytically continued as $f(x+iy)$ for $|y|<\delta$, the Fourier coefficients are multiplied by e^{ky}. When the Euler equations with analytic initial data are integrated numerically, it is thus important to follow the time evolution of δ, both to control the accuracy of the simulation and also to discriminate between singular solutions for which δ vanishes in a finite time, and globally smooth solutions for which δ remains strictly positive. Indeed, a solution of the Euler equation which is initially analytic remains so as long as the vorticity does not blow up ([1]). Monitoring δ is a much more stronger constraint on the numerics accuracy than monitoring $sup|\omega|$ ([2])–([4]), although mathematically, the behavior of the latter quantity controls the regularity of the solution ([5]).

2. Results of spectral simulations

In a first series of numerical simulations of the Euler equations, we considered the Taylor Green vortex ([6]), ([7]) which is an incompressible three–dimensional flow developing from the single mode initial conditions

$$(\ v_x = \sin x \cos y \cos z\ ;\ v_y = -\cos x \sin y \cos z\ ;\ v_z = 0\ )\ . \tag{3}$$

This flow displays several symmetries which are implemented in the numerical simulation in order to reduce memory requirement and operation number. With the same code as the one used to simulate viscous flows ([8]), inviscid calculations at resolutions up to 864^3 collocation points were performed. In this algorithm, the solution is dealiased by suppressing at each time–step, the Fourier modes for which at least one wavevector component exceeds two thirds of the wavenumber cut–off.

We also considered (2π)–periodic random initial conditions with a spectrum of the form $E(k) = ck^2 e^{-(k/k_0)^2}$ with $k_0 = 1$ and $c = 0.03$. With the same code as the one described in ([9]), dealiased simulations at resolutions up to 256^3 were performed. In order not to reduce significantly the effective

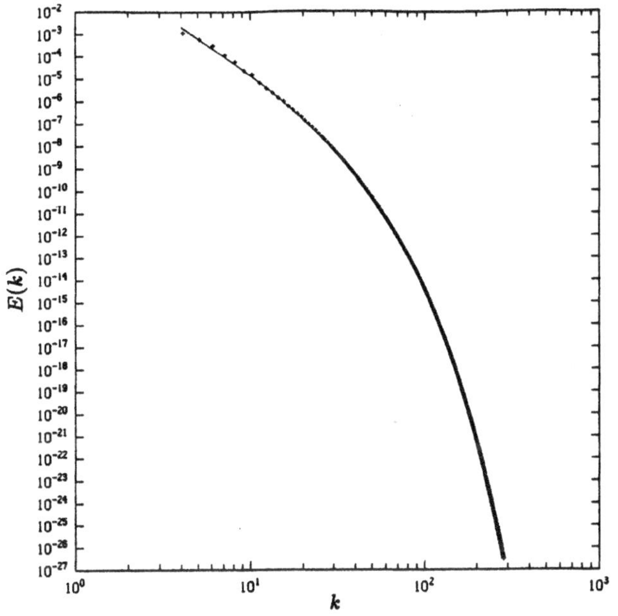

Fig. 1. Energy spectrum $E(k)$ (crosses) together with a mean square fit (solid line) of the form $ck^{-n}e^{-2\delta k}$ for the Taylor Green vortex with a resolution of 864^3, at time $t = 2.1$. The fit is performed in the range $5 < k < 279$.

resolution, aliasing was suppressed as in ([10]), a method which retains 94% of the maximum wavenumber but doubles the operation number.

At the maximum resolutions, our codes require all the machine memory. They are multitasked on the four CPU's in order to maximize the computing power.

Investigation of the regularity properties of the solution, by using the analyticity strip method ([11]), consists in analyzing the time evolution of the energy spectrum

$$E(k,t) = \frac{1}{2} \sum_{k-\frac{\Delta k}{2} < |\mathbf{k}'| \leq k+\frac{\Delta k}{2}} |\hat{\mathbf{v}}(\mathbf{k}',t)|^2 \tag{4}$$

obtained by averaging on spherical shells of width $\Delta k = 1$. In eq.(4), $\hat{\mathbf{v}}(\mathbf{k}',t)$ denotes the spatial Fourier transform of the solution. As long as the velocity field is analytic, the energy spectrum $E(k,t)$ decays exponentially at large k (with a possible algebraic prefactor). The logarithmic decrement is twice the width $\delta(t)$ of the analyticity strip of the solution when continued to complex spatial variables. The temporal evolution of the analyticity strip is

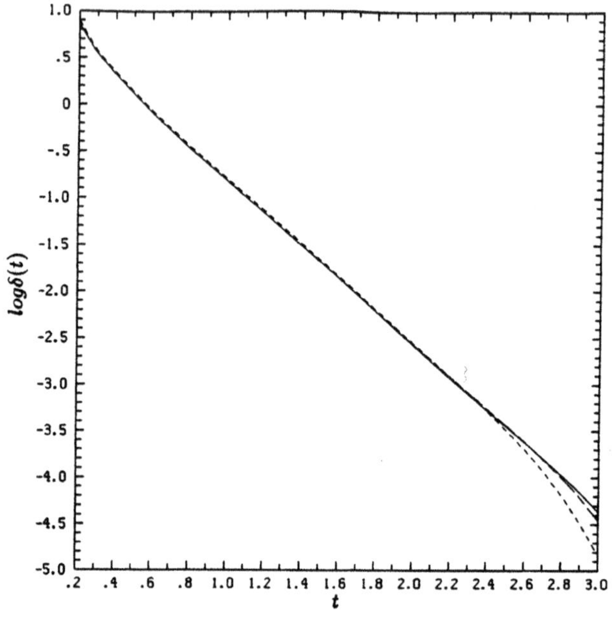

Fig. 2. Time evolution of the logarithmic decrement $\delta(t)$ obtained by fitting the energy spectrum on the ranges $5 < k < 71$, $5 < k < 161$ and $5 < k < 279$ for the inviscid Taylor Green vortex integrated with resolutions 256^3 (dotted line), 512^3 (dashed line) and 864^3 (solid line) respectively.

obtained by fitting the computed energy spectrum $E(k,t)$ with the assumed functional form

$$c(t)k^{-n(t)}e^{-2\delta(t)k} \; . \tag{5}$$

An example of a mean square fit of this form is presented in Fig.1 for the Taylor Green vortex at time $t = 2.1$ when integrated at a resolution of 864^3 collocation points. The crosses give the computed energy spectrum while the solid line is the result of the fit performed in the range $5 < k < 279$. In order to avoid spectral oscillations resulting from the symmetries of the Taylor Green flow, the energy spectrum was in this case averaged on shells of width $\Delta k = 2$. Except at the lowest wavenumbers where no universal behavior is expected, the fit appears to be very good.

The measure of $\delta(t)$ is reliable as long as it remains larger than a few mesh sizes, a condition required for the smallest scales to be accurately resolved and spectral convergence ensured. Figure. 2 shows $\delta(t)$ versus time for the inviscid Taylor Green vortex integrated at resolutions 256^3, 512^3 and 864^3. The fits of the energy spectrum were performed on ranges $5 < k < 71$,

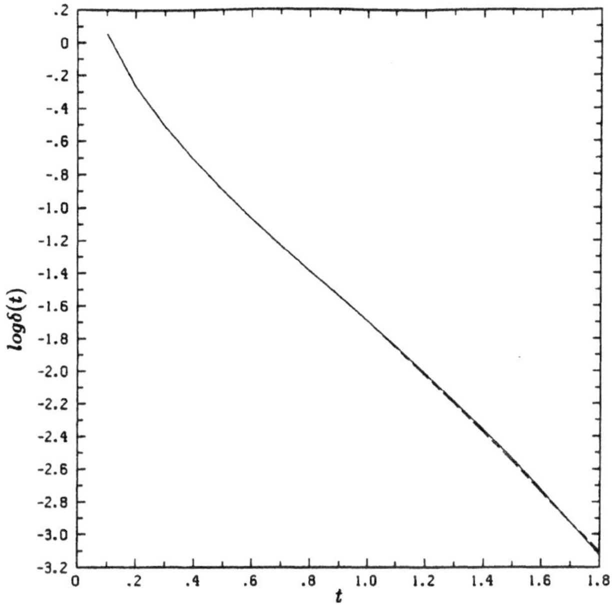

Fig. 3. Time evolution of the logarithmic decrement $\delta(t)$ for the general periodic flow, obtained by fitting the energy spectrum in the ranges $5 < k < 90$ and $5 < k < 110$, for simulations at resolution 200^3 (dashed line) and 256^3 (solid line) respectively.

$5 < k < 161$ and $5 < k < 279$ respectively. We clearly see that after a short transient, $\delta(t)$ decays like

$$\delta(t) = \delta_0 e^{-t/T} \tag{6}$$

with a characteristic shrinking time $T = 0.57$ and $\delta_0 = 2.5$, up to a time $t \simeq 3$ when it becomes comparable to twice the smallest resolved scale. If this exponential behavior can be extrapolated indefinitely, this would indicate that this the Taylor–Green vortex remains analytic for all times.

We now turn to simulations of general periodic flows with random initial conditions, in order to check the genericity of the results obtained with the Taylor Green vortex. The latter indeed displays specific features, like symmetries and lines of zero vorticity. The initial conditions we chosed, correspond to an energy spectrum of the form $E_0(k) = ck^2 e^{-(k/k_0)^2}$ with $k_0 = 1$ and $c = 0.03$. In the case of the realization we used, the initial energy $E_0 = 6.5\ 10^{-2}$ and the initial enstrophy $\Omega_0 = 0.29$. Dealiased numerical simulations were performed at resolution 200^3 and 256^3 collocation points. Again the computed energy spectrum (not shown) is well fitted with the functional form (5). An exponential decay is obtained for $\delta(t)$, up to the moment when it becomes of the order of a few mesh sizes and the accuracy of the simulation deteriorates (Fig. 3). This result strongly supports the genericity of the behavior observed for the Taylor Green vortex.

In addition to the logarithmic decrement, the representation of the energy spectrum by the functional form (5) provides the exponent $n(t)$. For both the Taylor Green vortex and the random flow, we observed that $n(t)$ varies between 4 and 5, and appears to approach 4 at the end of the runs. A similar observation is reported in ([4]). A consequence of this spectral behavior is that the enstrophy is not dominated by the smallest scales.

3. Formation of vortex layers

The decay in time of the logarithmic decrement measured on the energy spectrum has a counter part in physical space where it is associated with the generation of small–scale structures. In order to characterize this process, we choose to visualize the high vorticity regions of the flow. For both for the Taylor Green vortex and the general periodic flow, we observe the formation of pancake like structures whose thickness decays in time.

In the case of the Taylor Green vortex, the pancake is formed near the center of the $y = 0$ face which is a mirror symmetry plane of the flow. Figure 4 displays the vorticity in this plane at times $t = 3, 3.25, 3.50$ and 4. Shorter times ($t = 0$ and 2) were given in Fig. 1 of reference ([7]). We clearly observe the formation of a vorticity layer which shrinks in time without significant distortion. This process is observed even when the layer becomes so thin that its inner structure is no longer accurately resolved.

The formation of vorticity layers seems generic for a three dimensional ideal flow. In Fig. 5, high vorticity regions are represented for the general periodic flow at times $t = 0.05, 1.10, 1.65$ and 2.95. The visualizations were done on a Silicon Graphic machine, using the (three–dimensional) interactive VFFS software developed at CERFACS. We see that initially isotropic blobs of vorticity contract in one direction, while in the two others, the dimensions remain finite.

In order to reproduce the dynamics of an isolated pancake structure, we construct a simple self similar model by looking for a velocity field of the form $v_i = M_i(t)w_i(\xi_1, \xi_2, \xi_3)$ and a pressure $p = N(t)P(\xi_1, \xi_2, \xi_3)$ where $\xi_i = \frac{x_i}{\ell_i(t)}$.

Assuming that the pancake structure shrinks in the ξ_3–direction, while the characteristic scales in the transverse directions remain of order unity, a solution of eqs. (1), (2) consistent with this evolution, corresponds to the scaling factors $M_1 = M_2 = 1$,$\ell_1 = \ell_2 = 1$, $M_3 = \ell_3 = e^{-t/T}$ and $N = 1$.

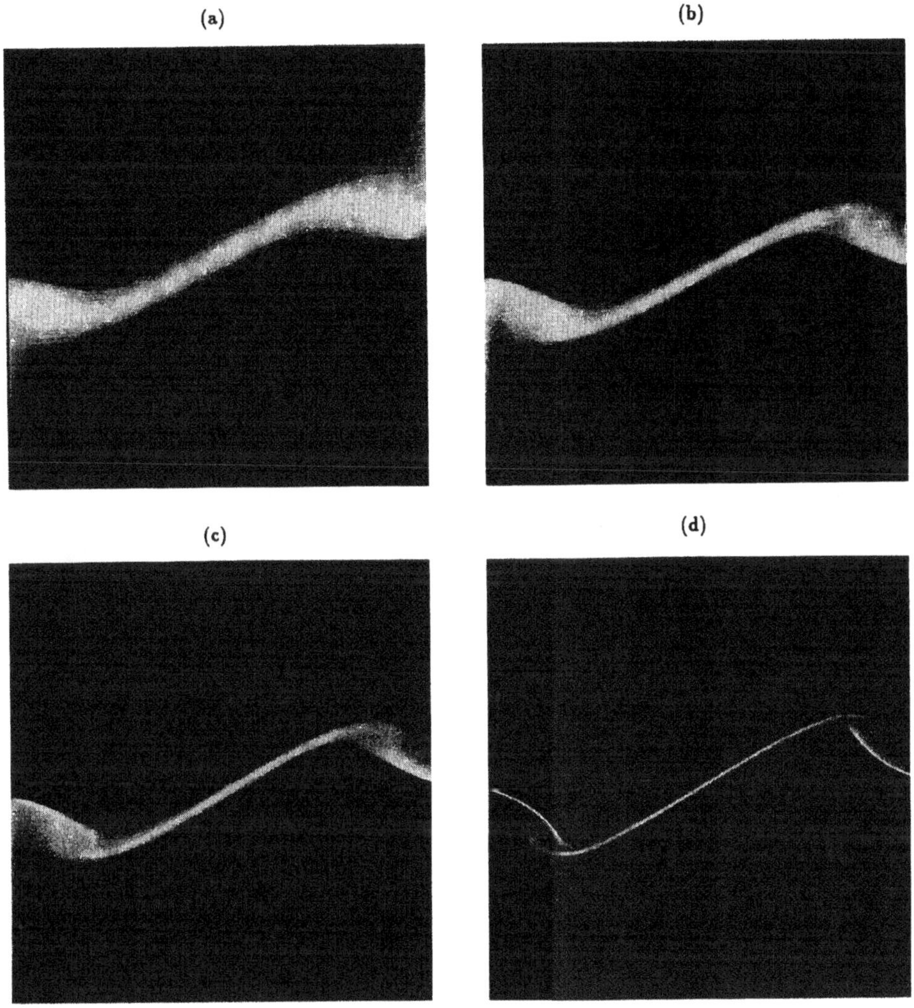

Fig. 4. Visualizations in physical space of high vorticity regions for the Taylor Green vortex at $t = 3$, $t = 3.25$, $t = 3.50$ and $t = 4$. Note that the thickness of the pancake decays in time.

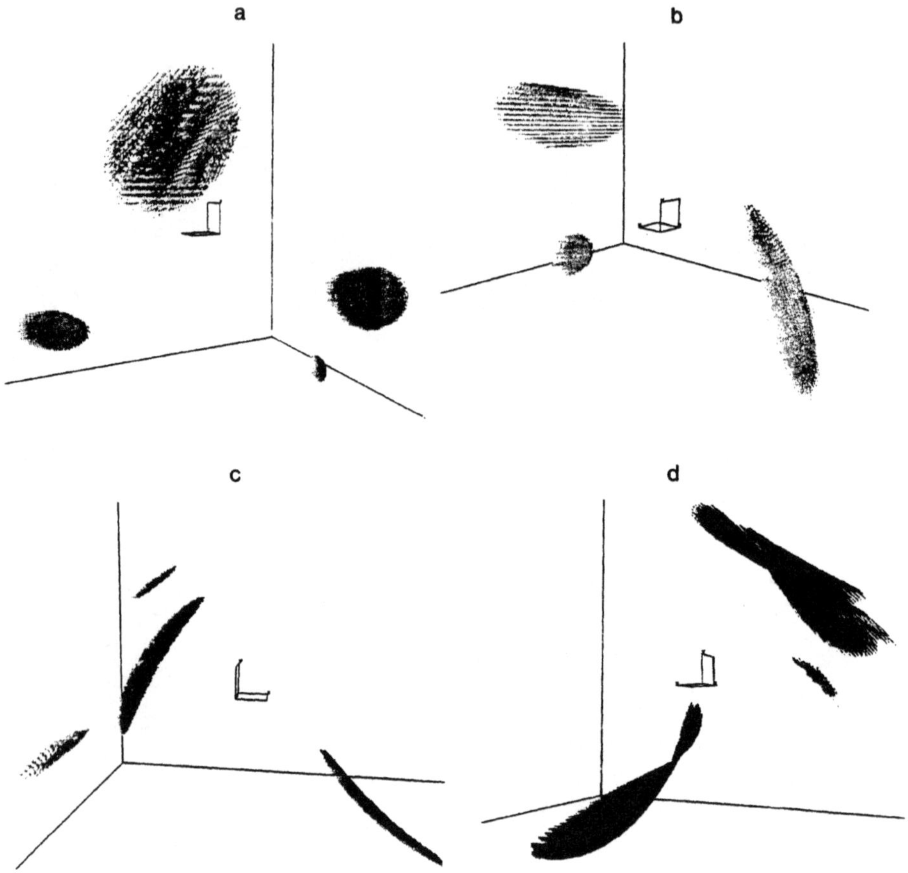

Fig. 5. High vorticity regions for the random flow at times $t = 0.05, 1.10, 1.65$, and 2.95.

The profile of the solution is given by the system

$$\nabla_\perp \cdot w_\perp + \frac{\partial w_3}{\partial \xi_3} = 0 \tag{7}$$

$$\frac{\xi_3}{T}\frac{\partial w_\perp}{\partial \xi_3} + (w_\perp \cdot \nabla_\perp + w_3 \frac{\partial}{\partial \xi_3}) w_\perp = -\nabla_\perp P \tag{8}$$

$$\frac{\partial P}{\partial \xi_3} = 0 \; . \tag{9}$$

where $\perp$ denotes the projection in the plane perpendicular to the ξ_3-direction. We notice that, as is usually the case for a boundary layer, the pressure does not change significantly when crossing the pancake. Note that, although space dimension does not explicitly enter in the above scaling, no non-zero solutions of eqs. (7)–(9) can be considered in the context of two-dimensional flows since the vorticity

$$\omega = \text{Curl } \mathbf{v} = \begin{pmatrix} -e^{t/T}\frac{\partial w_2}{\partial \xi_3} \\ e^{t/T}\frac{\partial w_1}{\partial \xi_3} \\ 0 \end{pmatrix} + O(1) \tag{10}$$

grows exponentially in time, while it is constant in two dimensions.

The eigenvectors associated to the eigenvalues

$$\lambda_1 = \frac{1}{2}e^{t/T}\sqrt{(\frac{\partial w_1}{\partial \xi_3})^2 + (\frac{\partial w_2}{\partial \xi_3})^2} + O(1) \tag{11}$$

$$\lambda_2 = O(1) \tag{12}$$

$$\lambda_3 = -\lambda_1 \tag{13}$$

of the strain matrix $\mathbf{S} = \frac{1}{2}(\nabla \mathbf{v} + \nabla \mathbf{v}^{\text{tr}})$ read

$$\mathbf{V}_1 = e^{t/T}\left(\frac{\partial w_1}{\partial \xi_3}, \frac{\partial w_2}{\partial \xi_3}, \sqrt{(\frac{\partial w_1}{\partial \xi_3})^2 + (\frac{\partial w_2}{\partial \xi_3})^2}\right) + O(1) \tag{14}$$

$$\mathbf{V}_2 = e^{t/T}\left(\frac{\partial w_2}{\partial \xi_3}, \frac{-\partial w_1}{\partial \xi_3}, 0\right) + O(1) \tag{15}$$

$$\mathbf{V}_3 = e^{t/T}\left(\frac{\partial w_1}{\partial \xi_3}, \frac{\partial w_2}{\partial \xi_3}, -\sqrt{(\frac{\partial w_1}{\partial \xi_3})^2 + (\frac{\partial w_2}{\partial \xi_3})^2}\right) + O(1)\ . \tag{16}$$

Our model thus predicts that the vorticity tends to be aligned with the strain eigenvector associated to the (finite) intermediate eigenvalue. Furthermore, the two other eigenvectors make an angle of $\pi/4$ with the ξ_3-axis.

The alignment of vorticity with the intermediate strain eigenvector was conjectured by Vieillefosse ([12]) on the basis of a simplified analysis where the pressure is supposed to act in a completely isotropic way. As discussed in ([13]), and ([14]), this assumption which leads to a finite time blow up of the vorticity, is likely to become inconsistent as the singularity is approached and to change the dynamical picture drastically.

Coming back to our self-similar model, an explicit solution of eqs.(7)–(9) which explains some of the features seen in the visualizations of our numerical simulations (see below) is obtained in the form

$$(\ w_1 = f(\xi_3)\ ,\ w_2 = \frac{\xi_2}{T}\ , w_3 = -\frac{\xi_3}{T}\ ) \tag{17}$$

or equivalently, when coming back to the primitive variables

$$(v_1 = f(ze^{t/T}) \ , \ v_2 = \frac{y}{T} \ , \ v_3 = -\frac{z}{T} \ ) \ . \tag{18}$$

Note that this velocity is in fact an exact solution of the Euler equation. It corresponds to the superposition of a shear flow (associated to a vortex layer in the (x, y)–plane) and a uniform strain field in the (z, y)–plane.

Computing strain and vorticity for this flow, one observes that, as time elapses, the eigenvector $\mathbf{V_2}$ remains aligned with the vorticity, while $\mathbf{V_1}$ and $\mathbf{V_3}$ rotate in time in the (x, z)–plane. The angle between the shrinking direction $\mathbf{V_3}$ and the (x, y)–vorticity plane varies from $\theta = \pi/2$ at $\alpha = 0$ to $\theta = \pi/4$ for large α.

In order to validate the dynamics as exemplified by the above model, we have simultaneously visualized the vorticity and the strain eigenvectors in regions of strong vorticity. These visualizations, done for the general periodic flow, support both the alignment of vorticity with the intermediate eigenvector and the rotation in time of the two other eigenvectors towards a direction making an angle of $\pi/4$ with the vortex layer. The interactive change of view point provided by VFFS software enable us to get a rather clear representation of the direction of the vectors. In contrast, a fixed picture is ambiguous in this respect and is thus not shown.

To get a quantitative insight, we have computed the histograms of the cosines of the angles between the vorticity and the three eigenvectors, in the regions where the former exceeds 50% of its maximum. The results displayed in Fig. 6 confirm the alignment of the vorticity with the intermediate eigenvector.

4. Collapse of vortex layers and generation of vortex tubes

The stability of a stretched vortex layer of the type described by (18) when slowly modulated in the x–direction, was studied in the context of free shear layers to model the formation of streamwise vortices ([15]), ([16]). This analysis sheds light on the dynamics of the vorticity pancakes generated in our numerical simulations, beyond the time at which the integration had to be stopped due to loss of accuracy. The most relevant results for our problem concern the existence of an instability of the vortex layer that yields the vorticity to focus in a finite time into vortex tubes.

The external strain will continue to exponentially shrink such vortex tubes ([17]). However, self–interactions and/or interactions among vortex filaments may possibly lead to a more singular behavior ([18])–([20]). Still, the conclusions of numerical simulations of interacting vortex tubes are controversial about the existence of a finite time singularity ([22]), ([2]), ([4]).

In the context of Navier–Stokes flows at large Reynolds numbers, the shrinking of a vorticity pancake is stopped by viscosity. The resulting vortex

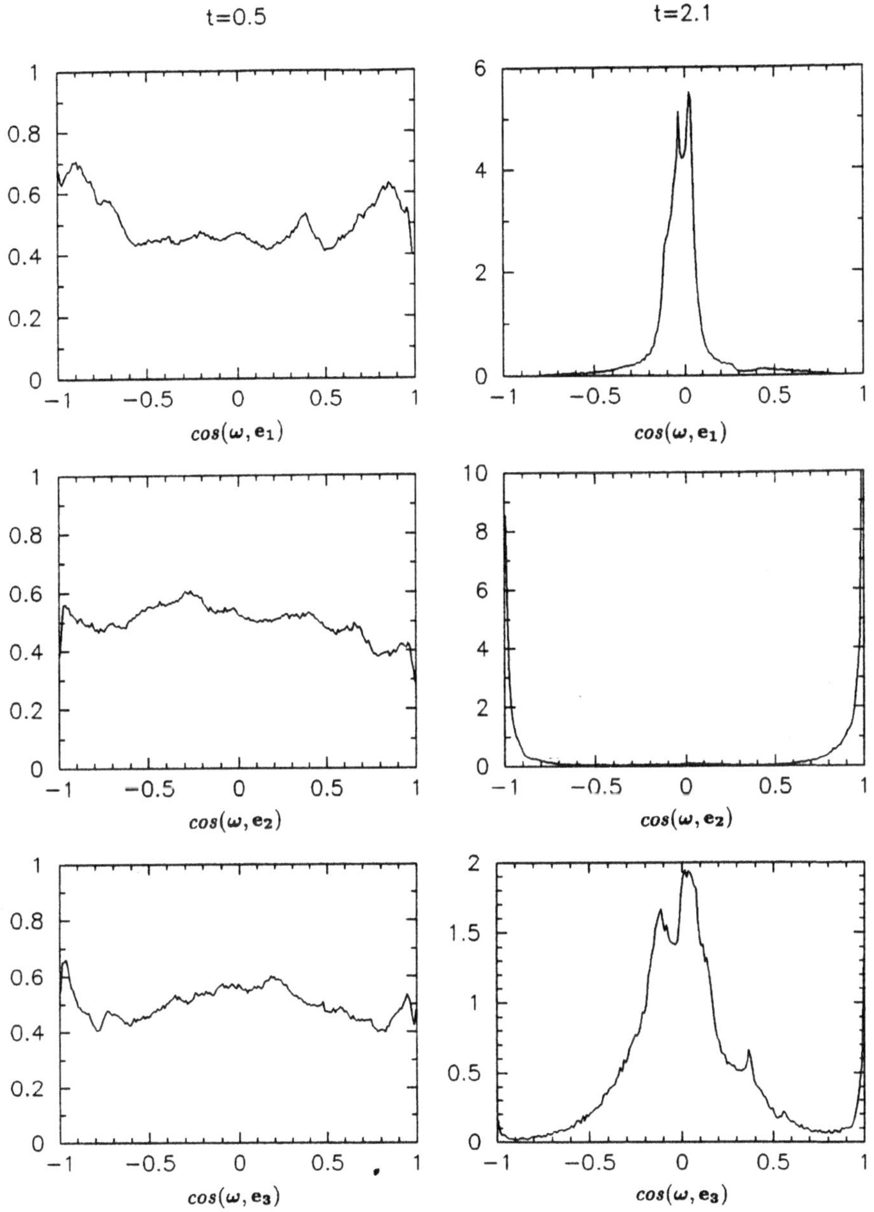

Fig. 6. Histograms of the cosines of the angles between the vorticity ω and the three strain matrix eigenvectors $\mathbf{e_1}$, $\mathbf{e_2}$ and $\mathbf{e_3}$, associated to the eigenvalues in ascending order, for the random periodic flow at times $t = 0.5$ and $t = 2.1$.

layer (with a small but finite thickness) collapses like in an ideal fluid. This process is interrupted when the width of the structure becomes comparable to its thickness. A vortex filament aligned with the intermediate strain eigenvector have then been formed, similar to those observed in numerical simulations of fully developed turbulence ([21])–([25]), ([8]), ([9]). Furthermore, the time for filament formation has a finite limit as viscosity tends to zero. This is a consequence of the catastrophic collapse of an infinitely thin vortex sheet ([15]). This quasi–singular behavior provides an interpretation to the abrupt formation of vorticity filaments observed in a recent experiment ([26]).

5. Conclusion

We have simulated numerically the short-time dynamics of space periodic three-dimensional ideal flows whose energy is initially concentrated in the largest scales. Both highly symmetric and randomly chosen initial conditions were considered. Since the nonlinearities lead to the development of arbitrarily small scales, the computation was limited to the time during which all the scales of motion were accurately resolved. After an early transient, we observed an exponential decay in time for the logarithmic decrement of the energy spectrum, or equivalently for the width of the analyticity strip of the velocity field. A simple model assuming a self-similar dynamics was also presented and satisfactory compared with numerical observations.

We suggest that, like in free shear layers, the subsequent dynamics consists in focusing instabilities which concentrate the vorticity leading to abrupt generation of isolated vortex filaments. The possible existence of a finite time singularity for the Euler equation may strongly depend on the details of the interactions between these structures. In this picture, the elementary processes leading to the formation of small scale structures in turbulent flows are of a purely inviscid origin. Viscosity prescribes minimal scales and permits reconnection.

Acknowledgements

We are grateful to R. Krasny for pointing out reference ([15]) to our attention. The calculations were done on the Cray–2 of the Centre de Calcul Vectoriel Pour la Recherche (Palaiseau, France). The main routine used in all our calculations is the Fast Fourier Transform of the Cray–2 library, which is a C. Temperton FFT adapted for this machine by Cray–Research. One of us (MEB) acknowledges the support of grant 90–173 from the Direction des Recherches Etudes et Techniques.

References

[1]C. Bardos and S. Benachour, Domaine d'analyticité des solutions de l'équation d'Euler dans un ouvert de R^n, Ann. Sc. Norm. Sup. Pisa, serie IV, **4**, 647 (1977).

[2]A. Pumir and E.D. Siggia, Collapsing solutions of the 3-D Euler equation, Phys. Fluids A **2**, 220 (1990).

[3]A. Pumir and E.D. Siggia, Development of singular solutions to the axisymmetric Euler equation, Phys. Fluids A **4**, 1472 (1992).

[4]R.M. Kerr, Evidence for a singularity of the three-dimensional incompressible Euler equation, Preprint (1991).

[5]J.T. Beale, T. Kato and A. Majda, Remarks on the breakdown of smooth solutions for 3-D Euler equations, Comm. Math. Phys. **94**, 61 (1984); G. Ponce, Remarks on a paper by J.T. Beale, T. Kato and A. Majda, Comm. Math. Phys. **98**, 349 (1985).

[6]G.I. Taylor and A. E. Green, Mechanism of the production of small eddies from large one, Proc. Roy. Soc. London **A 158**, 499 (1937).

[7]M. Brachet, D. Meiron, S.A. Orszag, B. Nickel, R. Morf and U. Frisch, Small-scale structure of the Taylor-Green vortex, J. Fluid Mech. **130**, 411 (1983).

[8]M.E. Brachet, Direct simulation of three-dimensional turbulence in the Taylor-Green vortex, Fluid Dyn. Res. **8**, 1 (1991).

[9]A. Vincent and M. Meneguzzi, The spatial structure and statistical properties of homogeneous turbulence, J. Fluid Mech. **225**, 1 (1991).

[10]S.A. Orszag and G.S. Patterson, Numerical simulation of the three-dimensional homogeneous turbulence, Phys. Rev. Lett. **28**, 76 (1972).

[11]C. Sulem, P.L. Sulem and H. Frisch, Tracing complex singularities with spectral methods, J. Comp. Phys. **50**, 138 (1983).

[12]P. Vieillefosse, Local interaction between vorticity and shear in a perfect incompressible fluid, J. Phys. Paris **43**, 837 (1982).

[13]Z.S. She, E. Jackson and S.A. Orszag, Structure and dynamics of homogeneous turbulence: models and simulations, Proc. R. Soc. London A **434**, 101 (1991).

[14]E. Dresselhaus and M. Tabor, The kinematics of stretching and alignment of material elements in general flow fields, J. Fluid Mech. **236**, 415 (1992).

[15]J.C. Neu , The dynamics of stretched vortices, J. Fluid Mech. **143**, 253 (1984).

[16]S.J. Lin and G.M. Corcos, The mixing layer: deterministic models of a turbulent flow. Part 3. The effect of plane strain on the dynamics of streamwise vortices, J. Fluid Mech. **141**, 139 (1984).

[17]J.C. Neu, The dynamics of a columnar vortex in an imposed strain, Phys. Fluids **27**, 2397 (1984).

[18]A.J. Chorin, The evolution of a turbulent vortex, Comm. Math. Phys. **83**, 517 (1982).

[19]E.D. Siggia, Collapse and amplification of a vortex filament, Phys. Fluids **28**, 794 (1985); A. Pumir and E.D. Siggia, Vortex dynamics and the existence of solutions to the Navier-Stokes equations, Phys. Fluids **30**, 1606 (1987).

[20]A. Fukuyu and T. Arai, Singularity formation in three-dimensional inviscid flow, Fluid Dyn. Res. **7**, 229 (1991).

[21]E.D. Siggia, Numerical study of small-scale intermittency in three-dimensional turbulence, J. Fluid Mech. **107**, 375 (1981).

[22]R.M. Kerr and F. Hussain, Simulation of vortex reconnection, Physica D **37**, 474 (1989).

[23]R.M. Kerr, Higher-order derivative correlations and the alignment of small-scale structures in isotropic numerical turbulence, J. Fluid Mech. **153**, 31 (1985).

[24]W.T. Ashurst, A.R. Kerstein, R.M. Kerr and C.H. Gibson, Alignment of vorticity and scalar gradient with strain rate in simulated Navier-Stokes turbulence, Phys. Fluids **30**, 2343 (1987).

[25]Z.S. She, E. Jackson and S.A. Orszag, Intermittent vortex structures in homogeneous isotropic turbulence, Nature **344**, 226 (1990).

[26]S. Douady, Y. Couder and M.E. Brachet, Direct observation of the intermittency of intense vorticity filaments in turbulence, Phys. Rev. Lett. **67**, 982 (1991).

Singularity Formation for Models of Axi-Symmetric Swirling Flow

Russel E. Caflisch and Peter Smereka
Department of Mathematics
UCLA
Los Angeles
CA 90024-1555

ABSTRACT. Singularities are found in solutions of three systems that model the equations for axi-symmetric swirling flow. The first example is a simple first order system in "Jordan form." The second is a one dimensional analogue of the 2D Boussinesq system. In the third example, a complex solution of the axi-symmetric swirling flow equations is numerically constructed. This solution is a traveling wave with a complex wave speed that brings a singularity from the complex plane to a real position at a finite real time. The complex-valued flow can be understood as coming from a solution Moore's approximation for axi-symmetric swirling flow.

1. Introduction

A principal outstanding question in the mathematical theory of fluid dynamics is whether singularities can form in finite time for solutions of the 3D incompressible, inviscid Euler equations. The significance of this problem lies first in its relation to existence theory for these equations. Singularities, if they exist for even a small set of possible initial data, would probably also cause significant problems for numerical computations and could be a source of development of small scales in the onset of turbulence.

The purpose of this paper is to present several singular solutions of simple models equations for axi-symmetric flow with swirl. Some of the features of these examples may provide mathematical insight into the process of singularity formation: The first two examples show that, for a first order system in "Jordan form," the off-diagonal terms can cause singularity development. Since the swirling flow equations may be written as a first order system in Jordan form with the vortex stretching terms as off-diagonal elements, we conjecture that singularity formation for swirling flow (if it occurs) can be understood as caused by this Jordan structure.

R. E. Caflisch and G. C. Papanicolaou (eds.), Singularities in Fluids, Plasmas and Optics, 53–64.
© 1993 *Kluwer Academic Publishers.*

The third example presents a numerically determined solution of the full equations for axi-symmetric swirling flow. The solution is complex-valued and can be understood as coming from Moore's approximation for swirling flow. A singularity in this solution is clearly seen through an analysis of the asymptotics of the Fourier coefficients. The simplicity of the resulting singularity structure suggests that it may be a generic singularity form. This example also shows that singularities may move around in the complex space plane until they hit the real line at some finite time.

The main analytic result on singularity formation for 3D inviscid, incompressible flow is that of Beale, Kato & Majda 1989 , who showed that if an initially smooth flow develops a singularity at time t, then

$$\int_0^t \sup_{\boldsymbol{x}} |\omega(\boldsymbol{x}\ , t')| dt' = \infty. \tag{1.1}$$

A similar result, that loss of analyticity implies blow-up of vorticity, was proved by Bardos & Benachour 1977 . For inviscid, incompressible flow in two dimensions, an initially smooth velocity field $\boldsymbol{u}\ (\boldsymbol{x}\ , 0)$ will stay smooth for all time due to the conservation of vorticity.

The numerical search for singularities was started on the Taylor-Green flow by Brachet *et al.* 1983 , for which the results indicate a singularity for a complex value of t but not for real t. Experimental results showing vortex reconnection for high-Reynolds number flows inspired a study of singularities for a filament model of two interacting vortex tubes by Siggia 1985 . The singularities in this model equation are smoothed out however due to flattening of the vortex cores, as seen in computations of Anderson & Greengard 1989 , Pumir & Siggia 1990 and Shelley, Meiron & Orszag 1992 . More recently Kerr 1992 presented a new set of computations for this problem which exhibit intensification of vorticity that may indicate singularity formation. Several related studies of singularity formation are in Bhattacharjee & Wang 1992, Bell & Marcus 1992, Childress *et al.* 1989, Chorin 1981, Chorin 1982 and Stuart 1989 and in the contributions of Brachet *et al.*, Kerr and Pumir & Siggia in these Proceedings.

Swirl in axi-symmetric flows amplifies vorticity by stretching, and the axi-symmetry may prevent core flattening, so that such flows seem a likely candidate for singularities. Grauer & Sideris 1992 , Meiron & Shelley 1992 and Pumir & Siggia 1992 have performed computations of axi-symmetric flow with swirl that show significant vortex stretching but no singularities within the computational time. In Pumir & Siggia 1992 , however, an adaptive numerical method, with nonlinear scaling of $\boldsymbol{x}$ and t, was used on an asymptotically reduced equation to produce singularities for axi-symmetric flow with swirl. Some critical remarks on this results are presented in Bhattacharjee & Wang 1991 . A study of singularity formation for axi-symmetric, swirling vortex sheets was performed by Caflisch, Li & Shelley 1992 .

2. Axi-Symmetric Flow with Swirl

The Euler equations for incompressible, inviscid flow are

$$\begin{aligned} \boldsymbol{u}_t + \boldsymbol{u} \cdot \nabla \boldsymbol{u} + \nabla p &= 0 \\ \nabla \cdot \boldsymbol{u} &= 0. \end{aligned} \tag{2.1}$$

The simplest flow in which there is non-trivial vortex stretching is *axi-symmetric flow with swirl* for which the velocity $\boldsymbol{u} = (u_r, u_\theta, u_z)$ in cylindrical coordinates satisfies $\partial_\theta \boldsymbol{u} = 0$. The swirl is the azimuthal velocity u_θ. The Euler equations for axi-symmetric swirling flow are

$$\partial_z u_z + r^{-1}\partial_r(r u_r) = 0 \tag{2.2}$$
$$\partial_t u_z + \boldsymbol{u} \cdot \nabla u_z = -\partial_z p \tag{2.3}$$
$$\partial_t u_r + \boldsymbol{u} \cdot \nabla u_r - r^{-1} u_\theta^2 = -\partial_r p \tag{2.4}$$
$$\partial_t u_\theta + \boldsymbol{u} \cdot \nabla u_\theta + r^{-1} u_r u_\theta = 0 \tag{2.5}$$

in which $\boldsymbol{u} \cdot \nabla = u_z \partial_z + u_r \partial_r$. The vorticity is

$$\omega = (\omega_r, \omega_\theta, \omega_z) = (-\partial_z u_\theta,\ \partial_z u_r - \partial_r u_z,\ r^{-1}\partial_r(r u_\theta)). \tag{2.6}$$

A convenient reformulation of these equations can be made in terms of stream function and vorticity. Define circulation $\Omega = r u_\theta$, azimuthal vorticity $\zeta = -r\omega_\theta$ and stream function ψ with

$$(u_r, u_z) = r^{-1}(-\partial_z \psi, \partial_r \psi). \tag{2.7}$$

Denote

$$D^2 = r\partial_r(r^{-1}\partial_r) + \partial_z^2. \tag{2.8}$$

Then the Euler equations for axi-symmetric swirling flow are equivalent to

$$\begin{aligned} (\partial_t + \boldsymbol{u} \cdot \nabla)\Omega &= 0 \\ (\partial_t + \boldsymbol{u} \cdot \nabla)(r^{-2}\zeta) &= -r^{-4}\partial_z(\Omega^2) \\ D^2\psi &= \zeta. \end{aligned} \tag{2.9}$$

The characteristic structure of these equations is most easily seen by transformation to first order form. Set

$$\begin{aligned} \rho &= \Omega^2 \\ \gamma &= r^{-2}\zeta \\ u &= u_r + i u_z \\ v &= u_r - i u_z \\ \alpha &= r + iz \\ \beta &= r - iz \end{aligned} \tag{2.10}$$

Then the axi-symmetric swirling flow equations (2.9) are equivalent to

$$\begin{aligned}(\partial_t + u\partial_\alpha + v\partial_\beta)\rho &= 0\\ (\partial_t + u\partial_\alpha + v\partial_\beta)\gamma &= -ir^{-4}(\partial_\alpha - \partial_\beta)\rho\\ \partial_\alpha u &= a\\ \partial_\beta v &= b\end{aligned} \tag{2.11}$$

in which

$$\begin{aligned}a &= -r^{-1}(u+v) + 2ir\gamma\\ b &= -r^{-1}(u+v) - 2ir\gamma.\end{aligned} \tag{2.12}$$

In this complex formulation of the Euler equations, the system has been diagonalized, with characteristic velocity vectors $(1,u,v)$, $(0,1,0)$ and $(0,0,1)$ in (t,α,β). The first of these velocities, representing the particle speed, is a double characteristic, and there is an irremovable "off-diagonal" term $ir^{-4}(\partial_\alpha - \partial_\beta)\rho$ in the second equation. This term is analogous to the off-diagonal element in a Jordan block, since it consists of a derivative applied to the function ρ, a characteristic variable for the same speed We will refer to this form of the Euler equations as their *Jordan form.* Note that the off-diagonal term is exactly the vortex stretching term.

The next two sections present simple examples of singularity formation for first order systems in Jordan form. On this basis, we conjecture that singularity formation for the Euler equations (if it occurs) can be understood as caused by the Jordan form of the equations.

3. A Simple System in Jordan Form

The simplest first order system in Jordan form is just Burgers equation with a second equation such as

$$\begin{aligned}f_t + ff_x &= 0\\ g_t + (fg)_x &= 0.\end{aligned} \tag{3.1}$$

This system was studied earlier by Forestier & LeFloch 1992 who noted that the Riemann problem for this system does not have a general solution. As they noted, the solution of this system can be written in a simple implicit form as

$$f(x,t) = f_0(x_0) \tag{3.2}$$

$$g(x,t) = \frac{g_0(x_0)}{1+tf_0'(x_0)} \tag{3.3}$$

$$x = x_0 + tf_0(x_0). \tag{3.4}$$

Since

$$\begin{aligned}\frac{\partial}{\partial x}f &= \frac{\partial f_0}{\partial x_0}\left(\frac{\partial x}{\partial x_0}\right)^{-1}\\ &= \frac{f_0'(x_0)}{1+tf_0'(x_0)}\end{aligned} \tag{3.5}$$

then singularities in the derivative of f occur where $1+tf_0'(x_0)=0$, and also cause g to blowup. In particular $g\to\infty$ when a shock forms in f.

A simple variation of this system is also exactly solvable but has singularities of much worse form. Consider the system

$$\begin{aligned}f_t+ff_x&=0\\ g_t+fg_x+g^nf_x&=0.\end{aligned} \tag{3.6}$$

The solution of this equation for $n\neq 1$ is

$$f(x,t) = f_0(x_0) \tag{3.7}$$

$$g(x,t) = ((1-n)\log(1+tf_0'(x_0))+g_0(x_0)^{1-n})^{\frac{1}{1-n}} \tag{3.8}$$

$$x = x_0+tf_0(x_0). \tag{3.9}$$

Note that the form of the singularity is now a power of a logarithm.

4. A 1D Model of the Boussinesq System

As pointed out by Pumir & Siggia 1992 , a singularity for axi-symmetric swirling flow should be approximately described by the slightly simpler Boussinesq system

$$\begin{aligned}&(\partial_t+u\partial_r+v\partial_z)\Omega=0\\ &(\partial_t+u\partial_r+v\partial_z)\zeta=\partial_z(\Omega^2)\\ &(\partial_r^2+\partial_z^2)\psi=\zeta.\\ &(u,v)=(-\partial_r\psi,\partial_z\psi)\end{aligned} \tag{4.1}$$

A simple one dimensional model for this system is the system

$$\begin{aligned}f_t+uf_x&=0\\ g_t+ug_x&=f_x\\ \psi_{xx}&=g\\ u&=\psi_x\end{aligned} \tag{4.2}$$

The modified system (4.2) has a Jordan form similar to the of (4.1); since it is only one-dimensional, however, it does not include incompressibility.

The "stream function" ψ can be eliminated by replacing the last two equation of (4.2) with

$$u_x = g \tag{4.3}$$

This system can be solved using a Lagrangian coordinate ξ, defined so that

$$\frac{\partial}{\partial t}x(\xi, t) = u. \tag{4.4}$$

Then

$$\begin{aligned} \frac{\partial}{\partial t}f(\xi, t) &= 0 \\ \frac{\partial}{\partial t}g(\xi, t) &= f_x = f_\xi x_\xi^{-1} \\ u_\xi &= g x_\xi \end{aligned} \tag{4.5}$$

with initial data

$$(f, g, x)(\xi, 0) = (f_0, g_0, x_0)(\xi). \tag{4.6}$$

Elimination of f and x yields the equation

$$g_{tt} = -g g_t \tag{4.7}$$

so that

$$g_t = -\frac{1}{2}g^2 + \frac{1}{2}g_0^2 + f_0'/x_0'. \tag{4.8}$$

The solution is

$$g = \sqrt{2d}\cot[(t - t_0)\sqrt{d/2}] \tag{4.9}$$

in which

$$t_0 = \sqrt{2/d}\left(\frac{\pi}{2} + \tan^{-1}(\frac{g_0}{\sqrt{2d}})\right) \tag{4.10}$$

and

$$d = -\frac{1}{2}g_0^2 - f_0'/x_0'. \tag{4.11}$$

We also find

$$u_\xi = -\frac{f_\xi}{\sqrt{2d}}\sin[(t - t_0)\sqrt{2d}]. \tag{4.12}$$

Therefore we see that the solution of (4.8) blows up with the asymptotic form

$$\begin{aligned} g &\approx \frac{2}{t - t_0} \\ x_\xi &\approx -\frac{1}{2}f_\xi(t - t_0)^2 \\ u_\xi &\approx -f_\xi(t - t_0) \end{aligned} \tag{4.13}$$

near the singularity. In particular this shows that the "vorticity" g blows up, while the "velocity" u does not blow up at the singularity.

5. Moore's Approximation for Axi-Symmetric Flow with Swirl

Moore's approximation keeps nonlinear terms that transfer energy to higher wavenumbers, which is expected to be the dominant process in singularity formation, but neglects terms that transfer energy back to smaller wavenumbers. First formulated by Moore 1979 and 1985 for singularity formation on vortex sheets during Kelvin-Helmholtz instability, this method was generalized by Caflisch, Orellana & Siegel 1989 and applied to singularity formation in the Rayleigh-Taylor problem by Baker, Caflisch & Siegel 1992 . We will describe here the application of Moore's approximation to axi-symmetric flow with swirl, as carried out in Caflisch 1992 .

First decompose the velocity $\boldsymbol{u}$ into three pieces, as

$$\boldsymbol{u} = \boldsymbol{u}_+ + \boldsymbol{u}_- + \boldsymbol{u}_0 \tag{5.1}$$

in which $\boldsymbol{u}_+$ contains the positive z-wavenumbers, $\boldsymbol{u}_-$ contains the negative wavenumbers and $\boldsymbol{u}_0$ is constant in z; i.e.,

$$\begin{aligned} \boldsymbol{u}_+ &= \sum_{k=1}^{\infty} \hat{\boldsymbol{u}}_k(r,t)e^{ikz} \\ \boldsymbol{u}_- &= \sum_{k=-1}^{-\infty} \hat{\boldsymbol{u}}_k(r,t)e^{ikz} \\ \boldsymbol{u}_0 &= \boldsymbol{u}_0(r). \end{aligned} \tag{5.2}$$

Moore's approximation splits the Euler equations (2.2-2.5) into two sets of Euler equations, each with the same form as (2.2-2.5). The first system is for $\boldsymbol{u}_+ + \boldsymbol{u}_0$, and the second, which is the conjugate of the first, is for $\boldsymbol{u}_- + \boldsymbol{u}_0$.

We now look for solutions $\boldsymbol{u}$ for (2.2-2.5) of the form

$$\boldsymbol{u} = \sum_{k=0}^{\infty} \hat{\boldsymbol{u}}_k(r,t)e^{ikz}. \tag{5.3}$$

As discovered by Siegel 1989 (and further described in Baker, Caflisch & Siegel 1992) for the Rayleigh-Taylor problem, solutions of this form may be conveniently sought as traveling waves with an imaginary wave speed $i\sigma$ in the z direction. Thus we look for $\boldsymbol{u} = \boldsymbol{u}(r, z - i\sigma t)$; i.e.,

$$\boldsymbol{u} = \sum_{k=0}^{\infty} \hat{\boldsymbol{u}}_k(r)e^{ikz+k\sigma t}. \tag{5.4}$$

This solution can also be interpreted as consisting only of purely growing modes.

After substitution into the system (2.2-2.5) and some manipulation, the equations for $f_k(r) = \hat{u}_r(k,r)$ are

$$\partial_r(r^{-1}\partial_r(rf_k)) - (k^2+\kappa)f_k = D_k \tag{5.5}$$

in which $D_k = D_k(f_0, ..., f_k - 1)$ depends only on f_l for $l < k$ and

$$\kappa(r) = \frac{2\bar{u}_\theta\bar{\omega}_z}{\sigma^2 r} \tag{5.6}$$

with $\bar{u}_\theta, \bar{\omega}_z$ the azimuthal velocity and vorticity for the zeroth mode $\boldsymbol{u}_0$. The zeroth component $\boldsymbol{u}_0 = \bar{u}_\theta(r)\boldsymbol{\theta}$ is a simple steady swirling flow. The first term f_1 represents a linearly unstable mode with growth rate σ for the steady flow $\boldsymbol{u}_0$, satisfying the homogeneous equation

$$\partial_r(r^{-1}\partial_r(rf_1)) - (1+\kappa)f_1 = 0. \tag{5.7}$$

The remarkable property of this solution, as found by Siegel 1989 , is that the traveling wave speed is determined by the linear eigenvalue problem (5.7), and that the equations (5.7) for f_k are "lower triangular." This makes their numerical solution fairly straightforward.

This problem was numerically solved for flow in an annulus $1 < r < 3$ with boundary conditions $u_r(r=1) = u_r(r=3) = 0$; i.e.,

$$f_k(1) = f_k(3) = 0 \tag{5.8}$$

for all k. First a basic state $\bar{u}_\theta(r)$ and a linearly unstable mode $f_1(r)$ were determined as a smoothed out version of the explicit steady flow and unstable mode for a cylindrical vortex sheet in an annulus, as determined in (Caflisch 1992). This smoothing was performed in a way that did not require solving a new eigenvalue problem, so that numerical error was minimized.

Once $\bar{u}_\theta(r)$ and $f_1(r)$ were determined, the equations (5.5) for f_k with $k > 1$ were solved without any ambiguity. Solution of the iteration equations (5.5) was delicate, however, because of an instability with respect to increase of the wavenumber k. This numerical instability amplified round-off error, as in the problem of singularity formation for vortex sheets. Krasny was able to remove the spurious growth of roundoff error in the Kelvin-Helmholtz problem through use of either a nonlinear filter (Krasny 1986a) or a smoothing parameter (Krasny 1986b) . In the present problem, the filtering method did not work since the problem is not periodic in r and the smoothing method was found to be too crude to yield information on the singularity formation. Instead, for this simple problem the effect of roundoff error was adequately minimized by the use of ultra-high precision arithmetic. Precision levels of 10^{-64} and 10^{-128} were implemented using the package MPFUN developed by David Bailey at NASA Ames (Bailey 1991a and 1991b) . This package comes with an automated translator that translates a standard FORTRAN program into a program that calls special high precision arithmetic subroutines for every arithmetic statement.

The forcing terms D_k in equations (5.5) were evaluated through a hybrid spectral-pseudospectral method with no aliasing or closure error. The computational complexity was $O(N_r N_z^{3/2})$ for N_r points in r and N_z wavenumbers. Since there was no closure error in the computation of the first N_z modes, we chose a fixed number of modes in z, and then refined the discretization of r until the results converged. The most refined computation for $N_z = 64$ used $N_r = 2048$ and agreed well with the results for $N_r = 1024$.

Singularities were analyzed through the asymptotics of the Fourier components f_k. Since the solution computed here has a symmetry with respect to reflection about $z = 0$, two singularities were expected to occur at positions

$$z_{\pm}(r) = -i\rho_1 \pm \rho_2. \tag{5.9}$$

Near each of these singularities the structure of the radial velocity u_r was sought in the form

$$u_r(z, r) \simeq c_1 e^{ic_2}(z - z_{\pm})^{-1-(\alpha_1 - i\alpha_2)} \tag{5.10}$$

for z near $z_{\pm}$. As shown in Sulem *et al.* 1983 , this implies that the Fourier coefficients f_k will have the asymptotic form

$$f_k \cong c_1 k^{-\alpha_1} e^{-\rho_1 k} \sin(c_2 + \alpha_2 \log k + \rho_2 k) \tag{5.11}$$

for $k >> 1$. In (5.9- 5.11) the parameters $(c_1, c_2, \alpha_1, \alpha_2, \rho_1, \rho_2)$ depend on r, i.e.,

$$(c_1, c_2, \alpha_1, \alpha_2, \rho_1, \rho_2) = (c_1, c_2, \alpha_1, \alpha_2, \rho_1, \rho_2)(r). \tag{5.12}$$

Following Baker, Caflisch & Siegel 1992, Pugh 1989 and Pugh & Cowley 1992 , $(c_1, c_2, \alpha_1, \alpha_2, \rho_1, \rho_2)$ were determined through a sliding fit; i.e., for each k, the 6 parameters were chosen to exactly fit the 6 values $f_k, f_{k+1}, \ldots, f_{k+5}$. The asymptotic fit was successful since the values $(c_1, c_2, \alpha_1, \alpha_2, \rho_1, \rho_2)$ were found to be (nearly) independent of the starting index k, as well as independent of the discretization size $dr = 2\pi/N_r$.

The parameters α_1 and α_2 were found to be $\alpha_1 = -\frac{2}{3}\pm.01$ and $\alpha_2 = 0.\pm.01$ independent of r. The amplitude term c_1 is also nearly constant; while the phase c_2 and the real position ρ_2 are approximately linear in r. Since the singularity moves toward the real axis at speed σ in the imaginary z direction, the first singularity in time corresponds to the minimum value of the complex position $\rho_1(r)$ for $r = r_{\min}$. This was seen to occur at approximately $r = 2$, near which ρ_1 was found to be approximately quadratic in $r - r_{\min}$.

Now the time dependence is included by replacing z with $z - i\sigma t$. Using the above results, for r near $r_{\min}$ and z near $z_+ + i\sigma t$ we approximate

$$\begin{aligned}
c_1 &= \bar{c}_1, \quad c_2 = \bar{c}_2 \\
\rho_1 &= \bar{\rho}_1 + \bar{\rho}_{11}(r - r_{\min})^2 \\
\rho_2 &= \bar{\rho}_2 + \bar{\rho}_{21}(r - r_{\min}) \\
\alpha_1 &- i\alpha_2 = -2/3.
\end{aligned} \tag{5.13}$$

Define new orthogonal variables centered at the real singularity position and time by

$$\begin{aligned} \tau &= \sigma t - \bar{\rho}_1 \\ \zeta &= (z - \bar{\rho}_2) - \bar{\rho}_{21}(r - r_{\min}) \\ \eta &= \sqrt{\bar{\rho}_{11}}((r - r_{\min}) + \bar{\rho}_{21}(z - \bar{\rho}_2)) \end{aligned} \tag{5.14}$$

The form of the singularity in (5.10) is then

$$u_r(z, r, t) \cong c\xi^{-1/3} \tag{5.15}$$

in which the singularity variable ξ is

$$\xi = \eta^2 - \tau - i\zeta. \tag{5.16}$$

The corresponding stream function is

$$\psi(z, r, t) \cong c\xi^{2/3}. \tag{5.17}$$

This shows that the singularity occurs at the centers of oval-shaped rolls which are flattening as the singularity forms. Further details of this solution, as well as for the numerical method, can be found in Caflisch 1992 .

6. Conclusions

The examples above provide some relatively simple contexts in which singularity formation can be seen more or less explicitly. Two principles seem to be at work in these examples: First the off diagonal elements in the Jordan form for a first order system can lead to singularity formation. Second, singularities can be understood as moving in the complex plane, until the real singularity time when they hit the real axis. This same point of view is espoused by the contributions of Baker & Tanveer and Cowley in these proceedings.

The ultimate goal of this investigation is to describe singularity formation for the 3D Euler equations (or to show that singularities do not form). A second goal is to identify the generic forms of singularities for 3D Euler, as performed for first order systems in diagonal form with two speeds in Caflisch, Ercolani, Hou & Landis 1992 . The simplicity of the singularity found in Section 5 suggests that such a form may be generic.

Acknowledgements

Research supported in part by URI grant #N00014092-J-1890 from DARPA.

References

C.R. Anderson and C. Greengard. The vortex ring merger problem at infinite Reynolds number. *Communications on Pure and Applied Mathematics*, 42:1123–1139, 1989.

David H. Bailey. Mpfun: A portable high performance multiprecision package. *RNR Technical Report RNR-90-022*, 1991a.

David H. Bailey. Automatic translation of fortran programs to multiprecision. *RNR Technical Report RNR-91-025*, 1991b.

G.R. Baker, Russel E. Caflisch, and Michael Siegel. Singularity formation during Rayleigh-Taylor instability. 1992.

Claude Bardos and S. Benachour. Domaine d'analycite des solutions de l'equation d'Euler dans un ouvert de R^n. *Annali della Scuola Normale Superiore di Pisa*, IV,4:647–687, 1977.

J.T. Beale, T. Kato, and A. Majda. Remarks on the breakdown of smooth solutions for the 3-D Euler equations. *Communications on Mathematical Physics*, 94:61–66, 1984.

John Bell and Daniel Marcus. *Communications on Mathematical Physics.*

A. Bhattacharjee and Xiaogang Wang. Finite-time vortex singularity in a model of three-dimensional Euler flows. 1992.

M.E. Brachet, D. Meiron, S. Orszag, B. Nickel, R. Morf, and U. Frisch. Small-scale structure of the Taylor-Green vortex. *Journal of Fluid Mechanics*, 130:411–452, 1983.

R.E. Caflisch. Singularity formation for complex solutions of the 3D incompressible Euler equations. *preprint*, 1992.

R.E. Caflisch, N. Ercolani, T.Y. Hou, and Yelena Landis. Multi-valued solutions and branch point singularities for nonlinear hyperbolic and elliptic systems. *Communications on Pure and Applied Mathematics*, 1992.

R.E. Caflisch, Xiaofan Li, and M.J. Shelley. The collapse of an axi-symmetric swirling vortex sheet. 1992.

R.E. Caflisch, O.F. Orellana, and M. Siegel. Singularity formulation and ill-posedness for vortex sheets. *SIAM J. Math. Anal.*, 20:293–307, 1989.

S. Childress, G.R. Ierley, E.A. Spiegel, and W.R. Young. Blow-up of unsteady two-dimensional Euler and Navier-Stokes solutions having stagnation point form. *Journal of Fluid Mechanics*, 203:1–22, 1989.

A. Chorin. Estimates of intermittency, spectra and blowup in developing turbulence. *Communications on Pure and Applied Mathematics*, 24:853–856, 1981.

A. Chorin. The evolution of a turbulent vortex. *Communications on Mathematical Physics*, 83:517–535, 1982.

A. Forestier and P. LeFloch. Multivalued solutions to some nonlinear and nonstrictly hyperbolic systems. *Japan J. Applied Math.*, 1992.

R. Grauer and T. Sideris. Numerical computation of 3d incompressible ideal fluids with swirl. *Physical Review Letters*, 25:3511–3514, 1991.

Robert M. Kerr. Evidence for a singularity of the three-dimensional, incompressible Euler equations. 1992.

Robert Krasny. A study of singularity formation in a vortex sheet by the point-vortex approximation. *Journal of Fluid Mechanics*, 167:65–93, 1986a.

Robert Krasny. Desingularization of periodic vortex sheet roll-up. *Journal of Computational Physics*, 65:292–313, 1986b.

D.I. Meiron and M.J. Shelley. personal communication. 1992.

D.W. Moore. The spontaneous appearance of a singularity in the shape of an evolving vortex sheet. *Proc Roy. Soc. London A*, 365:105–119, 1979.

D.W. Moore. Numerical and analytical aspects of Helmholtz instability. In Niordson and Olhoff, editors, *Theoretical and Applied Mechanics*, pages 629–633. Proc. XVI ICTAM, North-Holland, 1985.

David A. Pugh. *Development of vortex sheets in Boussinesq flows - Formation of singularities.* PhD thesis, Imperial College, 1989.

David A. Pugh and Stephen J. Cowley. On the formation of an interface singularity in the rising 2-d boussinesq bubble. 1992.

Alain Pumir and Eric D. Siggia. Collapsing solutions to the 3-d Euler equations. *Physics of Fluids*, A2:220, 1990.

Alain Pumir and Eric D. Siggia. Development of singular solutions to the axisymmetric Euler equations. 1992.

M.J. Shelley, D.I. Meiron, and S.A. Orszag. Dynamical aspects of vortex reconnection. 1992.

Michael Siegel. *An analytical and numerical study of singularity formation in the Rayleigh-Taylor problem*. PhD thesis, NYU, 1989.

Eric D. Siggia. Collapse and amplification of a vortex filament. *Physics of Fluids*, 28:794–805, 1985.

J.T. Stuart. Nonlinear Euler partial differential equations: Singularities in their solution. *Symposium in honor of C.C. Lin*, 1989.

C. Sulem, P.L. Sulem, and H. Frisch. Tracing complex singularities with spectral methods. *Journal of Computational Physics*, 50:138–161, 1983.

Xiaogang Wang and A. Bhattacharjee. Is there a finite-time singularity in axisymmetric Euler flows? 1991.

Complex Analytic Branching Structures in Porous Media Convection

Thierry Dombre
Centre de Recherches sur les Très Basses Températures
CNRS
BP166
38042 Grenoble Cedex 09
France

ABSTRACT. We apply the Schwarz function approach to the problem of convection in two-dimensional porous media, in the limit where a sharp interface develops between the hot and cold fluids. We show how this formalism provides a natural language to describe the proliferation of singularities in the complex plane which was found previously in a joint work with A. Pumir and E.D. Siggia (ref. [2] of this paper).

1. Introduction

Convection in porous media consists in the advection of an active scalar (the temperature) in an incompressible velocity field. The velocity is proportional to the buoyancy force (i.e. linearly related to the advected scalar) but a non local pressure term has to be added, in order to enforce the constraint of incompressibility [1],[2]. In the limit where a sharp interface develops between the hot and cold fluids, the problem becomes in two dimensions essentially equivalent to an Hele-Shaw flow with two fluids of different density and the same viscosity in the presence of gravity. In this idealized limit, the vorticity is concentrated on the interface with a density per unit length equal to $sin\alpha$ where α is the angle between the local tangent to the curve and the horizontal. The vortex sheet evolution is governed by the equation (1) given below, which is rather close to the Birkhoff-Rott equation.

In ref. [2], we showed that the presence of the Poisson kernel in the r.h.s. of (1) leads to the proliferation of singularities in the complex plane at short times, when one starts from an initial distorted shape different from a parabola (the parabola is a steady state of (1) in open geometries). In contrast to other vortex sheet problems like the Kelvin Helmholtz instability for instance, the Lagrangian coordinate drops out in the limit of a zero

R. E. Caflisch and G. C. Papanicolaou (eds.), Singularities in Fluids, Plasmas and Optics, 65–74.
© 1993 Kluwer Academic Publishers.

thickness interface. One is left with a purely geometric problem, where the "stretching" aspects become secondary. In ref. [2], we chose the horizontal coordinate x as the independent variable and tried to solve for $z(x)$ and $\bar{z}(x)$, where $z = x + iy$ and $\bar{z} = x - iy$ denote the position of the interface in complex notations. Therefore we had to work with a set of three variables $\{x, z, \bar{z}\}$, a circumstance which made the discussion rather cumbersome. Another annoying feature of this first parametrization is that it does not put on the same footing folds (which make the map $x(y)$ multivalued) and overhangs (which make the map $y(x)$ multivalued). Although this might be not too serious a limitation since gravity distinguishes the horizontal and vertical directions, we would like to suggest in this paper a more symmetrical choice of parametrization, namely solve for $\bar{z}(z,t)$. At all steps of the analysis one deals only with a set of two variables $\{z, \bar{z}\}$, and this is the least one can do to treat a problem involving two fluids.

In Section 2, we set up the formalism, derive the equation of motion for the so-called Schwarz function of the interface [3] and show how it should be analytically continued. A peculiarity of vortex sheet dynamics is that the evolutions in the z and $\bar{z}$ planes always interact in an intricate way and as we shall argue, this interaction is primarily responsible for the spontaneous generation of singularities. In Section 3, we reinvestigate in some details the example of a cubic curve at the initial time. Our purpose is not to find anything new at this level but we want to see how things really work in this approach. We get a picture very similar to what we had in [2], although we are now in a different manifold. Square root singularities form on the physical sheets of both planes and move towards the interface contour at early times of evolution. It should be noted that the physical contour itself is moving in the z and $\bar{z}$ planes but it has to avoid branch points on the first Riemann sheet (at least, it cannot be crossed by any of them). Hence the clustering of such singularities near the physical contour, if it turned out to occur at later times, could well be associated to a meandering shape of the interface and offer an explanation of the numerical observations reported in ref. [1].

2. The Interface Equation in the Schwarz Function Formalism

We start from the expression of the Lagrangian velocity for convection in porous media in complex notations [2]

$$\partial_t \bar{z}(\gamma, t) = \frac{1}{2\pi i} PV \int_{-\infty}^{+\infty} \frac{\frac{\partial y}{\partial \gamma}(\gamma', t)}{z(\gamma, t) - z(\gamma', t)} d\gamma' \tag{1}$$

where the integral is a principal value integral. Once finite thickness corrections are added to eq. (1), a natural choice for the Lagrangian coordinate γ is the area of the vortex layer between a reference point and the point of observation but in the limit of an ideal interface of zero thickness, γ becomes a dumb variable. Up to a trivial boundary term, (1) can be recast as

$$\partial_t \bar{z}(\gamma, t) = \frac{1}{4\pi} PV \int_{C(t)} \frac{\frac{\partial \bar{z}}{\partial z}(z', t)}{z - z'} dz' \tag{2}$$

where $C(t)$ is the interface contour in the z plane. It is tempting at this level to choose z as the new independent variable and try to solve for $\bar{z}$ in the form $\bar{z} = g(z, t)$. The map g is the Schwarz function of $C(t)$ [3]. Naturally, to go from the Lagrangian velocity written above to $\frac{\partial \bar{z}}{\partial t}|_z = \frac{\partial g}{\partial t}$, one has to subtract some tangential contribution due to reparametrization and the result reads

$$\frac{\partial g}{\partial t} = \frac{i}{4} Hg' + \frac{i}{4} g' \times \overline{Hg'} \tag{3}$$

In this equation g' is for $\frac{\partial g}{\partial z}$, while Hg' is the Hilbert transform of g' in the z plane. We define the Hilbert transform of a function $f(z)$ with respect to the instantaneous position of $C(t)$ as

$$Hf(z) = \frac{1}{\pi i} PV \int_{C(t)} \frac{f(z')}{z - z'} dz' \tag{4}$$

Note that by construction $g' = \frac{\partial \bar{z}}{\partial z} = e^{-2i\alpha}$, where α is the angle between the local tangent to the curve and the horizontal axis. With this in mind, it is straightforward to check that the r.h.s. of (3) is a purely normal velocity as it should in this formulation.

To make progress, we must understand how to analytically continue eq. (3) in the whole z plane, since at the moment it is defined only on $C(t)$. To do this, we shall consider C with the appropriate orientation as defining the "real axis" in the z plane (from now on, we shall drop the time dependence of C to make notations simpler). The region to the left (resp. right) of C, to be noted henceforth $\mathcal{U}$ (resp. $\mathcal{L}$), can then be seen as the upper (resp. lower) half complex planes in the z plane. The basic property of the Hilbert transform, as defined in (4), is that for any function $f(z)$, $(f + Hf)(z)$ is analytic in $\mathcal{L}$, while $(f - Hf)(z)$ is analytic in $\mathcal{U}$.

Similar notions can be introduced in the $\bar{z}$ plane, which we are forced to consider, because of the presence of $\overline{Hg'}$ in the r.h.s. of (3). First, the inverse function of g is nothing but its complex conjugate on the interface contour; i.e. g satisfies for z belonging to C

$$\bar{g}(g(z, t), t) = z \tag{5}$$

In the following, we shall denote $\bar{z}$ and $\bar{g}$ by z^* and g^* to recall that they are not anymore complex conjugates of z and g, once away from the contours C or $\overline{C}$. We define the Hilbert transform of a function $f(z^*)$ in the z^* plane as

$$H^{*}f(z^{*}) = \frac{1}{\pi i} PV \int_{\overline{C}} \frac{f(z'^{*})}{z^{*} - z'^{*}} dz'^{*} \tag{6}$$

Note that $\bar{\mathcal{U}} = \mathcal{L}^{*}$ and $\bar{\mathcal{L}} = \mathcal{U}^{*}$ if $\mathcal{U}^{*}$ and $\mathcal{L}^{*}$ are the upper and lower half complex planes with respect to $\overline{C}$ in the z^{*} plane.

Once equipped with these tools, we can rewrite eq. (3) as

$$\frac{\partial g}{\partial t} = \frac{i}{4} Hg' - \frac{i}{4} g' \times H^{*} g^{*\prime}(g) \tag{7}$$

which has now a meaning in the whole z plane. In the last term of this equation, the argument of $H^{*}g^{*\prime}$, which lives in the z^{*} plane, is $g(z,t)$, so that the quantity as a whole is indeed a function of z. The inverse Schwarz function satisfies a similar equation in the z^{*} plane, which is found to be

$$\frac{\partial g^{*}}{\partial t} = \frac{i}{4} H^{*} g^{*\prime} - \frac{i}{4} g^{*\prime} \times Hg'(g^{*}) \tag{8}$$

In principle, (8) is redundant with (7) but in some cases it is helpful to jump from one to the other equation, in order to identify more easily singular behaviours.

Now we can try to address our initial problem: does the dynamics embodied in eqs. (7),(8) give rise as time elapses to new singularities of g or g^{*}? It should be realized that, provided the contour C itself remains analytic, generic singular points of g are zeros of the derivative map g' and square root singularities. Zeros of g' are related to square root singularities of g^{*} and vice versa. Indeed, one has

$$z^{*} - z_c^{*} \sim (z - z_c)^2 \Leftrightarrow z - z_c \sim (z^{*} - z_c^{*})^{\frac{1}{2}} \tag{9}$$

From the relation $\frac{\partial \bar{z}}{\partial z} = e^{-2i\alpha}$, we infer that the analytic continuation of the angle α is logarithmically divergent at both kinds of points. It follows that these points are where the curvature $\kappa = \partial\alpha/\partial s$ diverges and are therefore the physically interesting objects. It is easier to ask about the possible nucleation of square root singularities. Let us concentrate on the equation (7) for $g(z,t)$. The effect of the first term in the r.h.s. of (7) is rather simple. Since by construction Hg' cannot be more singular than $\pm g'$ (where the sign depends on the position of the singularity we are looking at with respect to C), it follows that the first term cannot generate new singularities. At most does it advect the existing ones towards the physical contour C. So we come to the conclusion that the potential source of trouble lies in the second term of eq. (7). More precisely,"dangerous" points are where $H^{*}g^{*\prime}(g(z))$ diverges while $g'(z)$ does not vanish (the last condition is necessary to preclude a cancellation which would result in a regular behaviour). If such points exist, we expect to observe in their neighborhood the formation of daughter square root singularities. It would appear that the two preceding requirements exclude each other, because singularities of $g^{*\prime}$, and therefore the ones of $H^* g^{*\prime}$, are closely connected to zeros of g' according to (9). What happens is that the analytic continuation process in the z plane may be accompanied by a non trivial excursion of $g(z)$ in the z^* plane, during which the "real axis" $\overline{C}$ is crossed. Whenever this occurs, the link between the singularities of $g^{*\prime}$ and those of $H^* g^{*\prime}$ is broken. We shall demonstrate this clearly in the next Section by considering the example of a cubic curve at the initial time. The only justification for this choice is that it is the simplest one after the parabola and calculations can still be done by hand. However, we believe that the conclusions are rather general. With respect to our previous formulation of the problem in the x coordinate [2], the present approach, besides its aesthetic appeal, yields interesting information on the position of the new singularities in the various Riemann sheets which was missing before or at least would have been more difficult to obtain.

3. The Case of the Cubic

We take at the initial time

$$y = \frac{x^3}{6} \tag{10}$$

To simplify the calculations to be presented below, we assume a zero slope at the origin . The general case was discussed in [2] and could be addressed in the present formalism without any difficulty . The Schwarz function of the cubic is deduced from eq. (10) by replacing x, y by $\frac{z+\bar{z}}{2}$, $\frac{z-\bar{z}}{2i}$ and solving for $\bar{z}$. We quote, for further use, the equation obeyed by z and $\tilde{z}^* = z + \bar{z} = z + z^*$

$$\tilde{z}^{*3} - 24i\tilde{z}^* + 48iz = 0 \tag{11}$$

The resulting expression of g is

$$g(z) = -z + 2 \times 3^{\frac{1}{3}} \{V - \frac{z_c^{\frac{2}{3}}}{V}\} \tag{12}$$

where $z_c = \frac{2\sqrt{2}}{3} e^{-\frac{3i\pi}{4}}$ and $V = (-iz - i\sqrt{z^2 - z_c^2})^{\frac{1}{3}}$. Hence g has two square root singularities at $\pm z_c$. To get the right definition of g on the physical sheet, we attach to each of these points, branch cuts along the vertical direction in such a way that they do not cross the physical contour. The phase of $z - z_c$ varies from $-\pi/2$ to $3\pi/2$ as z_c is encircled anticlockwise , the one of $z + z_c$ from $-3\pi/2$ to $\pi/2$ as $-z_c$ is encircled in the same way, so that $\sqrt{z^2 - z_c^2}$ behaves as z for large $|z|$ in $\mathcal{L}$ and as $-z$ in $\mathcal{U}$. With these conventions, it can be checked that g as defined above is the only root of eq. (11) satisfying the conditions

$$\begin{aligned} &g(0) = 0 \\ &g(z) \sim -\ z + 2(-6iz)^{\frac{1}{3}} \text{ as } |z| \to \infty \text{ in } \mathcal{L} \\ &g(z) \sim -\ z - 2(+6iz)^{\frac{1}{3}} \text{ as } |z| \to \infty \text{ in } \mathcal{U} \end{aligned} \tag{13}$$

Note that $\pm(\mp 6iz)^{\frac{1}{3}} \sim x$ for large values of $|z|$ on respectively the positive and negative sides of the physical contour C, so that the last two equations in (13) are in fact expressing the obvious relation $\bar{z} = -z + 2x$ in the large.

Let us denote the Schwarz function g on the physical sheet as h_0. As branch cuts are crossed, h_0 gets permuted with the two other roots $h_\pm$ of eq. (11) whose expression is

$$-e^{\mp\frac{2i\pi}{3}}\frac{z_c^{\frac{2}{3}}}{V}\} \tag{14}$$

We find the following monodromy properties of the set $\{h_i; i = 0, \pm\}$ upon successive rotations by 2π around $-z_c$

$$h_0 \to h_+ \to h_0, \quad h_- \to h_- \tag{15}$$

The root h_- is in fact analytic at $-z_c$ and thereby upper analytic. We get just the opposite picture at z_c

$$h_0 \to h_- \to h_0, \quad h_+ \to h_+ \tag{16}$$

From eq. (11) follows the additional relation, which is also obvious from (12) and (14)

$$h_0 + h_+ + h_- = -3z \tag{17}$$

All these remarks allow us to get the expression of Hg on the physical sheet without any computation

$$Hh_0 = -(h_+ - h_-) \tag{18}$$

This is true because $h_0 - (h_+ - h_-) = -3z - 2h_+$ is lower analytic, while $h_0 + (h_+ - h_-) = -3z - 2h_-$ is upper analytic: hence $-(h_+ - h_-)$ has the expected properties of the Hilbert transform of h_0.

Corresponding relations hold for the inverse Schwarz function and are obtained by complex conjugation or putting "*". For instance $g^*(z^*) = h_0^*(z^*)$ on the physical sheet and

$$H^*h_0^* = +(h_+^* - h_-^*) \tag{19}$$

It is only for z and $g(z)$ both on the first Riemann sheet in their respective manifolds that the relation (5) between g and g^* translates into

$$h_0^*(h_0(z)) = z \tag{20}$$

The analytic continuation of eq. (20) in the z plane takes the form

$$h_j^*(h_i(z)) = z \tag{21}$$

where the index i depends solely on the path of analytical continuation towards z in the z plane and the index j solely on the path followed by $g(z)$ in the z^* plane. A closer inspection of eqs. (12),(14) and their complex conjugates reveals for instance that h_+^* (resp. h_-^*) is the inverse function of h_0 at large values of z in $\mathcal{L}$ (resp. $\mathcal{U}$).

We are now in position of asking whether the scenario outlined in Section 2 for the formation of new singularities can work in this case. First, the Hilbert transform $H^*g^{*\prime}$ diverges only at $z^* = \pm z_c^*$ at the initial time. Let us consider more particularly $-z_c^*$ (to get what happens with z_c^*, just change in what follows + and − signs). By taking the complex conjugate of eq. (11) and differentiating with respect to z^* one finds that $h_0^*(-z_c^*) = h_+^*(-z_c^*) = -2z_c^* = z_1$, which together with the complex conjugate of eq. (17) leads to the value of the third root $h_-^*(-z_c^*) = +7z_c^* = z_2$. On the other hand, it is easily proven by differentiating eq. (11), this time with respect to z, that g' vanishes only at $z = \pm 2z_c^* = \mp z_1$ where it takes the value $\pm z_c^*$, in other words the critical value of g^*. By definition there must exist a couple of indices i and j such that $h_i(z_1) = h_j(z_2) = -z_c^*$. Since z_1 and z_2 do not belong to the set of singular points of g, which are characterized by the merging of two of their images, the indices i and j are unique. In fact, it is not difficult to show that $i = j = 0$. Indeed, on the straight line $\mathcal{D}$ joining the origin to $-z_c^*$ in the z plane, we parametrize z as $e^{-\frac{i\pi}{4}}u$ with u real and look for g in the form $e^{-\frac{i\pi}{4}}(-u+\phi(u))$. Then eq. (11) yields upon differentiation with respect to z

$$\frac{\partial\phi}{\partial u} = \frac{16}{8+\phi^2} \tag{22}$$

Since ϕ vanishes at the origin, we deduce from (22) that ϕ remains real on the whole line $\mathcal{D}$. A graphical solution of (22) shows that h_0 equals $-z_c^*$ at two distinct points on $\mathcal{D}$. The closest point to the origin is easily identified with z_1 where h_0' vanishes. The furthest one is z_2. This is the interesting point because h_0' does not vanish there.

So we have what we needed from the beginning : a point z_2 which lies on the physical sheet ($g(z) = h_0(z)$), is such that $g(z_2)$ is a singularity of $H^*g^{*\prime}$, but at the same time is not a zero of g'. It can be checked that on the way from the origin to z_2, z passes by $-z_1$ where $h_0(z)$ equals $+z_c^*$. Beyond that point, the inverse map g^* turns into h_-^* and the analytic continuation of $H^*g^{*\prime}$ into $h_+^{*\prime} - h_0^{*\prime}$. In the neighborhood of z_2, the most singular part of the equation of motion reduces to

$$\frac{\partial g}{\partial t} = \frac{i}{4} g' h_0^{*\prime}(g) \tag{23}$$

We have implicitely assumed that the dynamics of $H^* g^{*\prime}$ will remain smooth at the short time scales we are interested in and replaced it by its most singular part *at the initial time*. The validity of this hypothesis will be checked *a posteriori*. Explicit computation of $h_0^{*\prime}$ near $-z_c^*$ gives $h_0^{*\prime} \sim 2^{\frac{1}{4}} e^{\frac{3i\pi}{8}} (z^* + z_c^*)^{-\frac{1}{2}}$ which turns eq. (23) into

$$\frac{\partial g}{\partial t} = \frac{i}{4} g' \frac{2^{\frac{1}{4}} e^{\frac{3i\pi}{8}}}{\sqrt{g + z_c^*}} \tag{24}$$

with the initial time condition $g + z_c^* = -\frac{3}{5}(z - z_2)$. Not too surprisingly, this equation has precisely the same structure as the one obtained in [2] and its solution reads

$$z - z_2 == -\frac{i}{4} \frac{2^{\frac{1}{4}} e^{\frac{3i\pi}{8}} t}{\sqrt{g + z_c^*}} - \frac{5}{3}(g + z_c^*) \tag{25}$$

By defining $z - z_2 = a e^{-\frac{i\pi}{4}} \tilde{z}$ and $g + z_c^* = b e^{\frac{3i\pi}{4}} \tilde{g}$, eq. (25) can be recast, with an appropriate choice of the real scaling coefficients a and b, as

$$\tilde{z} = \frac{e^{-\frac{i\pi}{4}} t}{\sqrt{\tilde{g}}} + \tilde{g} \tag{26}$$

For large $\tilde{z}$, $\tilde{g}$ behaves as

$$\tilde{g} \sim \tilde{z} - \frac{e^{-\frac{i\pi}{4}} t}{\sqrt{\tilde{z}}} \tag{27}$$

Within an inner region of width $O(t^{\frac{2}{3}})$, $\tilde{g}$ deviates strongly from its form at the initial time. We can borrow from [2] its exact expression on the physical sheet

$$\tilde{g} = i\sqrt{\frac{\tilde{z}}{3}} \{W - \frac{1}{W}\} \tag{28}$$

where

$$W = e^{-\frac{2i\pi}{3}} (\sqrt{1 - \frac{1}{Z^3}} - \frac{1}{Z^{\frac{3}{2}}})^{\frac{1}{3}} \tag{29}$$

and

$$Z = (\frac{2}{t})^{\frac{2}{3}} \times \frac{\tilde{z}}{3i} \tag{30}$$

With the phase of Z varying between 0 and 2π, it can be checked that the solution (28) exhibits the right asymptotic behaviour (27) everywhere in the complex plane.

According to (28)-(30), three square root singularities form at $Z^3 = 1$. However, not all of them reside on the first Riemann sheet. It turns out that the physical root (28) is indeed multivalued at the point $\tilde{z} = 3(\frac{t}{2})^{\frac{2}{3}}e^{-\frac{i\pi}{6}}$, which is moving towards the interface contour. We noticed in [2] that the first derivative of $\tilde{g}$ as given by (28), has no new zeros in the inner region. This remark is very helpful here since it means that $g^{*\prime}$ does not develop singularities in the neighborhood of $g(z_2)$ which would contradict our initial assumption of a smooth dynamics for $H^*g^{*\prime}$. Naturally, as the inverse function of g, $g^* \sim h_-^*$ acquires some fine structure in the neighborhood of $g(z_2)$. But this does not show up in $H^*g^{*\prime}$ which is related to the two other roots h_+^* and h_0^*. Therefore our whole approximation scheme is coherent.

The nucleation process does not stop here, as was already explained in [2]. Again this is easily understood in our new formulation. In the same way $h_0(z)$ gets daughter square root singularities in the neighborhood of z_2, $h_0^*(z^*)$ gets three of them in the neighborhood of z_2^* (just by complex conjugation). The question is then whether there exist points z_3 such that $h_0(z_3) = z_2^*$. For a cubic of zero slope at the initial time, the answer is yes, z_3 is unique and given by $h_+^*(z_2^*)$. The same reasoning we used near z_2 can now be repeated near z_3, where again $g'(z_3) \sim h_0'(z_3) \neq 0$ but $H^*g^{*\prime} \sim h_0^{*\prime} - h_-^{*\prime}$ is singular because of the new singularities of h_0^* close to z_2^*. We shall not pursue the analysis here which should resemble the one provided in [2]. In any case, a second generation of singularities forms and some of them are definitely on the first Riemann sheet, which means they may be physically relevant at later times.

4. Conclusion

Spontaneous generation of singularities at early times in the complex plane is a rather ubiquitous phenomenon, which appears for instance also in the contributions of Cowley and Tanveer to this Workshop [4]. It occurs everytime the initial condition contains some features like an isolated zero of the derivative map or some singularity, that the equation of motion does not "like". The initial singularity is then dressed at later times and this process may well give rise to a cloud of new singularities. The peculiarity of the mechanism described here, in an elaboration of previous work [2], is that it is non local and, as far as we can see, sensitive to rather subtle geometric properties of the initial interface shape, in a way that we do not yet fully apprehend.

The most interesting question from a physical point of view concerns the fate of this myriad of new algebraic singularities as time goes on: do they keep on moving towards the physical contour so that they can be eventually observed, or are they by nature hidden by the original singularity which gave them birth in the first place? On that difficult problem, we have nothing to say at the moment. Accurate numerical methods for tracing singularities in the complex plane, as recently developed by Baker and collaborators [5], might be very useful to settle this issue.

Finally, some participants in the Workshop raised the possibility that the effects discussed here would be mostly due to the topology of the cubic. Indeed, the cubic is not a very plausible shape for a vortex sheet: in particular, it cannot be obtained by a continuous deformation of the flat interface. To answer to some extent this objection, we should say that the same phenomena are found when considering bell-shaped interfaces like $y = \frac{a}{x^2+b^2}$ at the initial time.

References

[1] Pumir, A., Shraiman, B., and Siggia, E.D. : 1992, 'Vortex Morphology and Kelvin's Theorem', *Phys. Rev. A* **45**, 5351-5354.

[2] Dombre, T., Pumir, A., and Siggia, E.D. : 1992, 'On the Interface Dynamics for Convection in Porous Media', *PhysicaD* **57**, 311-329.

[3] See for instance Howison, S.D. : 1986, 'Fingering in Hele-Shaw cell', *J. Fluid Mechanics* **167**, 439 and references therein.

[4] Cowley, S.J. : 1992, 'Singularities in Vortex Sheets and Related Flows'; Tanveer, S. : 1992, 'Singularity Dynamics in Hele-Shaw Flow for Small Surface Tension' in the Proceedings of this Workshop.

[5] Baker, G. : 1992, 'Tracing Singularities Numerically in the Complex Physical Plane' in the Proceedings of this Workshop.

SINGULARITIES OF THE VELOCITY FIELD AND INTERACTION LOCALITY OF TURBULENCE

G.E. FALKOVICH
Dep. of Nuclear Physics
Weizmann Institute of Science
Rehovot 76100 Israel

ABSTRACT. A new exact relation found for the triple correlation function of velocities in the framework of Euler equation allows us to describe a balance of interactions between motions with distant scales for any spectrum of developed turbulence. For a steady spectrum carrying constant energy flux, the interactions of the given eddy with large and small eddies are shown to be counterbalanced i.e. these interactions give the opposite signs for energy change and they decrease by the same law with the distance in k-space. The locality properties of possible spectra carrying other motion integrals are studied.
Considering, in particular, the multiscaling picture of developed turbulence, we analyze the range of singularity exponents h of the velocity field $\left(\delta v(r) \propto r^h\right)$ that are compatible with the locality. It is shown that the condition of the infrared locality of interaction (with largest eddies) could give only the upper restriction for the exponent fixing thus the weakest singularity. The upper limit thus found ($h_{\max} = 1$) coincides with the boundary exponent of singularity of the first velocity derivative (i.e. energy dissipation). As far as an ultraviolet interaction is concerned, we prove that there is no lower restriction for the exponent h provided the probability to find h-subset decreases with h quickly enough for the mean exponent of the double correlator to be less than two which is certainly so. Therefore, a local cascade picture of developed turbulence is compatible, in principle, with arbitrary strong singularities of the velocity field. Such singularities define the behavior of high-order correlators though hardly influence low-order correlators (the double and the triple ones).

1. Introduction

1.1. FORMULATION OF THE PROBLEM

One could imagine two possible universal pictures of developed turbulence in the inertial range of scales that are intermediate between external and dissipation scales. The first picture presumes that the main contribution into the interaction providing energy redistribution between different scales stems from the interaction of motions with the spatial scales of the same order of magnitude. A relay energy transfer over scales from the pumping to the damping region is thus assumed to be local in k-space. The detailed information about the structure of the pumping is lost during many steps of the cascade so a universal spectrum of turbulence has to be defined only by the value of the energy flux and by the current wavenumber [1]. The second picture presumes that some universal spatio-temporal patterns are present in turbulent flow. Such a picture implies strong phase correlation between harmonics with different wavenumbers that is necessary to form a pattern. Among such spatio-temporal structures, those having singularity are of most importance

R. E. Caflisch and G. C. Papanicolaou (eds.), Singularities in Fluids, Plasmas and Optics, 75–91.
© 1993 *Kluwer Academic Publishers.*

since the singularity of the velocity or its derivatives should define the small-scale asymptotics of the Fourier image of the velocity field.

Which class (cascade or structural) the turbulence of an incompressible fluid belongs to? The interaction between different spatial harmonics is strong which should provide essential phase correlation. Some (though contradictory) evidence for singularity formation could be found in other contributions of this volume. On the other hand, numerous experiments show that the energy spectrum is satisfactory described by the Kolmogorov cascade picture [2]. From the viewpoint of a turbulence theory, the crucial point for cascade picture to be valid is the locality of interaction.

The locality analysis of developed turbulence in incompressible ideal fluid is the subject of this paper, its goal is two-fold. First, this paper is an attempt to show that the presence of rather arbitrary singularities could be compatible with the interaction locality which is the basic feature of the cascade picture. (Here locality should be considered in a statistical sense i.e. as being averaged over different velocity field configurations). To do this, one should develop a mathematical technique that allows one to analyze the energy exchange between different scales for quite an arbitrary turbulent distribution. Such a technique is based on the Wyld diagram formalism [3] improved by the elimination of sweeping by Kraichnan [4] and Lvov [5]. This technique allows us to obtain an exact relation for the asymptotic behavior of the triple velocity correlation function governing the energy exchange between eddies of strongly different scales [6]. This asymptotics expressed via the double velocity correlation function giving the energy spectrum of turbulence (Sect.2). Having such a relation, one could pursue the second goal of this paper: to provide a detailed analysis of the interaction balance in a stationary turbulence. As one can see in Sect.3, quite fundamental physical statement could be proved for any steady turbulence cascade with one flux of a motion integral (energy): the contribution into the energy exchange of small scales has the same law of decrease with the ratio of scales as that of large scales. This statement of interaction counterbalance is valid for the turbulence of waves with a decay dispersion law (capillary waves, sound etc) as well as for the turbulence of an incompressible fluid.

1.2. PHENOMENOLOGY OF DEVELOPED TURBULENCE

Here we briefly remind the main statements of the phenomenological approach to the turbulence of an incompressible fluid. The theory of developed hydrodynamic turbulence deals with flows for which the external scale of motion L is much larger than the scale η of effectively damping modes. This phenomenon occurs when the Reynolds number $R = LU/\nu$ is much larger than unity. Here U is some large-scale external velocity (for example, the mean velocity of the flow past a body) and ν is the kinematic viscosity. The statistical properties of the turbulent flow in the inertial range of scales l ($L \gg l \gg \eta$) are usually supposed to be self-similar, which means that, for example, the velocity increment over a distance l should behave as follows

$$\delta v(r,l) \equiv v(r+l) - v(r) = \delta v(l) \propto l^h \ . \tag{1}$$

Here equality means identical statistical properties. Power behavior with l is also assumed for high-order momenta of the velocity field:

$$S_p(l) \equiv \langle (\delta v(l))^p \rangle \sim U^p \left(\frac{l}{L}\right)^{\zeta_p} . \tag{2}$$

The first attempt to find exponent h was done by Kolmogorov [1] who assumed the energy dissipation rate ϵ to stay finite in the limit of infinite Reynolds numbers. Proceeding from this, Kolmogorov obtained the following expression for the third-order moment of the velocity field [1]

$$S_3 = \langle(\delta v(l))^3\rangle = -\frac{4}{5}\epsilon l \ . \tag{3}$$

This formula is one of a few exact relations in turbulence theory. The triple correlator governs the behavior of the energy flux over the scales (see Sect.2 for details) and the expression (3) means a constancy of the flux in the inertial range. Indeed, for the scale l, the rate of energy transfer is the energy $(\delta v)^2$ divided by the typical turnover time $l/\delta v$. One should remember that the statement of flux constancy is meaningful only provided the locality of interaction in the k-space (which means that only motions of comparable scales effectively interact).

The simplest phenomenological approach implies single-scaling behavior: $\zeta_p = ph$, $h = 1/3$, which gives the famous Kolmogorov $\frac{5}{3}$-law. As far as the energy spectrum is concerned, experiments [2] confirm this law. However, an anomalous scaling (i.e., a nonlinear dependence of ζ on p) seems to be observed both in natural and numerical experiments [7,8]. To take this scaling into account, the second phenomenological approach, namely the multifractality hypothesis [9] was suggested. It presumes a turbulent flow to possess a range of scaling exponents $(h_{\min}, h_{\max})$. Each scaling law $\delta v(l) \propto l^h$ takes place on the subset of the three-dimensional space having Hausdorff dimension $D(h)$. Such a picture gives thus the following expression for correlation function:

$$S_p(l) \sim U^p \int_{h_{\min}}^{h_{\max}} d\mu(h) \left(\frac{l}{L}\right)^{ph+3-D(h)} , \tag{4}$$

Here the measure of the different scalings $d\mu(h)$ is supposed to be smooth. The exponent $3 - D(h)$ defines the probability of finding the scaling exponent h. Multifractality concept also gives a power behavior for all correlation functions: since in (4) the smallest exponent dominates in the limit of $l \to 0$, then the correlators behave by the power laws

$$S_p(l) \sim U^p \left(\frac{l}{L}\right)^{\zeta_p} , \qquad \zeta_p = \min_h (ph + 3 - D(h)) \ , \tag{5}$$

though with an anomalous scaling defining by an unknown function $D(h)$. The lesser the exponent h, the stronger the respective singularity of the velocity or its derivative. Note that the multifractal approach still remains mainly formal and underlying physics is unclear.

Both above pictures implies (3) and are based thus on the locality assumption. The locality problem was studied previously for single-scaling Kolmogorov spectrum only. It was done in the framework of Direct Interaction Approximation [10] and in the entire perturbation series without any cutoff [11]. Here we solve general locality problem without any assumptions about the steady spectrum except scale invariance.

To analyze the mechanism of interaction balance, one should know the triple moment for noncoinciding arguments like

$$S_3(r_{12}, r_{23}, r_{13}) = \langle \delta v(r_{12}) \delta v(r_{23}) \delta v(r_{13}) \rangle \ . \tag{6}$$

In the section 2, the asymptotic expression of (6) for one distance being much smaller than others is given. To find such asymptotic behavior, one should use the exact formulas for the correlation functions which are obtained by analyzing the entire perturbation series (without any cutoff) for Euler equation. Neither dimensional approach, nor finite closures could give the correct answer in this case. Such an analysis (explained in more details in [6]) gives the second [in addition to (3)] exact relation for the asymptotic behavior of the triple correlation function. This relation allows us to give in the section 3 the whole analysis of the interaction locality. In particular, we analyze the multifractal picture directly in the framework of Euler equation. Possible presence of different exponents gives rise to a new question in the locality consideration. Is it possible that locality is violated in the ultraviolet region for sufficiently small h and in the infrared region for large h? We show that locality analysis gives the upper restriction for exponent h.

2. Exact Relation for the Triple Correlation Function

A natural way to analyze an interaction locality is to consider the Euler equation for incompressible fluid flow in the $\boldsymbol{k}$-representation [2]:

$$\frac{\partial v_\alpha(t,\boldsymbol{k})}{\partial t} = \frac{1}{2}\int \Gamma_{\alpha,\beta,\gamma}(\boldsymbol{k}|\boldsymbol{q},\boldsymbol{p}) v^*_\beta(t,\boldsymbol{q}) v^*_\gamma(t,\boldsymbol{p})\, d\boldsymbol{q}d\boldsymbol{p}\,, \tag{7}$$

$$\begin{aligned}\Gamma_{\alpha,\beta,\gamma}(\boldsymbol{k}|\boldsymbol{q},\boldsymbol{p}) &= P_{\alpha\alpha'}(\boldsymbol{k})P_{\beta\beta'}(\boldsymbol{q})P_{\gamma\gamma'}(\boldsymbol{p}) i(k_\beta\delta_{\alpha\gamma} + k_\gamma\delta_{\alpha\beta})\delta(\boldsymbol{k}+\boldsymbol{q}+\boldsymbol{p})\,,\\ P_{\alpha\beta}(\boldsymbol{k}) &= \delta_{\alpha\beta} - k_\alpha k_\beta/k^2\end{aligned} \tag{8}$$

where Γ is *the Eulerian vertex* and P is *the transverse projector.*

Equation (7) is nonlinear. Multiplying it by different powers of velocity $v(\boldsymbol{k},t)$ and averaging, one could obtain the equations for the correlation functions of the velocity field. Time evolution of n-th correlation function is governed by some integral with the correlation function of order $n+1$. As a first step, one could consider the simultaneous double correlation function

$$F_{\alpha\beta}(\boldsymbol{k},t)\delta(\boldsymbol{k}-\boldsymbol{q}) = \langle v_\alpha(\boldsymbol{k},t) v^*_\beta(\boldsymbol{q},t)\rangle\;, \tag{9}$$

which describes the energy density over scales: $E(k,t) \propto k^2 F(k,t)$. Here angular brackets denote turbulent ensemble averaging. The equation for $F_{\alpha\beta}$ is as follows:

$$\frac{\partial F_{\alpha\alpha'}(\boldsymbol{k},t)}{\partial t} = \mathrm{Re}\int \Gamma^*_{\alpha,\beta,\gamma}(\boldsymbol{k}|\boldsymbol{q},\boldsymbol{p}) F^{(3)}_{\alpha'\beta\gamma}(\boldsymbol{k},\boldsymbol{q},\boldsymbol{p};t)\, d\boldsymbol{q}d\boldsymbol{p} = I_{\alpha\alpha'}(\boldsymbol{k},t)\;. \tag{10}$$

Here $F^{(3)}$ is the simultaneous triple correlator:

$$F^{(3)}_{\alpha\beta\gamma}(\boldsymbol{k},\boldsymbol{q},\boldsymbol{p};t)\delta(\boldsymbol{k}+\boldsymbol{q}+\boldsymbol{p}) = \langle v_\alpha(\boldsymbol{k},t) v_\beta(\boldsymbol{q},t) v_\gamma(\boldsymbol{p},t)\rangle \tag{11}$$

From the physical viewpoint, the interaction locality means that the main effect on the behavior of given k-eddies (eddies of spatial scale $1/k$) arises because of their interaction with eddies of the same order of size. Mathematically, it means that the main contribution to the integral (10) arises from the regions where q and p are of order of k which means the convergence of the integral (10) and nothing else. There exist different definitions of locality separating, for instance, the case with the main contribution from the region $(k/2, 2k)$ from the case with that of the region $(k/5, 5k)$ etc. Such a difference seems to be of no principal importance for turbulence theory. Whereas the difference between the case of a convergent integral in (10) (where one could define the energy flux in the k-space) and the case of divergency (where energy is transferred nonlocally) is a qualitative one. Our definition of locality is usual in the general theory of turbulence [12].

So the main point here is the analysis of integral convergence and, in particular, the question about the convergence reserve that is the law of how does the contribution of a distant region in the k-space turn into zero while the distance increases. Interaction of k-eddies with q-eddies of much larger scales is defined by the asymptotic behavior of $F^{(3)}(\boldsymbol{k}, \boldsymbol{q}, \boldsymbol{p}; t)$ at $q \ll k \simeq p$ and $p \ll k \simeq q$. On the other hand, the interaction of k- and q-eddies with $q \gg k$ is defined by the asymptotic region $k \ll q \simeq p$. Thus, in the both cases (infrared and ultraviolet ones), the nonlocal part of eddy interaction is described by the triad interaction $\boldsymbol{k} + \boldsymbol{q} + \boldsymbol{p} = 0$ with one of the wavenumbers being much smaller than two others. So, in order to study the problem of interaction locality, it is necessary to find the asymptotic behavior of the triple correlation function (11) at $k \ll q \simeq p$ to obtain the range of exponents h providing convergence of integral. It could be shown that the locality analysis gives the same results for the exponent interval whatever order $n \geq 2$ of correlation function is used.

Let us remind the relations between the scaling properties of second and third correlation functions $F^{(2)}$ and $F^{(3)}$ in k-representation and that of S_2 and S_3 in r-representation. Double correlators are related as follows

$$S_2(l) = \int |\, 1 - \exp(i\boldsymbol{k} \cdot \boldsymbol{l}) \,|^2 \, F^{(2)}(\boldsymbol{k})\, d\boldsymbol{k} \; . \tag{12}$$

For a power function $F^{(2)}(k) \propto k^{-y}$ with $3 < y < 5$, the main contribution to this integral is given by the region $k \simeq 1/l$. Therefore, in the inertial interval where $S_2(l) \propto l^{\zeta_2}$ we have $y = 3 - \zeta_2$.

Considering similar integral for the triple correlators $S_3(l)$ and $F^{(3)}(\boldsymbol{k}, \boldsymbol{q}, \boldsymbol{p})$ and assuming the main contribution to be provided by the region $k \simeq q \simeq p \simeq 1/l$ we obtain the following scaling behavior

$$F^{(3)}(\lambda \boldsymbol{k}, \lambda \boldsymbol{q}, \lambda \boldsymbol{p}) = \lambda^{-7} F^{(3)}(\boldsymbol{k}, \boldsymbol{q}, \boldsymbol{p}) \tag{13}$$

in this region. According to the scaling hypothesis for high-order correlators, we shall assume that relation (13) valids for any k, p, q from the inertial interval. For the third correlator, we thus know the total exponent only:

$$F^{(3)}(k, q, q) \propto k^a q^b, \qquad a + b = 7 \; , \tag{14}$$

but not the exponents a and b separately. Therefore, one should analyze the equation for $F^{(3)}$ following from (7) .

The main problem here is due to the fact that integrals over $\boldsymbol{k}$ diverge in terms of Eulerian velocity v even for Kolmogorov spectrum. According to Kraichnan [13], such a divergency is caused by the strong sweeping of the small eddies by the largest ones, so it is irrelevant to the problem of the locality of dynamic interaction providing energy cascade due to an energy exchange between the eddies.

Following paper [11], one could use *the quasi-Lagrangian* (qL) velocity $\boldsymbol{u}(\boldsymbol{r}_0, \boldsymbol{r}, t)$ in order to eliminate the sweeping in the vicinity of some space point $\boldsymbol{r}_0$:

$$\boldsymbol{v}(\boldsymbol{r}, t) = \boldsymbol{u}\left[\boldsymbol{r}_0, \boldsymbol{r} - \int^t u(\boldsymbol{r}_0, \boldsymbol{r}_0, \tau)\, d\tau, t\right] . \tag{15}$$

If the velocities of all particles of fluid are identical, the above qL velocity naturally coincides with the Lagrangian velocity of fluid particle along the actual trajectory. It should be noted that equation (15) contains no approximations, since it is just the definition of the function $\boldsymbol{v}(\boldsymbol{r}, t)$. Formula (15) itself represents a precise relation between the Eulerian and quasi-Lagrangian velocities. In terms of the quasi-Lagrangian velocity, one can adequately construct a theory of turbulence [5,11] and, in particular, to find the asymptotic behavior of the triple velocity correlation function.

The equation for $\boldsymbol{u}(\boldsymbol{r}_0; \boldsymbol{k}, t)$ may be derived by substituting (15) into (7) . The resulting equation has the form of the initial Euler equation (7) but with another *dynamic* vertex V describing only the dynamic interaction of eddies in the qL approach to the theory of turbulence [5,6,11]

$$\begin{aligned} V_{\alpha,\beta,\gamma}(\boldsymbol{k}|\boldsymbol{q}, \boldsymbol{p}; \boldsymbol{r}_0) &= P_{\alpha,\alpha'}(\boldsymbol{k}) P_{\beta,\beta'}(\boldsymbol{q}) P_{\gamma,\gamma'}(\boldsymbol{p}) U_{\alpha',\beta',\gamma'}(\boldsymbol{k}|\boldsymbol{q}, \boldsymbol{p}; \boldsymbol{r}_0) , \\ U_{\alpha,\beta,\gamma}(\boldsymbol{k}|\boldsymbol{q}, \boldsymbol{p}; \boldsymbol{r}_0) &= i\{k_\gamma \delta_{\alpha,\beta}[\delta(\boldsymbol{k}+\boldsymbol{q}+\boldsymbol{p}) - \delta(\boldsymbol{k}+\boldsymbol{q}) \exp(i\boldsymbol{p}\boldsymbol{r}_0)] , \\ &\quad + k_\beta \delta_{\alpha,\gamma}[\delta(\boldsymbol{k}+\boldsymbol{q}+\boldsymbol{p}) - \delta(\boldsymbol{k}+\boldsymbol{p}) \exp(i\boldsymbol{q}\boldsymbol{r}_0)] . \end{aligned} \tag{16}$$

The main difference between the quasi-Lagrangian and the conventional (in terms of the Eulerian velocity) description of turbulence is that the wave vector $\boldsymbol{k}$ is no longer preserved in the vertex V and the functions (9) , (11) become nondiagonal in $\boldsymbol{k}$ [i.e., the correlation functions are not not proportional to $\delta(\boldsymbol{k}_1 - \boldsymbol{k}_2)$]. This is the result of the absence of spatial uniformity of the theory due to the explicit dependence of the qL velocity (15) on the coordinate of the marked point $\boldsymbol{r}_0$ where sweeping is precisely eliminated.

It is convenient to develop the turbulence description in ω-representation supposing statistical properties to be invariant with respect to a time shift. We will study double and triple correlation functions of velocity field. By analogy with (9) and (11) , let us introduce these functions for the qL velocities $\boldsymbol{u}(\boldsymbol{r}_0; \boldsymbol{k}, \omega)$ in the $\omega, \boldsymbol{k}$-representation:

$$\begin{aligned} F_{\alpha\beta}(\boldsymbol{r}_0; \omega_1, \boldsymbol{k}_1, \boldsymbol{k}_2)\delta(\omega_1 - \omega_2) &= \langle u_\alpha(\boldsymbol{r}_0; \boldsymbol{k}_1, \omega_1) u^*_\beta(\boldsymbol{r}_0; \boldsymbol{k}_2, \omega_2)\rangle , \\ F^{(3)}_{\alpha\beta\gamma}(\boldsymbol{r}_0; \omega_1, \omega_2, \omega_3, \boldsymbol{k}, \boldsymbol{q}, \boldsymbol{p})\delta(\omega_1 + \omega_2 + \omega_3) & \\ = \langle u_\alpha(\boldsymbol{r}_0; \boldsymbol{k}, \omega_1) &u_\beta(\boldsymbol{r}_0; \boldsymbol{q}, \omega_2) u_\gamma(\boldsymbol{r}_0; \boldsymbol{p}, \omega_3)\rangle , \\ \boldsymbol{u}(\boldsymbol{r}_0; \boldsymbol{k}, \omega) = \int \boldsymbol{u}(\boldsymbol{r}_0; \boldsymbol{r}, t) &\exp[i(\omega t - \boldsymbol{k}\cdot\boldsymbol{r})]\, d\boldsymbol{r} dt . \end{aligned} \tag{17}$$

It is important to emphasize that simultaneous correlation functions of the order n for qL-velocities $F^{(n)}$ coincide with the same functions of Eulerian velocities $F_e^{(n)}$ and do not depend on $\boldsymbol{r}_0$ [5,11]. Therefore we will consider the problem of triad interactions in the qL approach with the sweeping being eliminated completely. Using then the above mentioned coincidence, we will obtain a conclusion about the behavior of simultaneous correlation functions of Euler velocities. In particular, we will use the following relations for the double and the triple velocity correlation functions:

$$\int F(\boldsymbol{r}_0;\boldsymbol{k},\boldsymbol{q},\omega)\,d\omega/2\pi = F_e(\boldsymbol{k},\boldsymbol{q}) = F_e(\boldsymbol{k})\delta(\boldsymbol{k}-\boldsymbol{q})\ . \tag{18}$$

$$\begin{aligned} &F^{(3)}_{\alpha\beta\gamma}(\boldsymbol{k},\boldsymbol{q},\boldsymbol{p})\delta(\boldsymbol{k}+\boldsymbol{q}+\boldsymbol{p}) \\ &= \int F^{(3)}_{\alpha\beta\gamma}(\boldsymbol{r}_0;\omega_1,\omega_2,\omega_3,\boldsymbol{k},\boldsymbol{q},\boldsymbol{p})\delta(\omega_1+\omega_2+\omega_3)\,d\omega_1 d\omega_2 d\omega_3/(2\pi)^3 \end{aligned} \tag{19}$$

To obtain the simultaneous correlation function, we have integrated the expression (18) and (19) over all of ω_j.

In order to find an asymptotic behavior of the simultaneous triple correlation function of Eulerian velocity Lvov and Falkovich [6] analyzed the entire diagram series for the triple moment of qL velocity. In the asymptotic limit $k \ll q \simeq p$ the main contribution into the triple correlator has been found. It is given by some principal subsequence of diagrams containing the double correlator of small wave vector $\boldsymbol{k}$. The reason for this could be readily understood: Any term in the perturbation series is expressed in terms of the double correlation function and Green's function $G(\vec{r}_0,\omega,\vec{k},\vec{k}_1)\delta(\omega-\omega_1) = \delta u(\vec{r}_0,\vec{k},\omega)/\delta f(\vec{r}_0,\vec{k}_1,\omega_1)$ giving the response of the velocity to extraneous force f (per unit mass). Having dimensionality of time, Green's function scales with k as an inverse frequency $G \propto \omega^{-1} \propto k^{-x}$. The double correlation function $F(\vec{k},\vec{k}_1,\omega)$ in ω-representation scales with k as k^{-y-x}. So for small k the contribution from the double correlator of small wavenumbers larger by a parameter $(q/k)^y$ than the contributions from Green's function if y is positive.

The result for the simultaneous triple correlation function of qL velocity (that coincides with the same function of Eulerian velocity!) is given by the following expression [6]:

$$\begin{aligned} F^{(3)}_{\alpha\beta\gamma}(\boldsymbol{k},\boldsymbol{q},\boldsymbol{p})\delta(\boldsymbol{k}+\boldsymbol{q}+\boldsymbol{p}) = \int & d\boldsymbol{k}' d\boldsymbol{q}' d\boldsymbol{p}' d\omega d\omega_1 d\omega_2 d\omega_1' d\omega_2' \\ & \times \delta(\omega_1+\omega_2+\omega_1'+\omega_2')\delta(\omega_1'+\omega_2'+\omega)F(\boldsymbol{k},\boldsymbol{k}';\omega) \\ & \times D_{\beta\gamma\beta'\gamma'}(\boldsymbol{q},\boldsymbol{p},\boldsymbol{q}',\boldsymbol{p}';\omega_1,\omega_2,\omega_1',\omega_2')V_{\beta'\gamma'\alpha}(\boldsymbol{q}'|\boldsymbol{p}',\boldsymbol{k}')\ . \end{aligned} \tag{20}$$

The dependence of $F^{(3)}(\boldsymbol{k},\boldsymbol{q},\boldsymbol{p})$ on small wave vector $\boldsymbol{k}$ is presented in (20) both in explicit form via $\boldsymbol{k}$-dependence of double correlation function $F(\boldsymbol{k},\boldsymbol{k}';\omega)$ and in implicit form via $(\boldsymbol{k},\omega)$-dependence of D and ω-dependence of F. Unknown function D is expressed in terms of some infinite diagram series. We do not know how to summarize it. However, it is not necessary for our treatment. It is enough for us to know that function D may be expanded at small κ and ω in a Taylor series with respect to the powers of $\boldsymbol{k}$ and ω since it depends on κ,ω only via sums like $\kappa+\boldsymbol{q}\simeq\boldsymbol{q}$ and $\omega+\omega_1\simeq\omega_0$ [6].

The main contribution to integral (20) over $\boldsymbol{k}',\omega$ is from the region $\boldsymbol{k}'\simeq\boldsymbol{k}$, $\omega\simeq k^x$ (x being the dynamic exponent that is the frequency scaling index) where function $F(\boldsymbol{k},\boldsymbol{k}',\omega)$ is not small. On the other hand, the main contribution into the integral over ω_1' and ω_2' is from the region $\omega_1'\simeq\omega_2'\simeq q^x \gg k^x \simeq \omega$. It allows us to neglect the value of ω and put $\omega_2' = -\omega_1'$ (instead of $\omega_2' = -\omega_1' - \omega$) in the argument of function D. As a result, only the function $F(\boldsymbol{k},\boldsymbol{k}',\omega)$ depends on ω in (20) . It allows to integrate (20) over ω with the help of (18) and then to integrate (20) over $\boldsymbol{k}'$, ω_2 and ω_2' using

δ-functions. One thus has:

$$
\begin{aligned}
F^{(3)}_{\alpha\beta\gamma}(\boldsymbol{k},\boldsymbol{q},\boldsymbol{p})\delta(\boldsymbol{k}+\boldsymbol{q}+\boldsymbol{p}) = F(\boldsymbol{k}) \int d\boldsymbol{q}'d\boldsymbol{p}'d\omega_1 d\omega_1' \\
\times D_{\beta\gamma\beta'\gamma'}(\boldsymbol{q},\boldsymbol{p},\boldsymbol{q}',\boldsymbol{p}';\omega_1,-\omega_1,\omega_1',-\omega_1')V_{\beta'\gamma'\alpha}(\boldsymbol{q}'|\boldsymbol{p}',\boldsymbol{k}) \ .
\end{aligned}
\tag{21}
$$

Here V is the *dynamic vertex* (16) . Substituting (16) into (21) and integrating over $\boldsymbol{p}'$ one has:

$$
\begin{aligned}
& F^{(3)}_{\alpha\beta\gamma}(\boldsymbol{k},\boldsymbol{q},\boldsymbol{p})\delta(\boldsymbol{k}+\boldsymbol{q}+\boldsymbol{p}) = -iP_{\alpha\alpha'}(\boldsymbol{k})F(\boldsymbol{k}) \int d\boldsymbol{q}'d\omega_1 d\omega_1' \{ k_{\gamma''}P_{\gamma''\gamma'}P_{\gamma'\beta'} \\
& \times D_{\beta\gamma\beta'\gamma'}(\boldsymbol{q},\boldsymbol{p},\boldsymbol{q}',-(\boldsymbol{q}'+\boldsymbol{k});\omega_1,-\omega_1,\omega_1',-\omega_1') + q'_{\alpha'}P_{\beta'\beta''}(\boldsymbol{q}') \\
& \times [P_{\gamma'\beta''}(\boldsymbol{q}'+\boldsymbol{k})D_{\beta\gamma\beta'\gamma'}(\boldsymbol{q},\boldsymbol{p},\boldsymbol{q}',-(\boldsymbol{q}'+\boldsymbol{k});\omega_1,-\omega_1,\omega_1',-\omega_1') \\
& - P_{\gamma'\beta''}(\boldsymbol{q}')D_{\beta\gamma\beta'\gamma'}(\boldsymbol{q},\boldsymbol{p},\boldsymbol{q}',-\boldsymbol{q}';\omega_1,-\omega_1,\omega_1',-\omega_1')]\} \ .
\end{aligned}
\tag{22}
$$

Now let us expand D in the Taylor series with respect to $\boldsymbol{k}$: $D(\boldsymbol{k}) \simeq D(0) + D'_\delta k_\delta + D''_{\delta\sigma}k_\delta k_\sigma/2$. The following expansion of the triple correlator (22) thus arises in the limit of small $\boldsymbol{k}$

$$
F_{\alpha\beta\gamma}(\boldsymbol{k},\boldsymbol{q},\boldsymbol{p}) = \left[\Phi_{1,\alpha\beta\gamma\delta}(\boldsymbol{q})\frac{k_\delta}{q} + \Phi_{2,\alpha\beta\gamma\delta\sigma}(\boldsymbol{q})\frac{k_\delta k_\sigma}{q^2} + \dots\right] F(k) \ .
\tag{23}
$$

Here $F(k) \propto k^{-y}$ is the simultaneous double correlation function of qL velocity that coincides with the same function of Eulerian velocity. So, the scaling exponent y is determined by the scaling exponent ζ_2 of double velocity moment (2) by the relation $y = \zeta_2 + 3$ which follows from (12) . It is worth to remind that in (21) we replaced $\omega_2' = -\omega_1' - \omega$ by $\omega_2' = -\omega_1'$. In order to take into account the dependence of D on ω, let us expand $D(\omega)$ at small ω, since the main contribution in the integral (20) stems from the region of small ω if k is chosen to be small. Now we substitute the expansion $D(\omega) \simeq D(0) + \omega\partial D/\partial\omega + \dots$ into (20) and integrate the additional terms arisen in (21) in the same manner. Finally, we obtain the following corrections to (23)

$$
\begin{aligned}
F_{\alpha\beta\gamma}(\boldsymbol{k},\boldsymbol{q},\boldsymbol{p}) = \Bigg\{ & \Phi_{1,\alpha\beta\gamma\delta}(\boldsymbol{q})\frac{k_\delta}{q}\left[1 + a_1\left(\frac{k}{q}\right)^x\right] \\
& + \Phi_{2,\alpha\beta\gamma\delta\sigma}(\boldsymbol{q})\frac{k_\delta k_\sigma}{q^2}\left[1 + a_2\left(\frac{k}{q}\right)^x\right] + \dots \Bigg\} F(k) \ .
\end{aligned}
\tag{24}
$$

Here a_1 and a_2 are some dimensionless constants, x being dynamic scaling exponent (for Kolmogorov solution, $y = \frac{11}{3}$, $x = \frac{2}{3}$). For multifractal subsets with an arbitrary h, one could obtain $x = 1 - h$ since the typical frequency scales with l as $\delta v(l)/l$. We shall demonstrate below that $h \le 1$, so $x > 0$. Account of the frequency dependence gives indeed the small corrections in each order of the expansion in k/q. Therefore, we shall use (23) in the further calculations. It is important to emphasize here that δ-functions of the frequences in (20) are the same in both terms arising in (20) after substituting the vertex V. This leads to the absence of a term proportional to $k^x F(k)$ in the expansion (23) .

So, the main term in the asymptotic expansion of the triple correlation function (24) is proportional to $k^{1-y}\Phi_1(q)/q$. The total scaling exponent of the triple correlation function depends on the distribution in question. For the spectrum carrying constant energy flux, it is $y_3 = 7$ according to (14) . The scaling exponent of $\Phi_{1,2}(q)$ may be thus expressed in terms of the scaling exponent y of the double correlator

$$\Phi(q) \propto q^z, \quad z = 7 - y \tag{25}$$

Let us allow a small digression. It is interesting to compare z which one would obtain in the Direct Interaction Approximation [10] for triple correlation function with the above correct answer. In that approximation (see [6] for details)

$$D_{\alpha\beta\gamma\delta}(\mathbf{q},\mathbf{p},\mathbf{q}',\mathbf{p}';\omega_1,\omega_2,\omega_1',\omega_2') = G_{\alpha\gamma}(\mathbf{q},\mathbf{q}';\omega_1)F_{\beta,\delta}(\mathbf{p},\mathbf{p}';\omega_2)\delta(\omega_1+\omega_1') \ . \tag{26}$$

Using (20) one has in the same way the estimation for functions Φ:

$$\Phi(q) \sim qF(q)G(q,0) \propto q^{z_0}, \quad z_0 = x + y - 1 \ . \tag{27}$$

Due to the scaling relation $2x + y = 5$ [5,11] the same result for z_0 could be easily obtained in any (but finite!) order of diagrammatic perturbation theory or under assumption that dynamic vertex V is not renormalized. The value of z_0 coincides with the correct result $z = 7 - y$ only on the Kolmogorov solution $x = \frac{2}{3}$, $y = \frac{11}{3}$. For example, in the multifractality approach, $y > \frac{11}{3}$ which reflects the renormalization of the vertexes for a multifractal turbulence. We do not know now how it occurs but we know the total exponent of the triple correlators. So, in order to find the scaling exponent z of large wave vectors in the triple velocity correlator it is necessary to take into account the entire diagrammatic series which leads to a renormalization of the vertexes.

The difference between the initial result of Kraichnan [10] and present approach is due to the fact that the Direct Interaction Approximation is an uncontrolled procedure. On the contrary, analyzing the whole series, one can find the main contribution into the triple correlator for $q \gg k$ and prove that neglected terms are small by parameter $(k/q)^{11/3}$. Our approach are thus controlled with a really small parameter [6]. And below we get the result different from [10]: interactions with distant scales are counterbalanced for a real spectrum with a constant energy flux (whatever ζ_2 it has) but not only for Kolmogorov solution as in the Direct Interaction Approximation.

Now let us briefly explain the physical meaning of the final expressions (23) – (24) . The triple correlation function (11) with one wavenumber being small in comparison with two others corresponds to the triple correlation function (26) of velocity increments between three points in the r-space with one distance being smaller than two others: $r_{23} = r \ll R = r_{13} \approx r_{12}$. The points 2 and 3 are thus close being distant from the point 1. The main contribution in the correlation of velocity increments is due to the pair correlator between distant points 1 and 2 (or 3). It gives $F(k)$ in (20) – (23) . The small factor k/q in (23) turns into r/R in the r-representation. It describes a decorrelating influence of the spatial structure of a small r-eddy on a large R-eddy. The terms $(k/q)^x$ describes a decorrelating influence of the motion of the small eddy. The rest dependence of S_3 on r could be restored with the help of (3) [or (14)] using the knowledge of the total exponent:

$$S_3(r,R,R) \propto S_2(R)\frac{r}{R}\frac{S_3(r)}{S_2(r)} \propto R^{\zeta_2-1}r^{2-\zeta_2} \ . \tag{28}$$

This formula corresponds to the first term of the expansion (24) . The diagrammatic approach allows us to obtain also the next three terms having different causes and giving different contributions into the locality analysis.

The final relations (24) or (28) are exact since we made no uncontrolled approximation while analyzing Euler equation. Of course, using diagrammatic series (as well as functional integral) is a poorly defined procedure from the viewpoint of rigorous mathematics. Nevertheless, in theoretical physics it could give exact results in fluid mechanics as well as in quantum field theory or statistical mechanics if one analyze series as a whole (or, equivalently, functional integral). For example, Dyson equation or Ward identities are usually obtained in that way. Note that the relation obtained is some analog of Ward identities in the field theory.

3. Locality Analysis

3.1 TURBULENCE OF INCOMPRESSIBLE FLUID

Now we are ready to substitute the triple correlation function (23) into the equation (10) and to check the integral convergence. The procedure is divided into two parts: infrared ($q \ll k$) and ultraviolet ($q \gg k$) ones. In the first case, the integral in (10) behaves as follows (we drop all constant factors):

$$I_{\mathrm{IR}} \propto \int_{1/L} dq \int d\Omega V(q|k,k) F^{(3)}(q,k,k) q^2 , \tag{29}$$

while in the second case it is

$$I_{\mathrm{UV}} \propto \int^{1/\eta} dq \int d\Omega V(k|q,q) F^{(3)}(k,q,q) q^2 , \tag{30}$$

Here L and η are external and internal scales respectively, Ω is a solid angle of the vector q.

In the triple correlation function (23) , the first term is odd with respect to the change of the sign of the small wavenumber. Therefore, due to the integration over $d\Omega$, this term does not contribute into infrared region of energy exchange (29) and the next term becomes essential. This term is even under $k \to -k$ and is proportional to $\Phi_2(k) q^{2-y}$. Therefore, the infrared convergence is defined by the integral

$$I_{\mathrm{IR}} \propto \Phi_2(k) k^{-1} \int_{1/L} F(q) q^4 \, dq , \tag{31}$$

Infrared locality is thus defined by the double correlation function only.

However, for ultraviolet integral (30) , the first term in (23) gives a nonvanishing value:

$$I_{\mathrm{UV}} \propto F(k) k^2 \int^{1/\eta} \Phi_1(q) q \, dq . \tag{32}$$

The vertex $\Gamma(k,q,q) \propto k$ was substituted in both (31) and (32) .

Dividing these integrals by $F(k)$ one obtains the estimate for a typical time of the nonlinear interaction of k-eddy with the largest eddies

$$\tau_{\mathrm{IR}}^{-1}(k, 1/L) \simeq \frac{\Phi_2}{kF(k)} \int_{1/L} F(q) q^4 \, dq , \tag{33}$$

and a typical time of the interaction with the smallest eddies

$$\tau_{UV}^{-1}(k,1/\eta) \simeq k^2 \int^{1/\eta} \Phi_1(q) q\, dq \ . \tag{34}$$

As one can see, the interaction with small scales produce some turbulent viscosity since $\tau_{UV} \propto k^2$.

Locality of the spectrum carrying energy flux. Firstly, we substitute the simple single-scaling Kolmogorov solution into (33)–(34) to demonstrate how it converges. In this case, $F(k) \propto k^{-11/3}$ and $\Phi(q) \propto q^{-10/3}$ by virtue of (25). Calculating (33) and (34), we obtain that the contributions of distant regions in the k-space into the integral (10) governing energy exchange behave as $(kL)^{-4/3}$ and as $(k\eta)^{-4/3}$. That law was previously obtained by Kraichnan [10], it demonstrates the locality of the simple Kolmogorov solution. It is worth to note that the contributions of small and large scales decrease with the distance by the same law. The same is true for a general local steady spectrum (not necessarily with the Kolmogorov exponent), carrying constant energy flux. Indeed, the exponent of Φ is defined by (25) as $z = 7 - y = 4 - \zeta_2$ due to the flux constancy which fixes the exponent of the triple correlation function. Therefore, $F(q) \propto q^{\zeta_2+3}$ while $\Phi(q) \propto q^{4-\zeta_2}$. Substituting these into (33) – (34), we obtain integrals to behave in the same way independently of the value of ζ_2: $(kL)^{\zeta_2-2}$ and $(k\eta)^{2-\zeta_2}$. The coincidence of the laws of decreasing the contributions from largest and smallest scales was firstly pointed out in [14] for Kolmogorov-like spectra of wave turbulence. Such a coincidence means that in a stationary spectrum, the contributions to interaction of all scales, from small to large ones, are balanced [12,14]. It also means that for a spectrum with a constant flux, the dynamical interaction either converges at $k \to 0$ and $k \to \infty$ or diverges at both limits. It is worth to note that such a balance does not necessary hold for any cascade steady spectra. For example, it is not the case for a turbulence system with two (or more) motion integrals giving different steady spectra as in two-dimensional turbulence.

The locality condition is thus

$$\zeta_2 < 2 \ , \tag{35}$$

which takes place for any reliable model of turbulence (the exponents of low-order correlators are close to those of single-scaling Kolmogorov picture $\zeta_p \approx p/3$).

Since we could not yet explicitly summarize the subsequence of diagrams giving function D (and Φ) then the numerical factor in I_{IR} and I_{UV} is unknown. Though the signs of these integrals could be readily established. Looking for an arbitrary term in the series for D (see Appendix at [6]), one could find that $\Phi(q)$ is positive [the first term is given by (27)]. Therefore,

$$I_{IR} \propto \frac{(kL)^{\zeta_2-2}}{2-\zeta_2} > 0 \,, \qquad I_{UV} \propto \frac{(k\eta)^{2-\zeta_2}}{\zeta_2 - 2} < 0 \ .$$

These formulas show the counterbalanced locality of the turbulence spectrum carrying energy flux from large to small scales.

Locality investigation for other hypothetical spectra of turbulence of an incompressible fluid. In principle, one could try to suggest spectra of hydrodynamic turbulence carrying motion integrals other than energy. There exist two other motion integrals of isotropic turbulence: averaged squared linear momentum, suggested by Saffman [15]:

$$M = \int \langle \vec{v}(\vec{r})\vec{v}(\vec{r}+\vec{R})\rangle \, d\vec{R} \; ,$$

and squared angular momentum

$$L = \int R^2 \langle \vec{v}(\vec{r})\vec{v}(\vec{r}+\vec{R})\rangle \, d\vec{R}$$

suggested by Loitsyanskii (see [16]). Its actual conservation which depends on the behavior of $E(k)$ at $k \to 0$ is not considered here. Proceeding from the dimensional argument, one may suggest the spectra with the fluxes of either M ($F_M(k) \propto k^{-4/3}$) or L ($F_L(k) \propto k^{-1/3}$ [17]. Both these distributions satisfy infrared locality. To check the convergence of the ultraviolet integral (32) one should find the exponent z of the function Φ_1. According to (23) - (25) this exponent is equal to $z = y_3 - y$ that is the difference between the exponents of the triple and the second correlators [the exponents of the latter are positive for both distributions in question so (23) still be valid]. Note that y_3 is defined now by the constancy of other (than energy) fluxes so $y_3 \neq 7$. Requiring the flux of M to be constant we find $y_3 = 4$ which gives $z = 8/3$ and (32) diverges. The same story is for L. So both distributions correspond to a strong ultraviolet nonlocality which means that the cascade turbulence with the fluxes of these integrals is impossible.

Other motion integral of ideal incompressible fluid could be obtained in the following way [19]. The Euler equation for incompressible fluid ($\rho = 1$)

$$\frac{\partial \vec{v}}{\partial t} + (\vec{v}\vec{\nabla})\vec{v} = -\vec{\nabla}p$$

could be written in the Hamiltonian form [18]:

$$i\frac{\partial a}{\partial t} = \frac{\delta H}{\delta a^*} \; ,$$

$$H = \frac{1}{2}\int v^2 \, d\vec{r} = \frac{1}{4}\int T_{1234} a_1^* a_2^* a_3 a_4 \delta(\vec{k}_1 + \vec{k}_2 - \vec{k}_3 - \vec{k}_4) \, d\vec{k}_1 d\vec{k}_2 d\vec{k}_3 d\vec{k}_4 \; .$$

Here $T_{1234} = (\vec{\psi}_{13}\vec{\psi}_{24}) + (\vec{\psi}_{14}\vec{\psi}_{23})$ with

$$\vec{\psi}_{ij} = \frac{(2\pi)^{-3/2}}{2}\left[\vec{k}_i + \vec{k}_j - (\vec{k}_i - \vec{k}_j)\frac{k_i^2 - k_j^2}{|\vec{k}_i - \vec{k}_j|^2}\right] .$$

Fourier image of the fluid velocity is expressed in terms of the new (Clebsch) variables as follows:

$$\vec{v}(\vec{k}) = -i \int \vec{\psi}_{12} a_1^* a_2 \delta(\vec{k} + \vec{k}_1 - \vec{k}_2) \, d\vec{k}_1 d\vec{k}_2 \; .$$

Besides the energy, Euler equation in the Hamiltonian form preserve also the action $\int |a_k|^2 \, d\vec{k}$. For a homogeneous turbulence, one could introduce pair correlator $n(\vec{k},t)\delta(\vec{k}-\vec{k}') = \langle a(\vec{k},t)a^*(\vec{k}',t)\rangle$ so that "the total number of waves" (action density per unit volume) is conserved. It follows from the fact that nonlinearity is of a kind of wave scattering $2 \to 2$ in terms of a_k. The spectrum supporting a constant action flux could be obtained using formal analogy of the Hamiltonian with that of a wave turbulence under four-wave scattering [18,12]:

$$n(k,Q) = \lambda Q^{1/3} k^{-4} \ . \tag{36}$$

After recalculation into the energy spectrum, it gives

$$F(k,Q) = C_1 Q^{2/3} k^{-3} \ .$$

Unlike the cases of M and L, the spectrum carrying constant flux of N is local that could be shown as follows. Since the exponent of the double correlator $y = 3$, then the infrared contribution is $I_{\mathrm{IR}} \propto (kL)^{-2}$. Assuming this spectrum to be single-scaling (i.e. the scaling index of n-th correlator is proportional to n), we find from the dimensional consideration $y_3 = 6$ and $z = 3$. Substituting this into (32), one could find that the collision term converges also at the ultraviolet limit and $I_{\mathrm{UV}} \propto (k\eta)$. As one can see, there is no counterbalance for the spectrum (36).

The locality of (36) could be obtained directly in the Clebsch variables (also in a quasi-Lagrangian form [5]) by deriving the relation between the asymptotics at $k \to 0$ of the fourth correlator $J(\vec{k},\vec{k}_1,\vec{p},\vec{p}_1))\delta(\vec{k}+\vec{k}_1-\vec{p}-\vec{p}_1) = \langle a(\vec{k})a(\vec{k}_1)a^*(\vec{p})a^*(\vec{p}_1)\rangle$ and the double correlator $n(k)$. The notion of a balance is a bit more complicated for a quadruplet interaction than for a triplet one (see the remark at the end of the next subsection).

To avoid misunderstanding, note that unlike the Kolmogorov spectrum with the energy flux the spectrum (36) is not yet shown to be a solution of the equations for the correlators. The possible physical meaning of this hypothetical spectrum is discussed in [20].

3.2. COUNTERBALANCED LOCALITY OF WEAK WAVE TURBULENCE

To put our locality investigation in a more general context, let us briefly consider the turbulence of waves with some dispersion law $\omega(k)$. If the amplitudes $a(\vec{k},t)$ of the waves are sufficiently small, the nonlinear interaction between the waves could be assumed to be weak in comparison with linear phenomenon of wave dispersion. Closed kinetic equation for the wave density in k-space (that is the double correlation function of the wave amplitudes $n(\vec{k},t)\delta(\vec{k}-\vec{k}') = \langle a(\vec{k},t)a^*(\vec{k}',t)\rangle$) can be obtained in this case since the triple correlation function is expressed via the double ones for any wavevectors (not only in the asymptotical limit) [12]. If the dispersion relation $\omega(k)$ allows the validity of the equality $\omega(k_1)+\omega(k_2) = \omega(\vec{k}_1+\vec{k}_2)$ for some $\vec{k}_1, \vec{k}_2$ (so-called decay case), then the kinetic equation is as follows [12]:

$$\begin{aligned}\frac{\partial n(\vec{k},t)}{\partial t} &= \int U(k,k_1,k_2)\delta(\vec{k}-\vec{k}_1-\vec{k}_2)\delta(\omega_k-\omega_1-\omega_2)(n_1n_2-n_kn_1-n_kn_2)\,d\vec{k}_1d\vec{k}_2 \\ &- \int U(k_1,k,k_2)\delta(\vec{k}_1-\vec{k}-\vec{k}_2)\delta(\omega_1-\omega_k-\omega_2)(n_kn_2-n_kn_1-n_1n_2)\,d\vec{k}_1d\vec{k}_2 \\ &- \int U(k_2,k_1,k)\delta(\vec{k}_2-\vec{k}_1-\vec{k})\delta(\omega_2-\omega_1-\omega_k)(n_1n_k-n_2n_1-n_kn_2)\,d\vec{k}_1d\vec{k}_2 \ .\end{aligned} \tag{37}$$

Here interaction probability U is assumed to be scale invariant as well as ω:

$$\omega(\lambda k) = \lambda^{\alpha}\omega(k), \quad U(\lambda\vec{k},\lambda\vec{k}_1,\lambda\vec{k}_2) = \lambda^m U(\vec{k},\vec{k}_1,\vec{k}_2) \ .$$

Let us discuss the convergence of collision integral (37) on power distributions $n(k) \propto k^{-s}$. We have a sum of integrals of power functions, of which every one clearly diverges either at $\omega_1 \to 0$ or at $\omega_1 \to \infty$. However, the structure of the kinetic equation provides the reduction of divergences by one power of ω_1 at $\omega_1 \to \infty$

$$n(\omega_1 - \omega) - n(\omega_1) \approx \omega \frac{\partial n(\omega_1)}{\partial \omega_1} \propto n(\omega_1) \frac{\omega}{\omega_1}$$

and by two powers (i.e. by ω_1^2) at $\omega_1 \to 0$.

Indeed, let at $k \ll k_1$ the function U has the following asymptotics $U(k, k_1, k_2) \propto k^{m_1} k^{m-m_1}$. Then the second and third terms in (37) converge at $\omega_1 \to \infty$ if the condition

$$s > s_2 = m - m_1 + d + 1 - 2\alpha$$

is satisfied. Such a reduction takes place at $\omega_1 \to 0$

$$n(\omega - \omega_1) - n(\omega) \approx \omega_1 \frac{\partial n(\omega)}{\partial \omega} \propto n(\omega) \frac{\omega_1}{\omega} \ .$$

The divergences at $\omega_1 \to 0$ are also reduced by one more power of ω_1 with divergences at $\omega_1 \to \omega$ [here one should take into account the symmetry of interaction probability $U(k, k_1, k_2) = U(k, k_2, k_1)$]. As a result, the convergence criterion of the collision integral at small frequencies is:

$$s < s_1 = m_1 + d - 1 + 2\alpha \ .$$

Thus, if $s_1 > s_2$, i.e.,

$$2m_1 > m + 2 - 4\alpha \ ,$$

then there exists an interval of s indices providing locality of interaction. It is important to note that the exponent of a steady Kolmogorov-like spectrum with an energy flux $s_0 = m/2 + d$ always lies exactly in the middle of the "locality interval": $s_0 = (s_1 + s_2)/2$. This also gives the counterbalanced locality of energy cascade spectra of wave turbulence: The contribution that stems from ultraviolet region into $\partial n_k / \partial t$ is proportional to $k_{max}^{s_2 - s_0}/(s_2 - s_0) < 0$ while the infrared contribution is $k_{min}^{s_1 - s_0}/(s_1 - s_0) > 0$. The detailed analysis shows that such signs of the contributions take place under the condition $s_0 > \alpha$ which is quite natural since the last inequality is the condition for energy flux to be positive (directed to large wavenumbers). Such a counterbalance means that motions with intermediate scales take the same energy from very large-scale motions as it give to small-scale ones.

Note that if three-wave interactions are forbidden [i.e. $\omega(\vec{k}_1 + \vec{k}_2) \neq \omega(k_1) + \omega(k_2)$ for any $\vec{k}_1, \vec{k}_2$], then the rhs in (37) turns into zero and one should take into account four-wave interactions $\omega(k_1) + \omega(k_2) = \omega(k) + \omega(\vec{k}_1 + \vec{k}_2 - \vec{k})$. Here the balance of interactions is organized in a more complicated way for steady spectra of turbulence. Waves with the given frequency ω_k could interact with the waves having frequencies ω_1 and ω_2 that change independently in the region $\omega_1 + \omega_2 > \omega$. One can show that the subregions $\{\omega_2 < \omega, \omega_1 < \omega\}$ and $\{\omega_1 > \omega, \omega_2 > \omega\}$ give a positive contribution into collision integral while the subregions $\{\omega_1 < \omega, \omega_2 > \omega\}$ and $\{\omega_1 > \omega, \omega_2 < \omega\}$ give a negative one (again, if the exponent of the turbulence spectra s_0 larger than that of an equilibrium one: $s_0 > \alpha$).

4. Interaction Locality in the Multiscaling Model

The interaction is assumed to be local in k-space for the multifractal picture. It is necessary to check that it is true for each h-subset. Mathematically, the integral convergence means that one could change the order of integrations over dh and dq. One may represent in (33) the double correlation function as an integral over exponents h like (4) :

$$F(q) \sim \int_{h_{\min}}^{h_{\max}} d\mu(h) q^{-2h-3} (qL)^{D(h)-3} . \tag{38}$$

Substituting it into (33), we obtain that the integral is proportional to the negative power of L (i.e. converges) if $h_{\max} < 1$. It is interesting to note that the dimension $D(h)$ does not arise in the final expression since $qL = 1$ for the lower limit of integration. The probability to find the largest eddy is equal to unity for all h-subsets. Multifractality plays its role for small scales only. The condition $h < h_{\max} = 1$ means that the maximal velocity gradients $\delta v(\eta)/\eta \propto \eta^{h-1}$ become singular as $R \to \infty$ for all h-subsets.

For the ultraviolet integral (34), we could not represent $\Phi(q)$ as an integral over dh since it is proportional to the ratio of two correlation functions $F_3(q)/F(q)$ [see (25) and (28)]. So the locality analysis for the interaction of k-eddies with much smaller scales could give us only the condition (35) and nothing else. It gives no restrictions for the possible values of exponent h. Indeed, it deals only with the exponent of double correlation function which is defined as $\min[2h - D(h) + 3]$ and not by the minimal value of h. The point is that $D(h)$ decreases while h decreases, so whatever slow (for small h) the drop of the correlation function is in the k-space, the decrease of probability to find the small eddy from that h-subset provides the interaction locality.

One could explain the physical meaning of the above locality analysis in the following way. Aforementioned locality means that a typical interaction time $\tau(k, k')$ for the eddies of the same order of size should be much larger than $\tau_{\rm IR}(k, 1/L)$ and $\tau_{\rm UV}(k, 1/\eta)$. The first time could be estimated as a turnover time for the k-eddy from h-subset (we put $k \simeq k'$)

$$\tau(k, h) \simeq [k\delta v(k, h)]^{-1} \simeq \frac{L}{U}(kL)^{h-1} . \tag{39}$$

The time of an infrared interaction is estimated as necessary for the k-eddy to be distorted noticeably by the velocity difference $\Delta v_L(k)$ of the large-eddy flow on the scale of the small one. It is thus as follows

$$\tau_{\rm IR} \simeq [k\Delta v_L(k)]^{-1} \simeq \frac{L}{U} .$$

It is important that this time is defined by the scale of k-eddy only and is independent of its velocity (that is which h-subset the eddy belongs to). Comparing $\tau(k)$ and $\tau_{\rm IR}$, we obtain the above locality criterion $h < h_{\max} = 1$.

A bit more complicated is the consideration of an ultraviolet interaction. The interaction of the given k-eddy with small k_1-eddies (with $\eta \geq k_1 \gg k$) should provide some turbulent viscosity ν_T:

$$\tau_{\rm UV}(k, k_1, h) \simeq [k^2 \nu_T(k_1, h)]^{-1} .$$

The value of viscosity should depend on the turbulence level at the place and the time under consideration. It means that the viscosity depends on the velocity of k-eddy which generates the small eddies. The rate of energy dissipation

$$\nu_T k_1^2 (\delta v(k_1))^2 \simeq \nu_T \epsilon^{2/3} k_1^{4/3} \left(\frac{k_1}{k}\right)^{2/3-\zeta_2}$$

should be equal to the local energy flux $\epsilon \simeq k\delta v^3(k,h)$. It gives the viscosity as follows:

$$\nu_T(k_1) \simeq \epsilon^{1/3} k_1^{-4/3} \left(\frac{k_1}{k}\right)^{2/3-\zeta_2} \simeq k^{1/3} k_1^{-4/3} \delta v(k,h) \left(\frac{k_1}{k}\right)^{2/3-\zeta_2}$$

and the following estimate for τ_{UV}

$$\tau_{UV}(k,k_1,h) \simeq \frac{L}{U}\left(\frac{k_1}{k}\right)^{2-\zeta_2} (kL)^{h-1} .$$

Requiring the ultraviolet locality, we obtain $\tau/\tau_{UV} \simeq (k_1/k)^{2-\zeta_2} \ll 1$ which gives criterion (35) independent of h under consideration. The locality analysis thus gives only upper restriction for the values of exponents h.

Concluding, we would like to say that the range of exponents evidently should be restricted from below. I myself suppose that the high-order correlation functions behaves as if they are determined by the positive exponents only [7,8]. What we have shown here is that such a restriction should be given not by locality analysis but by the other conditions (e.g. like incompressibility [21]).

Acknowledgements. I am grateful to V.Lvov and V.Zakharov for valuable remarks.

References

1. Kolmogorov, A.N. (1941) "The local structure of turbulence in incompressible viscous fluid for very large Reynolds number", Dokl. Akad. Nauk SSSR **30**, pp.301–305
2. Monin, A.S and Yaglom, A.M. (1971) Statistical Fluid Mechanics. Vol. I, II. MIT Press, Cambridge Mass., London
3. Wyld, H. (1961) Ann. Phys. (N.Y.) 14, p.143
4. Kraichnan, R. (1965) "Lagrangian-history closure approximation for turbulence", Phys. Fluids **8**, p.575
5. L'vov, V. (1991) "Scale invariant theory of fully developed hydrodynamic turbulence — Hamiltonian Approach", Phys. Rep. 207 (1), pp.1–47
6. Lvov, V. and Falkovich, G. (1992) "Counterbalanced interaction locality of developed hydrodynamic turbulence", Phys. Rev. **A46** (in press)
7. Anselmet, F., Gagne, Y., Hopfinger, E. and Antonia, R., (1984) "High-order velocity structure functions in turbulent shear flowa", J. Fluid Mech. **140**, pp.63–89
8. Vincent, A. and Meneguzzi, M., (1991) "The spatial structure and statistical properties of homogeneous turbulence" J. Fluid Mech. **225** pp.1–15

9. Parisi, G. and Frisch, U. (1985) in Turbulence and Predictability in Geophysical Fluid Dynamics, Proc. Int. School "E. Fermi", M.Ghil, R.Benzi and G.Parisi (eds.) p.84
10. Kraichnan, R.H. (1966) "Isotropic turbulence and inertial range structures", Phys. Fluids **9**, p.1728
11. Belinicher, V. and L'vov, V. (1987) "Sweeping elimination in the theory of hydrodynamic turbulence" Sov. Phys. JETP **66**(2), pp.303–313
12. Zakharov, V. L'vov, V. and Falkovich, G. (1992) Kolmogorov Spectra of Turbulence, vol.1 Wave Turbulence, Springer Verlag, Heidelberg
13. Kraichnan, R.H. (1959) "The structure of isotropic turbulence at very high Reynolds numbers", J. Fluid Mech. **6**, pp.497–543
14. Falkovich, G. (1987) "On the stability of the Kolmogorov spectrum of wave turbulence", Sov. Phys. JETP **66**(1), p.97
15. Saffman, P. (1967) "The large scale structure of homogeneous turbulence", J. Fluid Mech. **27**, p.581
16. Landau, L. and Lifshits, E. (1982) Fluid Mechanics, Pergamon Press, Oxford.
17. Kuznetsov, E., Newell, A. and Zakharov, V. (1991) "Intermittency and Turbulence", Phys. Rev. Let. **67**, p.3243
18. Zakharov, V. and Lvov, V. (1975) "On the statistical description of nonlinear wave fields", Radiophys. Quan. Electronics **18**
19. Lamb, P. (1945) Hydrodynamics, Dover, New York
20. Falkovich, G. (1993) "Revised universality concept in the turbulence theory" in V.Zakharov and W.Woychinsky (eds.), Nonlinear Waves and Weak Turbulence, Kluwer Academic Publishers, Dordrecht
21. Frisch, U. (1991) "From global Kolmogorov scaling to multifractality" in "A.N.Kolmogorov" special issue of Proc. Royal Soc. London A434

Solitons, Euler's Equation, and the Geometry of Curve Motion

Raymond E. Goldstein and Dean M. Petrich
Department of Physics
Joseph Henry Laboratories
Princeton University
Princeton
NJ 08544

ABSTRACT. A review is presented of some recent developments in the study of geometrical aspects of the dynamics of curves in the plane. An examination of general conditions necessary for global geometric conservation laws such as those of length and enclosed area leads to a class of motions mathematically equivalent to a hierarchy of integrable systems related to the Korteweg-de Vries (KdV) equation. The KdV, modified KdV, and Harry Dym hierarchies are then seen to be three equivalent views of the same underlying dynamics. It is shown that the nonlinear transformations between these systems have straightforward geometric meaning. These dynamics are found to be local approximations to the motion of vortex patches in ideal two-dimensional fluids, a result paralleling the connection between the Nonlinear Schrödinger equation and the motion of a vortex filament in space. The Hamiltonian structure of these integrable systems is recast in a form emphasizing the geometric interpretation in the language of curve motion. Applications of these results to physical systems are suggested.

1. Introduction and Motivation

If one contemplates a theoretical description of the dynamics of shapes in nature, whether of the complex supercoiling of a closed strand of DNA, the remarkable deformations of a red blood cell in flow, or the motion of a drop of incompressible fluid, the existence of conservation laws is readily apparent. These may be as simple as the length of a nonstretching polymer or the surface area of a cell membrane, or, in the case of DNA, as complex as the "linking number," a topological quantity associated with the windings of one edge of the double helix around the other.

How can these conservation laws be embodied in the dynamics of such shapes? More generally, how do we think about describing shapes? What kinds of variational principles might pertain to their motion? When we speak of a shape in hydrodynamics, it is usually the boundary of some domain of

R. E. Caflisch and G. C. Papanicolaou (eds.), Singularities in Fluids, Plasmas and Optics, 93–109.
© 1993 *Kluwer Academic Publishers.*

fluid. An important feature of the dynamics of *incompressible* fluids is that while the constraint

$$\nabla \cdot v = 0 \tag{1}$$

on the velocity field v is local, the dynamics of the boundary is strongly nonlocal. Such nonlocality also arises elsewhere in problems of pattern formation, often associated with diffusion.

One notion which pervades many theoretical treatments of pattern formation is that essential aspects of the physics may be captured by a *geometric law of motion* by which a point on a curve or surface moves with a velocity dependent on the local geometry there. This arises perhaps most clearly in the well-studied "soap-bubble" problem (Gage, 1984) in which surface tension leads to motion proportional to the mean curvature. It also appears in the study of motion of a solid-liquid interface in phenomena such as dendritic growth and directional solidification (Brower, et al., 1984), particularly in velocity regimes where there is a separation of length scales between that of diffusion and the scale of the pattern itself.

Here we focus on some general aspects of geometric equations of motion for curves in summarizing recent work (Goldstein & Petrich, 1991; Goldstein & Petrich, 1992) which reveals how the differential geometry underlying such dynamics is intimately related to the structure of certain integrable systems and, in turn, to the hydrodynamics of ideal incompressible, inviscid fluids. This work has its origins in research initiated with S.A. Langer on the dissipative dynamics of shapes with global geometric conservation laws (Langer, Goldstein, & Jackson, 1992; Goldstein & Langer, 1992), and was directly motivated by an observation by P. Constantin (Constantin, 1990a) that the modified Korteweg-de Vries (mKdV) equation is equivalent to the area-conserving motion of a closed curve in the plane.

Connections between hydrodynamics and the geometry of curve motion have been known for some time in the context of the Nonlinear Schrödinger equation (Hasimoto, 1972), an integrable system (Zakharov & Shabat, 1972). More formal results involving the KdV equation and Eulerian hydrodynamics have also been obtained (Morrison & Greene, 1980; Olver, 1982; Marsden & Weinstein, 1983).

In what follows, we first review (Section 2) some basic geometric and dynamical considerations regarding curve motion. Section 3 describes the correspondence between certain geometric laws of motion and KdV dynamics, while the relationship of these results to Eulerian hydrodynamics is given in Section 4. We close in Section 5 with some conclusions and avenues for future investigation.

2. Geometry of Curve Motion in the Plane

We begin by establishing notation and considering general features of the motion of a *closed* curve in the plane (Brower, et al., 1984; Constantin, 1990b). While it later proves most convenient to imagine a parametrization of the curve in terms of arclength, it is pedagogically more useful to imagine first that there exists an *arbitrary* parametrization $\alpha \in [0,1]$ such that $\mathbf{r}(\alpha)$ is the vectorial location of a point on the curve (Fig. 1). Then, the metric factor $\sqrt{g} \equiv |\mathbf{r}_\alpha|$ (with subscripts indicating differentiation) is just the magnification factor relating distances in α to distances along the curve, so $ds = \sqrt{g}d\alpha$. The length L and area A enclosed by the curve are given by

$$L = \int d\alpha, \qquad A = \frac{1}{2}\int d\alpha\, \mathbf{r} \times \mathbf{r}_\alpha, \tag{2}$$

treating the cross product as a scalar in two dimensions.

At each point along the curve, there exists a local coordinate system defined by the unit tangent and normal vectors to the curve, $\hat{\mathbf{t}} = (1/\sqrt{g})\mathbf{r}_\alpha$ and $\hat{\mathbf{n}}$, and if the curve is traversed counterclockwise, then their components are related by $t_i = -\epsilon_{ij}n_j$ and $n_i = \epsilon_{ij}t_j$. The pair are related to the curvature κ through the Frenet-Serret equations

$$\partial_s \begin{pmatrix} \hat{\mathbf{t}} \\ \hat{\mathbf{n}} \end{pmatrix} = \begin{pmatrix} 0 & -\kappa \\ \kappa & 0 \end{pmatrix} \begin{pmatrix} \hat{\mathbf{t}} \\ \hat{\mathbf{n}} \end{pmatrix}. \tag{3}$$

A curve may also be described by the *tangent angle* $\theta(s)$, with $\kappa = \theta_s$.

A dynamics for the curve may specified by resolving the curve velocity $\mathbf{r}_t$ (at fixed α) into components normal and tangential to the curve,

$$\mathbf{r}_t = U\hat{\mathbf{n}} + W\hat{\mathbf{t}}, \tag{4}$$

where for a geometric law of motion U and W can be arbitrary functions of $\kappa, \kappa_s, \kappa_{ss}$, etc., although more generally they could be arbitrarily nonlocal as well. With the dynamics of the curve as in Eq. (4), we may compute the time derivatives of basic geometric quantities such as L and A. They are found to be

$$L_t = \oint ds\,(\kappa U + W_s) = \oint ds \kappa U, \qquad A_t = \oint ds U, \tag{5}$$

where in the second integral we have assumed that W is periodic. It is also straightforward, given Eq. (4), to obtain the evolution of the normal and tangent vectors,

$$\hat{\mathbf{n}}_t = -\left(U_s - \kappa W\right)\hat{\mathbf{t}}, \qquad \hat{\mathbf{t}}_t = \left(U_s - \kappa W\right)\hat{\mathbf{n}}, \tag{6}$$

as well as of the tangent angle and curvature,

$$\Theta_t = -U_s + \kappa W, \qquad \kappa_t = -\left(\partial_{ss} + \kappa^2\right)U + \kappa_s W. \tag{7}$$

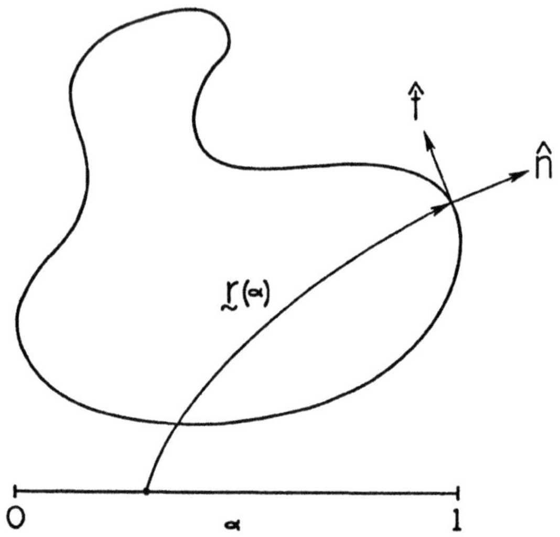

Fig. 1. A closed curve in the plane, parametrized by an arbitrary label α, with normal and tangent vectors indicated.

The two dynamics in Eq. (7) follow from the commutator of arclength and time derivatives,

$$[\partial_s, \partial_t] = (W_s + \kappa U)\,\partial_s. \tag{8}$$

When we speak of length conservation, we may imagine imposing it globally or locally, the former requiring only that the second integral in (5) vanish, the latter requiring that the integrand $W_s + \kappa U$ vanish for all s. A solution to this differential equation for W is unique up to an arbitrary function of time $c(t)$;

$$W(s) = -\int^s ds' \kappa U + c \equiv -\partial^{-1}\kappa U. \tag{9}$$

This choice of "gauge," which from Eq. (8) implies the commutation of arclength and time derivatives, also leads to a parametrization that is time independent: $s_t(\alpha, t) = 0$.

In the local arclength conserving gauge, the curvature evolution is completely given by U as (Brower, et al., 1984)

$$\kappa_t = -\Omega U, \qquad \Omega = \partial_{ss} + \kappa^2 + \kappa_s \partial^{-1}\kappa. \tag{10}$$

This relation will play an important role in the discussions below.

3. Integrable Hierarchies and Curve Motion

Let us now consider the class of velocity functions which conserve both area and perimeter. It is apparent from Eq. (5) that all functions U of the form $U = \partial_s f$ which also satisfy $\kappa U = \partial_s g$ for some local $f(\kappa, \kappa_s, \ldots)$ and $g(\kappa, \kappa_s, \ldots)$ will automatically conserve L and A. This follows from the fact that the integral of a total derivative vanishes for a closed curve. Clearly, the simplest choice is a reparametrization of the curve, with

$$U^{(1)} = 0, \qquad W^{(1)} = c, \tag{11}$$

with c an arbitrary constant. (The superscripts denote order in a sequence whose meaning will become clear below.) The dynamics then has the equivalent forms

$$\mathbf{r}_t = c\mathbf{r}_s, \qquad \kappa_t = c\kappa_s. \tag{12}$$

The first nontrivial choice of interest was suggested by Constantin (Constantin, 1990a) and also appears implicitly in earlier work (Lamb, 1977),

$$U^{(2)} = \kappa_s, \qquad W^{(2)} = -\frac{1}{2}\kappa^2, \tag{13}$$

with $c \equiv 0$. The curvature evolution is then the modified Korteweg-de Vries equation (Drazin & Johnson, 1989; Das, 1989),

$$\kappa_t = -\kappa_{sss} - \frac{3}{2}\kappa^2\kappa_s. \tag{14}$$

which is an integrable system. The particular numerical coefficients in this p.d.e. are irrelevant, as they can be changed by suitable rescalings. The relative signs, however, are important – Eq. (14) is the *focusing* version of the mKdV equation. After some study, one finds that the choices

$$U^{(3)} = \kappa_{sss} + \frac{3}{2}\kappa^2\kappa_s, \qquad W^{(3)} = -\frac{3}{8}\kappa^4 + \frac{1}{2}\kappa_s^2 - \kappa\kappa_{ss} \tag{15}$$

lead to another integrable system

$$\kappa_t = -\kappa_{5s} - \frac{15}{8}\kappa^4\kappa_s - \frac{5}{2}\kappa_s^3 - \frac{5}{2}\kappa^2\kappa_{sss} - 10\kappa\kappa_s\kappa_{ss}, \tag{16}$$

the third member of the mKdV hierarchy of integrable partial differential equations.

The successive members of the hierarchy have velocity functions determined by a recursion relation,

$$U^{(n)} = \Omega U^{(n-1)} = \Omega^{n-2}\kappa_s, \qquad (n \geq 2) \tag{17}$$

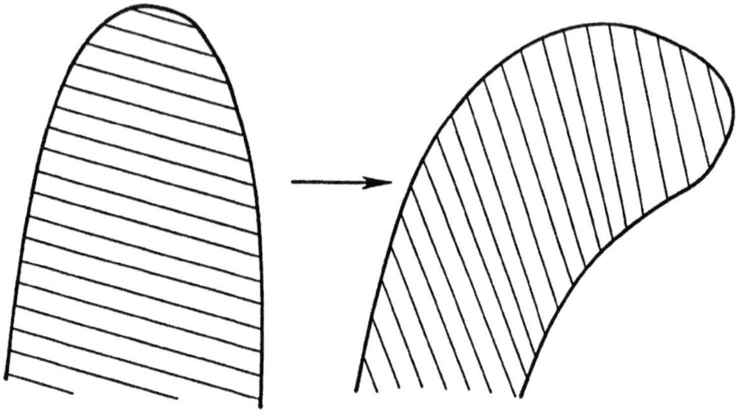

Fig. 2. Motion of a curve under chiral dynamics of the KdV type.

with $W^{(n)} = -\partial^{-1}\kappa U^{(n)}$, and all curvature evolution equations take the form $\kappa_t = -\Omega U^{(n)}$. From (10) we see that the operator Ω, which was known from previous work (Chern & Peng, 1979) to determine the hierarchy, has very general geometric significance as the connection between the vectorial curve dynamics and the curvature evolution.

Since the operator Ω is even under the transformation $s \to -s$, and the velocity functions U are odd, all of the curvature dynamics are intrinsically *chiral.* This feature leads to the kind of motion shown in Fig. 2.

These dynamics have been constructed to conserve the length and enclosed area of the moving curve, but they also possess the infinite number of constants of the motion of the mKdV hierarchy. One very important such constant is the "winding number"

$$N_w = \frac{1}{2\pi}\oint ds\kappa, \tag{18}$$

equal to the number of complete rotations through 2π the tangent vector makes upon complete traversals of the curve. This follows from the fact that the mKdV equation is already in conservative form,

$$\kappa_t = -\partial_s\left(\kappa_{ss} + \frac{1}{2}\kappa^3\right), \tag{19}$$

The conservation law in Eq. (18) has interesting consequences for the evolution of a closed curve. As illustrated in Figure. 2(a), the conservation of winding number forbids the creation of a single self-crossing, while the double crossing in Fig. 2(b) is not ruled out.

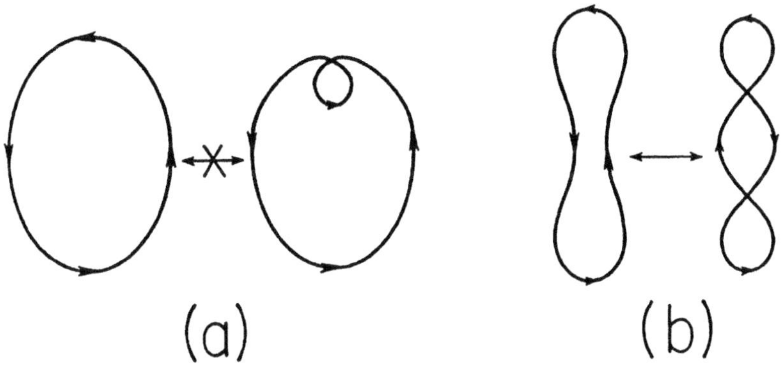

Fig. 3. Forbidden and allowed curve evolutions which follow from the modified Korteweg-de Vries equation for the curvature.

The additional conserved quantities of the mKdV hierarchy are polynomials in the curvature and its derivatives. It is a property of the hierarchy (Das, 1989) that these constants of the motion $\mathcal{H}^{(n)}$ serve also as Hamiltonians which generate the equations of the hierarchy by the relation

$$\kappa_t = \partial_s \left(\frac{\delta \mathcal{H}^{(n)}}{\delta \kappa} \right). \tag{20}$$

It may be verified that these Hamiltonians are just the integrated tangential velocities $\oint ds W^{(n)}$. This is just a restatement of the Kelvin circulation theorem for an ideal fluid (Landau & Lifshitz, 1959),

$$\oint_C d\mathbf{l} \cdot \mathbf{v} = \text{constant}, \tag{21}$$

where C is a contour in the fluid flow and $\mathbf{v}$ is the local velocity. We thus have two of the basic conserved quantities associated with inviscid, incompressible fluid flow: enclosed area and circulation.

Since it is known from the work of Miura (Miura, 1968) that each member of the KdV hierarchy is related to its partner in the mKdV hierarchy by a nonlinear mapping, it is natural to ask about the geometric meaning of that transformation. In our present notation, the Miura transformation connects solutions of the KdV equation

$$u_t = -u_{sss} + 3uu_s, \tag{22}$$

to those of mKdV by the substitution

$$u = -\frac{1}{2}\kappa^2 - i\kappa_s. \tag{23}$$

The significance of this relation is seen by considering the curve dynamics in the complex plane, where a point on the curve is given by $z(s,t) = x(s,t) + iy(s,t)$ and z_s and $-iz_s$ are, respectively, the unit tangent and normal. The general equation of motion (4) is then $z_t = (W - iU)z_s$. To express geometric quantities in terms of z and its derivatives, note that the coordinate may be expressed in terms of a nonlocal transformation of the tangent vector as

$$z(s,t) = \int^s ds' e^{i\theta(s',t)}, \tag{24}$$

so $\theta = -i\ln(z_s)$, $\kappa = -iz_{ss}/z_s$, and so on. Thus, the Miura function is

$$u = -\left[\left(\frac{z_{ss}}{z_s}\right)_s - \frac{1}{2}\left(\frac{z_{ss}}{z_s}\right)^2\right] \equiv -\{z, s\}, \tag{25}$$

which (apart from a minus sign) is the Schwarzian derivative (Nehari, 1975). Recall that this quantity is invariant under fractional linear transformations in the complex plane. That is, under

$$z \rightarrow w = \frac{az + b}{cz + d} \tag{26}$$

then $\{z, s\} = \{w, s\}$. The role of the Schwarzian derivative in KdV dynamics has been noted before in another context (Weiss, 1983, 1986). With these results, it is possible to write a complex version of the modified KdV dynamics in the compact form

$$z_t = -\{z, s\}z_s = -z_{sss} + \frac{3}{2}\frac{z_{ss}^2}{z_s}, \tag{27}$$

and one finds from this additional conserved quantities associated with the motion. These are discussed in Section 4.

Thus far, we have seen that two different geometric viewpoints on a particular set of local curve dynamics leads to two integrable hierarchies – KdV and mKdV. A third related hierarchy is known, based on the so-called Harry Dym equation (Kruskal, 1975; Hereman, et al., 1989) for a function $F(x,t)$

$$F_t = -F^3 F_{xxx}, \tag{28}$$

which has the intriguing property of both nonlinearity and dispersion in a single term. We now illustrate that it is equivalent to yet another mathematical description of the same curve dynamics. Note that this equation has arisen in recent studies of viscous fingering in Hele-Shaw dynamics (Kadanoff, 1990; Constantin & Kadanoff, 1991), further reinforcing the suspicion that these integrable systems are intimately related to curve motion in two-dimensional hydrodynamics.

The viewpoint of curve dynamics embodied in the Harry-Dym hierarchy is the *Eulerian* description in which a point on the curve is labelled by its position in the (complex) plane $z(s)$, as opposed to the *Lagrangian* description (in terms of arclength) we have been discussing so far. In particular, if we consider (Hereman, et al., 1989; Kawamoto, 1985; Kadanoff, 1991) the evolution of the tangent vector under a redefinition of the independent variable and define the new function $Y(z(s,t),t) = \exp(i\theta(z(s,t),t)$, then we find in general

$$Y_t = Y^2\partial_z (W - iU). \tag{29}$$

In this equation, we interpret the velocity functions U and W as functions of $Y(z,t)$ by expressing the curvature and its derivatives in terms of Y, with $\kappa = -iY_z$, etc. and $\partial_z = Y\partial_s$.

We may now simply substitute the velocity functions associated with the first few members of the mKdV hierarchy to arrive at the tangent vector evolution. The simple reparametrization in (11) does not involve actual motion of the curve, consistent with the result $Y_t = 0$. The Harry Dym equation (28) follows directly from the velocities in (13),

$$Y_t = -Y^3 Y_{zzz}, \tag{30}$$

and so on down the hierarchy. It should be remarked that this kind of redefinition is mainly formal at this point, since the variable z is constrained to lie on the curve. It remains an open problem to extend these results to pertain to a function defined over the entire complex plane.

As a final geometric interpretation of the mathematics of these integrable systems we turn to the Lax formulation of the hierarchy (Lax, 1968), a development which plays a central role in the inverse scattering transform (Ablowitz, et al., 1974). (As an aside, recent work (Nakayama, Segur, & Wadati, 1992) has shown an intriguing connection between the commutation relation in Eq. (8) and some aspects of the AKNS approach to inverse scattering theory.) This analysis proceeds as follows. The substitution $\kappa = 2i\psi_s/\psi$ into the Miura transformation (23) linearizes that Riccati equation and yields a Schrödinger equation for the as yet unknown function ψ,

$$L\psi = 0, \qquad L = 2\partial_{ss} + u. \tag{31}$$

From this, we deduce that the "wave function" is simply related to the tangent vector in the complex plane,

$$\psi(s,t) = \exp(-i\theta(s,t)/2). \tag{32}$$

In the Lax formulation, the members of the KdV hierarchy are written as $L_t + [L, M^{(n)}] = 0$, for suitable operators $M^{(n)}$, and the wave function evolves in time according to

$$\psi_t = M^{(n)}\psi. \tag{33}$$

One then readily verifies that the hierarchy of dynamics (33) comprises the first integrals (with respect to arclength) of the curvature dynamics given by the mKdV hierarchy. For instance, the the reparametrization (12) is $\theta_t = c\theta_s$, and the second mKdV equation (14) corresponds to

$$\theta_t = -\theta_{sss} - \frac{1}{2}\theta_s^3. \tag{34}$$

4. Connection with Eulerian Hydrodynamics

To see the connection between the KdV dynamics of the previous section and Eulerian hydrodynamics, it is helpful to recall the connection between the NLS and the motion of a vortex filament. If the filament has circulation $\mathcal{K}$, and is described by the curve $\mathbf{r}(s)$, then the fluid velocity at some point $\mathbf{r}$ is given by a Biot-Savart integral,

$$\mathbf{v}(\mathbf{r}) = -\frac{\mathcal{K}}{4\pi}\oint ds' \frac{(\mathbf{r}-\mathbf{r}(s'))\times\hat{\mathbf{t}}(s')}{|\mathbf{r}-\mathbf{r}(s')|^3}. \tag{35}$$

In the localized induction approximation, the motion of the curve is obtained by considering the fluid velocity at some point $\mathbf{r}$ close to the filament at some arclength coordinate s (the distance from the filament serving as a small-scale cutoff), approximating the curve near that point by a parabola whose curvature is that of the filament, and truncating the integral with a large-scale cutoff.

By a suitable rescaling of time, the curve motion within this approximation may be expressed as a generalization of the planar form, Eq. (4),

$$\mathbf{r}_t = U\hat{\mathbf{n}} + V\hat{\mathbf{b}} + W\hat{\mathbf{t}}, \qquad (U = W = 0; V = \kappa). \tag{36}$$

The tangent, normal, and binormal vectors $\hat{\mathbf{t}}$, $\hat{\mathbf{n}}$ and $\hat{\mathbf{b}}$ obey the Frenet-Serret equations,

$$\partial_s \begin{pmatrix} \hat{\mathbf{t}} \\ \hat{\mathbf{n}} \\ \hat{\mathbf{b}} \end{pmatrix} = \begin{pmatrix} 0 & \kappa & 0 \\ -\kappa & 0 & \tau \\ 0 & -\tau & 0 \end{pmatrix} \begin{pmatrix} \hat{\mathbf{t}} \\ \hat{\mathbf{n}} \\ \hat{\mathbf{b}} \end{pmatrix}, \tag{37}$$

where τ is the torsion. (For space curves, it is conventional to define the curvature with opposite sign to the planar case). Hasimoto (Hasimoto, 1972) showed from the law of motion (36) that the geometric quantity

$$\psi(s,t) = \kappa(s,t)\exp\left(i\int^s ds'\tau(s',t)\right) \tag{38}$$

obeys the Nonlinear Schröding equation

$$i\psi_t = -\psi_{ss} - \frac{1}{2}|\psi|^2\psi, \tag{39}$$

which, like the KdV and related equations, is integrable (Zakharov & Shabat, 1972). The essence of this result is then that a local approximation to inviscid incompressible hydrodynamics leads, through a geometric description, to an integrable dynamics.

We have found that the KdV curve dynamics constitute a localized induction approximation to the motion of a vortex patch, a singly connected region of constant vorticity bounded by irrotational fluid. If ω_p is the value of the vorticity in the patch, then a point on the boundary moves with velocity

$$\mathbf{r}_t(s,t) = \frac{\omega_p}{2\pi} \oint ds' \ln\left(\frac{|\mathbf{r}(s,t) - \mathbf{r}(s',t)|}{r_0}\right) \hat{\mathbf{t}}(s',t) \tag{40}$$

(Zabusky, Hughes, & Roberts, 1979), a manifestly nonlocal equation of motion. For the purpose of completeness, we have introduced the (arbitrary) parameter r_0 which assures dimensional consistency but does not affect the dynamics because the integral of the tangent vector is zero for a closed curve.

The localized induction approximation for the patch consists first in expanding the terms in the integrand in powers of $\Delta = (s' - s)$ and using the Frenet-Serret equations (3) to obtain, for instance,

$$\mathbf{r}(s') \simeq \mathbf{r}(s) + \Delta\hat{\mathbf{t}}(s) - \frac{1}{2}\Delta^2\kappa\hat{\mathbf{n}}(s) + \cdots . \tag{41}$$

The integral over s' in (40) is then truncated with a cutoff Λ at $s' = s \pm \Lambda/2$. The resulting integrals require no short-range cutoff like that needed in the filament problem. The lowest order terms in the dynamics are then found to have the form in Eq. (4) with velocities

$$U = C_1\kappa_s + \cdots, \qquad W = C_2 - (1/2)C_3\kappa^2 + \cdots, \tag{42}$$

where $C_1(\Lambda, r_0)$, etc. are elementary integrals. A suitable choice of r_0 leads to the local arclength conservation condition $W_s = -\kappa U$, and then the dynamics is completely equivalent to the second member of the mKdV hierarchy, Eq. (13). We now recognize the chirality of the dynamics as a reflection of the underlying vorticity of the fluid. Recall that a circular vortex patch has a boundary with constant curvature (in space and time), while, for instance, the Kirchoff ellipse has a varying curvature and a moving boundary whose rate of rotation is related to the magnitude of the enclosed vorticity.

Given the central importance of the Hamiltonian formulation of the KdV dynamics (Gardner, 1971; Das, 1989), it is natural to develop its interpretation in the context of curve motion and to relate it to that of ideal fluids in two dimensions. The Hamiltonian $\mathcal{H}$ of an ideal fluid is simply its kinetic energy, expressible in terms of the vorticity $\omega(\mathbf{r})$ as

$$\mathcal{H}[\omega] = \frac{1}{2}\int d^2r \int d^2r' \omega(\mathbf{r})\omega(\mathbf{r}')\mathcal{G}(\mathbf{r},\mathbf{r}') . \tag{43}$$

The Green's function $\mathcal{G}(\mathbf{r},\mathbf{r}') = \log(|\mathbf{r}-\mathbf{r}'|/r_0)$ is that appropriate to the Poisson equation $\nabla^2\psi = -\omega$ obeyed by the stream function ψ. For the special case of a vortex patch, it is possible to re-express $\mathcal{H}$ as a functional of the boundary (Dritschel, 1986)

$$\mathcal{H}_p[\mathbf{r}] = -\frac{1}{8}\omega_p^2 \oint ds \oint ds' \,\hat{\mathbf{t}}(s)\cdot\hat{\mathbf{t}}(s')\Phi(\mathbf{R}), \tag{44}$$

where $\mathbf{R} = \mathbf{r}(s) - \mathbf{r}(s')$ and $\Phi(\xi) = \xi^2 \ln(|\xi|/er_0)$.

To reconcile the two viewpoints of the fluid Hamiltonian, one as a functional of the vorticity, the other as a functional of the boundary shape, we examine the Poisson bracket appropriate to Euler's equation. The vorticity equation $\omega_t = -\mathbf{v}\cdot\nabla\omega$ follows from $\omega_t = \{\omega, \mathcal{H}\}$ if the Poisson bracket of two functions of ω is (Morrison & Greene, 1980; Olver, 1982; Marsden & Weinstein, 1983)

$$\{F,G\} = \int d^2r'\omega(\mathbf{r}')\left(\nabla'\frac{\delta F}{\delta\omega}\right)\times\left(\nabla'\frac{\delta G}{\delta\omega}\right). \tag{45}$$

In considering functional variations of the vorticity for a patch, it is only necessary to imagine displacements of the boundary normal to itself, suggesting the modification

$$\frac{\delta}{\delta\omega(\mathbf{r})} \to \frac{1}{\omega_p}\hat{\mathbf{n}}(s)\cdot\frac{\delta}{\delta\mathbf{r}(s)}. \tag{46}$$

After some algebra, this leads to the bracket of two functions of the boundary,

$$\{F,G\} \equiv \oint ds'\left(\hat{\mathbf{n}}(s')\cdot\frac{\delta F}{\delta\mathbf{r}(s')}\right)\partial_{s'}\left(\hat{\mathbf{n}}(s')\cdot\frac{\delta G}{\delta\mathbf{r}(s')}\right). \tag{47}$$

It follows that the normal velocity of the boundary (the physically important motion) has the very simple variational form

$$\hat{\mathbf{n}}\cdot\mathbf{r}_t = \partial_s\left(\hat{\mathbf{n}}\cdot\frac{\delta\mathcal{H}}{\delta\mathbf{r}}\right). \tag{48}$$

The localized induction approximation used directly on the equation of motion to yield Eq. (42) may now be applied to the Hamiltonian. It leads to a curvature expansion for $\mathcal{H}$,

$$\mathcal{H} = \oint ds\left(a + b\kappa^2 + c\kappa^4 + d\kappa_s^2 + \cdots\right), \tag{49}$$

where again the coefficients $a, b, c, \ldots$ depend on the arbitrary large-scale cutoff, but require no short-distance regularization. Using the variational form in Eq. (48) we now see how the dynamics generated by successive terms in (49) are those of the first few members of the mKdV hierarchy. The Hamiltonian proportional to the length, $\mathcal{H} = a \oint ds$ gives $U = a\kappa_s$, which is that associated with the ordinary mKdV equation. The second term, $\mathcal{H} = b \oint ds\kappa^2$, is that associated with an elastic rod (Landau & Lifshitz, 1970) and leads to $U = 2b[\kappa_{sss} + (3/2)\kappa^2\kappa_s]$. This in turn generates the third member of the mKdV hierarchy as discussed following Eq. (15). The last two terms in (49) can generate the next member of the hierarchy by suitable tuning of the coefficients c and d in (49) (by choice of r_0 and/or Λ).

An important feature of integrable systems of the KdV type is the existence of a so-called "dual Poisson bracket" formulation by which a given equation of motion follows from two distinct Hamiltonians and two distinct Poisson brackets. In the present context, it is possible to give a simple geometric interpretation to this fact. The two Poisson brackets correspond to the two points of view of the Hamiltonian – that it is a functional of the vector $\mathbf{r}(s)$ or of the curvature $\kappa(s)$. The second view is legitimate since the Hamiltonians in (49) do not contain the vector $\mathbf{r}(s)$ explicitly. We saw this already in the original Hamiltonian formulation of mKdV, Eq. (20). Once again, the relation between these two views follows purely from differential geometry as the connection between functional derivatives with respect to $\mathbf{r}(s)$ and with respect to $\kappa(s)$;

$$\partial_s \left(\hat{\mathbf{n}}(s) \cdot \frac{\delta}{\delta \mathbf{r}(s)} \right) = \Omega \partial_s \left(\frac{\delta}{\delta \kappa(s)} \right), \tag{50}$$

where the operator Ω is defined in (10).

Thus, with $\mathcal{H}^{(0)} = \oint ds$ we obtain the simple reparametrization from

$$\kappa_t = \partial_s \left(\hat{\mathbf{n}} \cdot \frac{\delta \mathcal{H}^{(0)}}{\delta \mathbf{r}} \right), \tag{51}$$

whereas with $\mathcal{H}^{(1)} = (1/2) \oint ds\kappa^2$ the same dynamics arises from

$$\kappa_t = \partial_s \left(\frac{\delta \mathcal{H}^{(1)}}{\delta \kappa} \right). \tag{52}$$

Similarly, we obtain the mKdV dynamics from $\mathcal{H}^{(1)}$ and the variational form in Eq. (51), whereas with $\mathcal{H}^{(2)} = -\oint ds\,(\kappa^4/8 - \kappa_s^2/2)$ the same dynamics arises from the functional variational with respect to κ.

We remarked in Section 3 that additional conservation laws could be deduced from the complex form of the mKdV dynamics. Having seen that this equation is an approximation to vortex patch dynamics, we may now present these constants of the motion. Expressed in terms of the complex coordinate $z(s,t)$ these conserved quantities are

$$I_1 = \oint ds \frac{z}{z_s}, \quad I_2 = \oint ds \frac{z^2}{z_s}, \quad I_3 = \oint ds \frac{|z|^2 z}{z_s}. \tag{53}$$

They correspond (up to constants of proportionality) respectively to the quantities

$$J_1 = \int_{\mathcal{P}} d^2r, \quad \mathbf{J}_2 = \int_{\mathcal{P}} d^2r\,\mathbf{r} \quad J_3 = \int_{\mathcal{P}} d^2r(x^2 + y^2), \tag{54}$$

which are respectively the area, components of the center of mass, and the angular momentum, proper conserved quantities for a vortex patch (Batchelor, 1967).

The *soliton* solutions of the curvature dynamics are functions with the traveling-wave form $\kappa(s,t) = K(z)$, where $z = s - ct$ for some speed c. The variational form of the dynamics in Eq. (51) can be shown to imply that such solutions satisfy

$$\hat{\mathbf{n}} \cdot \frac{\delta}{\delta \mathbf{r}} (\mathcal{H} + cL + dA) = 0, \tag{55}$$

for some constant d. This appears as a constrained extremization problem for $\mathcal{H}$ with Lagrange multipliers c and d conjugate to the length and area. The general form (48) is naturally complementary to the dissipative relaxational dynamics recently proposed in the study of pattern formation in two dimensions (Langer, Goldstein, & Jackson, 1992; Goldstein & Langer, 1992),

$$\hat{\mathbf{n}} \cdot \mathbf{r}_t = -\hat{\mathbf{n}} \cdot \frac{\delta}{\delta \mathbf{r}} (\mathcal{H} + cL + dA). \tag{56}$$

The connection between traveling waves and constrained minimization is valid for any Hamiltonian. The mKdV solitons correspond to uniformly rotating shapes having, for a given perimeter L and n periods of the curvature, a one-parameter family of solutions specified by the area (See Fig. 4). These are analogs of n-fold periodic rotating shapes found numerically as exact solutions of the contour dynamics algorithm (Deem & Zabusky, 1978).

5. Conclusion and Future Prospects

The work reviewed here has revealed conceptual and mathematical links between two-dimensional inviscid, incompressible hydrodynamics, differential geometry, and integrable systems. These results parallel in many ways those known in three dimensions in the context of the Nonlinear Schrdinger equation. Moreover, they indicate that a number of apparently complex nonlinear transformations between different integrable systems may be understood from simple differential geometric considerations.

The particular class of curve motions considered here is of course a rather small set of those with interesting conservation laws. Whether there are integrable systems unrelated to the KdV equation which are also equivalent to planar curve motion remains to be seen. For instance, the choice of velocities $U = \kappa^n \kappa_s$ and $W = -\kappa^{n+2}/(n+2)$, for $n \geq 1$, conserves perimeter and

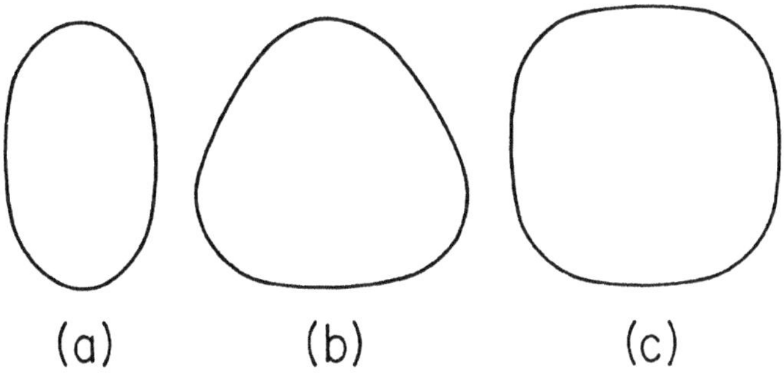

Fig. 4. Soliton solutions of the modified Korteweg-de Vries equation with 2–, 3–, and 4–fold symmetry.

enclosed area like the case KdV case $n = 0$, but it is not known if the resulting dynamics is integrable. There are also interesting topological aspects of the integrable motion of curves which remain incompletely understood, especially for curves in three spatial dimensions, although progress has been made recently in this direction (Ricca, 1992). In addition, the possibility that higher dimensional integrable systems reflect an underlying hydrodynamics appears worth pursuing.

The reformulation of the vortex patch Hamiltonian into a double contour integral, known also from the work of Dritschel (Dritschel, 1986), expands the range of energies one associates with shapes, and is reminiscent of the energy of a uniform dipolar (or magnetic) slab. There, the field energy involves an essentially Coulombic interaction between tangent vectors. Such domains formed by amphiphilic molecules at the air-water interface are known to form remarkable patterns, some with strong chirality (Weis & McConnell, 1984), reflecting the underlying chirality at the molecular level. Some aspects of this chiral motion may be contained in geometric descriptions akin to those discussed here. Moreover, the elastic-like Hamiltonian of the mKdV equation may be directly relevant to the dynamics of semiflexible macromolecules. Of course, in any real physical system there will viscous damping and other physical features which destroy the Hamiltonian structure of the dynamics. It is thus of interest to study solitonic shape dynamics with dissipative components (Kivshar & Malomed, 1989).

Finally, we observe that all of the dynamics discussed here, whether Hamiltonian or dissipative, are deterministic. Just as with the motion of a Brownian particle, it is of interest to investigate stochastic generalizations of curve dynamics in order to develop a *statistical mechanics of shapes.* At the same time, and especially in light of recent work on the chiral dynamics of Fermi surfaces in the Quantum Hall Effect (Wen, 1990), it is of interest to investigate quantum mechanical generalizations of curve motion.

Acknowledgements

We are indebted to P. Constantin for the crucial observations which initiated this line of inquiry, to J. Harvey for probing questions regarding the integrable hierarchies, to S.A. Langer and D.P. Jackson for ongoing collaborations conceptually related to this work, and to J. Distler, L.P. Kadanoff, D. McLaughlin, P. Nelson, A.I. Pesci, T. Powers, and M.J. Shelley for important discussions. This work was supported by the National Science Foundation, through a graduate fellowship (D.M.P.) and Grant CHE-9106240. Finally, it is a pleasure to thank R.J. Caflisch for organizing a most interesting conference in a most beautiful setting!

References

Ablowitz, M.J., Kaup, D.J., Newell, A.C., & Segur, H. 1974 The inverse scattering transform – Fourier analysis for nonlinear problems. *Stud. Appl. Math.* **53**, 249.

Batchelor, G.K. 1967 *An Introduction to Fluid Dynamics*. Cambridge University Press.

Brower, R.C., Kessler, D.A, Koplik, J. & Levine, H. 1984 Geometrical models of interface evolution. *Phys. Rev. A* **29**, 1335.

Chern, S.-S. & Peng, C.-K. 1979 Lie groups and KdV equations. *Man. Math.* **28**, 207.

Constantin, P. 1990a private communication.

Constantin, P. 1990b Dynamics of interfaces. unpublished.

Constantin, P. & Kadanoff, L.P. 1991 Dynamics of a complex interface. *Physica D* **47**, 450.

Das, A. 1989 *Integrable Models*. Singapore: World Scientific.

Deem, G.S. & Zabusky, N.J. 1978 Vortex waves: Stationary "V states," interactions, recurrence, and breaking. *Phys. Rev. Lett.* **40**, 859.

Drazin, P.G. & Johnson, R.S. 1989 *Solitons: an Introduction*. New York: Cambridge University Press.

Dritschel, D.G. 1986 The nonlinear evolution of rotating configurations of uniform vorticity. *J. Fluid. Mech.* **172**, 157.

Gage, M.E. 1984 Curve shortening makes convex curves circular. *Invent. Math.* **76**, 357.

Gardner, C.S. 1971 Korteweg-de Vries equation and generalizations. IV. The Korteweg-de Vries equation as a Hamiltonian system. *J. Math. Phys.* **12**, 1548.

Goldstein, R.E. & Petrich, D.M. 1991 The Korteweg-de Vries hierarchy as dynamics of closed curves in the plane. *Phys. Rev. Lett.* **67**, 3203.

Goldstein, R.E. & Petrich, D.M. 1992 Solitons, Euler's equation, and vortex patch dynamics. *Phys. Rev. Lett.* **69**, 555.

Goldstein, R.E. & Langer, S.A. 1992 Dissipative dynamics of closed curves in two dimensions. unpublished.

Hasimoto, H. 1972 A soliton on a vortex filament. *J. Fluid. Mech.* **51**, 477.

Hereman, W., Banerjee, P.P. & Chatterjee, M.R. 1989 Derivation and implicit solution of the Harry Dym equation and its connection with the Korteweg-de Vries equation. *J. Phys. A* **22**, 241.

Kadanoff, L.P. 1990 Exact solutions for the Saffman-Taylor problem with surface tension. *Phys. Rev. Lett.* **65**, 2986.

Kadanoff, L.P. 1991 private communication.

Kawamoto, S. 1985 An exact transformation from the Harry Dym equation to the modified KdV equation. *J. Phys. Soc. Jpn.* **54**, 2055.

Kivshar, Y.S. & Malomed, B.A. 1989 Dynamics of solitons in nearly integrable systems. *Rev. Mod. Phys.* **61**, 763.

Kruskal, M.D. 1975 Nonlinear wave equations. in *Dynamical Systems, Theory and Applications*, Lecture Notes in Physics, Vol. 38 New York: Springer-Verlag, p. 313.

Lamb, G.L., Jr. 1977 Solitons on moving space curves. *J. Math. Phys.* **18**, 1654.

Landau, L.D. & Lifshitz, I.M. 1959 *Fluid Mechanics*. Pergamon.

Landau, L.D. & Lifshitz, I.M. 1970 *Theory of Elasticity*. Pergamon.

Langer, S.A., Goldstein, R.E., & Jackson, D.P. 1992 Dynamics of labyrinthine pattern formation in magnetic fluids. *Phys. Rev. A*, in press.

Lax, P.D. 1968 Integrals of nonlinear equations of evolution and solitary waves. *Comm. Pure. Appl. Math.* **21**, 467.

Marsden, J. & Weinstein, A. 1983 Coadjoint orbits, vortices, and Clebsch variables for incompressible fluids. *Physica D* **7**, 305.

Miura, R.S. 1968 Korteweg-de Vries equation and generalizations. I. A remarkable explicit nonlinear transformation. *J. Math. Phys.* **9**, 1202.

Morrison, P.J. & Greene, J.M. 1980 Noncanonical Hamiltonian density formulation of hydrodynamics and ideal magnetohydrodynamics. *Phys. Rev. Lett.* **45**, 790.

Nakayama, K., Segur, H., & Wadati, M. 1992 On integrability and the motion of curves. preprint.

Nehari, Z. 1975 *Conformal Mapping*. Dover.

Olver, P.J. 1982 A nonlinear Hamiltonian structure for the Euler equations. *J. Math. Anal. Appl.* **89**, 233.

Ricca, R.L. 1992 Physical interpretation of certain invariants for vortex filament motion under LIA. *Phys. Fluids A* **4**, 938.

Weis, R.M. & McConnell, H.M. 1984 Two-dimensional chiral crystals of phospholipid. *Nature* **310**, 47.

Weiss, J. 1983 The Painlevé property for partial differential equations. II: Bäcklund transformation, Lax pairs, and the Schwarzian derivative. *J. Math. Phys.* **24**, 1405.

Wen, X.G. 1990 Electrodynamical properties of gapless edge excitations in the Fractional Quantum Hall states. *Phys. Rev. Lett.* **64**, 2206.

Zabusky, N.J., Hughes, M.H., & Roberts, K.V. 1979 Contour dynamics for the Euler equations in two dimensions. *J. Comp. Phys.* **30**, 96.

Zakharov, V.E. & Shabat, A.B. 1972 Exact theory of two-dimensional self-focusing and one-dimensional self-modulation of waves in nonlinear media. *Sov. Phys. JETP* **34**, 62.

Asymptotic and Numerical Results for Blowing-Up Solutions to Semilinear Heat Equations

Joseph.B. Keller
Stanford University

and

John S. Lowengrub
Stanford University and Institute for Advanced Study

ABSTRACT. Asymptotic and numerical results for blowing-up solutions to a class of semilinear heat equations are presented. They are obtained using the original space-time coordinate frame without any dynamical rescaling of space, time or solution. A new equivalent equation is derived using a nonlinear transformation. Under some assumptions, the asymptotic form, the time and the position of the singularity are determined. Blow-up in the original equation is equivalent to vanishing in the new equation. As a consequence of the transformation *and* the asymptotic form of the blow-up, all terms in the new equation are bounded. Thus, numerical solutions are easy to obtain right up to the blow-up time without any loss of resolution. Inverting the transformation yields the solution to the original problem.

1. Introduction

The study of solutions of partial differential equations which blow up is an important and challenging problem. Although physically, we might expect solutions to remain smooth and bounded, it is not uncommon to see physical quantities vary by several orders of magnitude over relatively short times. Therefore, we can think of them as nearly blowing-up. This is particularly true in combustion problems where semilinear heat equations are useful models (see [1]). A detailed study of blow-up in the mathematical model may then give some insight into the near blow-up of physical quantities and may also provide matching conditions for other, more realistic models that may be used near the blow-up time.

We shall present asymptotic and numerical results for blowing-up solutions to a class of semilinear heat equations. The results are obtained without any dynamical rescaling of space, time or solution, but are obtained using the original space-time coordinates instead. A new equivalent equation is derived

R. E. Caflisch and G. C. Papanicolaou (eds.), Singularities in Fluids, Plasmas and Optics, 111–129.
© 1993 *Kluwer Academic Publishers.*

by using a nonlinear transformation of the original unknown solution.

The transformation is derived by ignoring spatial dependence, which is appropriate if the solution is slowly varying. It gives an *equivalent* formulation of the problem and hence it does *not* restrict the type of blow-up that can occur. The blow-up is determined dynamically by solving the full reformulated problem. The transformation can be derived more generally, and this will be discussed elsewhere [11].

Under some reasonable assumptions, we determine the asymptotic form, time and position of the blow-up. The asymptotic form shows that in the reformulated problem, all the terms are bounded. The asymptotic results are compared with numerical simulations of the full reformulated equation.

Traditionally, the main computational tool has been a dynamical rescaling of time, space and solution that takes advantage of the scale invariant structure of the equations to keep the numerical solutions bounded (see [2,4,12]). In our method, since all the terms in the reformulated equation are bounded, we can use straightforward finite differences to solve the problem numerically. The numerical solutions are resolved right up to the singularity time. We find good agreement between the asymptotic and the numerical results for the structure, time, and position of the singularity.

In section 2 the asymptotic results are derived and in section 3 the numerical results are presented. Section 4 contains some conclusions.

2. Reformulation and Asymptotics

We consider the class of semilinear heat equations

$$u_t = \Delta u + f(u); \; u(x,0) = u_0(x) > 0; \; x \in R^n \tag{1}$$

where

$$\frac{f(u)}{u} \to \infty \text{ as } u \to \infty \tag{2}$$

Examples of this type of nonlinearity include $f(u) = u^p$ for $p > 1$ and $f(u) = e^u$, both of which occur in models of combustion [1]. We expect this class of problems to have blowing-up solutions where the blow-up is isolated and has a characteristic structure. For exact statements of theorems, see [1,5,6,9,10] for example.

Our interest is to study the asymptotic behavior of solutions which blow up at the time t_c and at the point x_c. To do so we introduce a new unknown v defined by

$$v(x,t) = \int_{u(x,t)}^{\infty} \frac{d\tilde{u}}{f(\tilde{u})} \tag{3}$$

Then we have the following *formal* asymptotic results:

Asymptotic Results

1. *Suppose that $v(x,t)$ is asymptotically spherically symmetric about x_c and is an even function of $|x-x_c|$. Then v has the following asymptotic form:*

$$v(x,t) \sim (t_c - t) - \frac{n}{2}\int_0^{t_c-t} |log f(u(s))|^{-1} ds + \frac{|x-x_c|^2}{4|log f(u(t_c-t))|} + \cdots \tag{4}$$

2. *Suppose that the initial function is slowly varying, i.e. that it is of the form $u_0(\epsilon x)$ with ϵ small, and that it attains its maximum at x_m. Let $T_0(x) = \int_{u_0(x)}^{\infty} \frac{d\bar{u}}{f(\bar{u})}$. Then*

$$t_c = T_0(x_m) + \epsilon^2 T_0(x_m)\Delta T_0(x_m) + O(\epsilon^4)$$

$$x_c = x_m - \epsilon^2 \frac{T_0(x_m)\nabla\Delta T_0(x_m)}{\Delta T_0(x_m)} + O(\epsilon^4) \tag{5}$$

A. Derivation of Transformation

Omit the Δu and let the nonlinearity be given by u^p in equation (1). This reduces (1) to the ODE problem

$$u_t = u^p, \quad u(x,0) = u_0(x) > 0, \quad x \epsilon R^n. \tag{6}$$

This is exactly solvable with the solution:

$$u(x,t) = (p-1)^{\frac{1}{1-p}} (T_o(x) - t)^{\frac{1}{1-p}}. \tag{7}$$

Here $T_0(x)$ is defined by

$$T_0(x) = \frac{u_0(x)^{1-p}}{p-1} \tag{8}$$

Clearly u blows up as $t \to \min T_0(x)$.

This example motivates the nonlinear transformation $u \to v$:

$$u(x,t) = [(p-1)v(x,t)]^{\frac{1}{1-p}} \tag{9}$$

Thus the blow-up of u is replaced by the vanishing of v in a special way. By (7), we expect v to vanish linearly in time as $t \to t_c$, the blow-up time. Reformulating (1) in terms of v gives:

$$v_t = -1 + \Delta v - \frac{p}{p-1}\frac{|\nabla v|^2}{v}, \; v(x,0) = (p-1)^{-1}[u_0(x)]^{1-p}. \tag{10}$$

Thus, the effect of the transformation is to replace the u^p term in (1) by -1 (this is the linear vanishing) but at the expense of adding the nonlinear gradient term. This additional term might be cause for worry as v appears in the denominator and is expected to vanish. However, we will show later that the asymptotic structure of v near the blow-up point is such that this term is in fact bounded.

Now, suppose that we have the general nonlinearity $f(u)$. Similar considerations suggest the following nonlinear transformation:

$$v(x,t) = \int_{u(x,t)}^{\infty} \frac{d\tilde{u}}{f(\tilde{u})} \tag{11}$$

which gives the reformulated problem

$$v_t = -1 + \Delta v - f'(u(v))|\nabla v|^2 \,, \; v(x,0) = \int_{u_0(x)}^{\infty} \frac{d\tilde{u}}{f(\tilde{u})}. \tag{12}$$

An advantage of the transformation is that the main effect of the nonlinearity is scaled out of the equation for v. The v equation depends only weakly on the nonlinearity $f(u)$. The original solution u is obtained by inverting the transformation (11), either analytically or numerically.

B. Asymptotic Structure in Time

Let us suppose that the first singularity appears at $x = 0$ (without loss of generality) and at $t = t_c$. Let v be given by (11) and satisfy equation (12). Further, *suppose* that near the singularity, v has an even, radial expansion:

$$v(x,t) = v_0(t) + \frac{r^2}{2!} v_2(t) + \frac{r^4}{4!} v_4(t) + \cdots \tag{13}$$

where $r = |x|$. Now, substituting the expansion (13) of v into equation (12) and equating coefficients of r, we obtain the coupled differential equations for the coefficients v_0 and v_2:

$$\begin{aligned} \frac{dv_0(t)}{dt} &= -1 + nv_2 \\ \frac{dv_2(t)}{dt} &= -2f'(u(v_0))v_2^2 + \frac{2+n}{3} v_4 \end{aligned} \tag{14}$$

Of course, the system is not yet closed as v_4 appears explicitly and thus v_0 and v_2 depend on the higher coefficients through v_4. However, let us *suppose* that $|v_4| << |f'(u(v_0))v_2^2|$ near the singularity time. Then omitting v_4 closes the system (14) near the singularity time as:

$$\begin{aligned} \frac{dv_0(t)}{dt} &= -1 + nv_2 \\ \frac{dv_2(t)}{dt} &= -2f'(u(v_0))v_2^2 \end{aligned} \tag{15}$$

This system can be solved in the following manner. First, we rewrite (15) as

$$
\begin{aligned}
v_0(t) &= (t_c - t) - n\int_t^{t_c} v_2(\tau)d\tau \\
v_2(t) &= \left[2\int_0^t f'(u(v_0(\tau)))d\tau + \frac{1}{v_2(0)}\right]^{-1}
\end{aligned}
\tag{16}
$$

by integrating in time. We can solve (16) by the iteration method:

$$
\begin{aligned}
v_0^{j+1}(t) &= (t_c - t) - n\int_t^{t_c}\left[2\int_0^{\tau} f'(u(v_0^j(\tilde{\tau})))d\tilde{\tau} + \frac{1}{v_2(0)}\right]^{-1} d\tau \\
v_2^{j+1}(t) &= \left[2\int_0^t f'(u(v_0^{j+1}(\tau)))d\tau + \frac{1}{v_2(0)}\right]^{-1}
\end{aligned}
\tag{17}
$$

We use the starting values

$$
\begin{aligned}
v_0^0(t) &= t_c - t \\
v_2^0(t) &= \frac{1}{2}|\log f(u(t_c - t))|
\end{aligned}
\tag{18}
$$

They are suggested by eliminating the v_2 term from the first equation in (16) to decouple the two equations. See the appendix for the evaluation of the integral in the second equation. That this method is convergent, for $t_c - t$ small enough, and asymptotic, in the sense that

$$
\frac{v_0^{j+1}}{v_0^j} = 1 + G_{j,j+1}(t), \text{ where } G_{j,j+1}(t) \to 0 \text{ as } t \to t_c
\tag{19}
$$

will be presented elsewhere [11].

By using (19), the starting conditions (18), and the expansion (13), we find that:

$$
\begin{aligned}
v(x,t) &= \int_{u(x,t)}^{\infty} \frac{d\tilde{u}}{f(\tilde{u})} \\
&\sim (t_c - t) - \frac{n}{2}\int_0^{t_c - t} |log f(u(s))|^{-1} ds + \frac{r^2}{4|log(f(u(t_c - t))|} + \cdots
\end{aligned}
\tag{20}
$$

In the case $f(u) = u^p$, (20) reduces to:

$$
v(x,t) \sim (t_c - t) - \frac{n}{2p}\int_0^{(p-1)(t_c - t)} \frac{dx}{|logx|} + \frac{p-1}{4p}\frac{r^2}{|log(t_c - t)|} + \cdots
\tag{21}
$$

In terms of u, (21) gives

$$u(x,t) \sim \left\{(p-1)\left[t_c - t - \frac{n}{2p}\int_0^{(p-1)(t_c-t)} \frac{dx}{|log x|} + \frac{p-1}{4p}\frac{r^2}{|log(t_c-t)|} + \cdots\right]\right\}^{\frac{1}{1-p}} \quad (22)$$

Alternatively, in the case $f(u) = e^u$, we get:

$$u(x,t) \sim -log\left[(t_c - t) - \frac{n}{2}\int_0^{t_c-t} \frac{dx}{|log x|} + \frac{r^2}{4|log(t_c-t)|} + \cdots\right], \quad (23)$$

since the inverse of the transformation is $u(x,t) = -log(v(x,t))$. By expanding (22), using the binomial theorem, we see that it agrees with the results in [5,7-10]. Similarly (23) agrees with those derived in [3]. An advantage of our approach is that it yields the more general result (20) for v. Inverting the transformation (11), either analytically or numerically, to find u determines its general asymptotic structure. This completes the derivation of (4).

C. Slowly Varying Initial Data

In the preceding analysis we assumed that blow-up occurs at some point t_c, x_c, but we did not determine that point. By considering slowly varying initial data we shall show that blow-up does occur at one point, and we shall determine it.

We suppose that $u(x,0) = u_0(\epsilon x)$. Then letting $x' = \epsilon x$ in (1) and then omitting the prime gives

$$u_t = \epsilon^2 \Delta u + f(u) \quad (24)$$

Defining v by (11) as before, and transforming (24) correspondingly gives

$$v_t = -1 + \epsilon^2 \Delta v - \epsilon^2 f'(u(v))|\nabla v|^2 \quad (25)$$

The initial data for v are given by

$$v(x,0) = T_0(x) = \int_{u_0(x)}^{\infty} \frac{d\tilde{u}}{f(\tilde{u})} \quad (26)$$

Thus (1) has been transformed to (25) and (26) for v.

We now suppose that v has the asymptotic form

$$v(x,t,\epsilon) = v_0(x,t) + \epsilon^2 v_2(x,t) + O(\epsilon^4) \quad (27)$$

Only even powers of ϵ appear since (25) depends on ϵ^2. Substituting (27) into (25) and equating powers of ϵ gives the system

$$\begin{aligned} \frac{dv_0(t)}{dt} &= -1; \quad v_0(x,0) = T_0(x) \\ \frac{dv_2(t)}{dt} &= \Delta v_0 - f'(u(v_0))|\nabla v_0|^2; \quad v_2(x,0) = 0 \end{aligned} \quad (28)$$

These equations (28) for the first two coefficients can be solved in closed form. Substituting the results into the expansion (27) gives

$$v(x,t) = T_0(x) - t + \epsilon^2 \left[t\Delta T_0(x) - |\nabla T_0(x)|^2 log \frac{f(u(T_0(x)-t))}{f(u_0(x))} \right] + O(\epsilon^4) \tag{29}$$

We now seek the first time at which $v(x,t) = 0$ for some x. To do so we evaluate (29) at x_c, t_c and equate it to zero:

$$\begin{aligned} v(x_c,t_c) = 0 &= T_0(x_c) - t_c \\ &+\epsilon^2 \left[t_c\Delta T_0(x_c) - |\nabla T_0(x_c)|^2 log \frac{f(u(T_0(x_c)-t_c))}{f(u_0(x_c))} \right] + O(\epsilon^4) \end{aligned} \tag{30}$$

Without loss of generality, we assume that $T_0(x)$ attains its minimum value at $x = 0$. We also assume that $\nabla T_0(0) = 0$. Now we write the following expansions for t_c and x_c:

$$\begin{aligned} t_c &= T_0(0) + \epsilon^2 t_{c,2} + \cdots \\ x_c &= 0 + \epsilon^2 x_{c,2} + \cdots \end{aligned} \tag{31}$$

We use the expansions (31) in (30) and equate powers of ϵ. At the second order, this yields only the correction to the blow-up time:

$$t_{c,2} = T_0(0)\Delta T_0(0) \tag{32}$$

The second order correction to the blow-up position is still undetermined. To find it, we take the gradient of (29) and suppose further that (x_c, t_c) is a critical point of v. Then

$$\begin{aligned} \nabla v(x_c,t_c) &= 0 \\ &= \nabla T_0(x_c) + \epsilon^2 t_c \nabla\Delta T_0(x_c) \\ &\quad -\epsilon^2 \nabla \left\{ |\nabla T_0(x_c)|^2 log \frac{f(u(T_0(x_c)-t_c))}{f(u_0(x_c))} \right\} + O(\epsilon^4) \end{aligned} \tag{33}$$

Now, using the expansions (31) in (33) determines the second order correction to the position:

$$x_{c,2} = \frac{-T_0(0)\nabla\Delta T_0(0)}{\Delta T_0(0)} \tag{34}$$

This completes the derivation of (5).

3. Numerical Results

We now investigate, numerically, the behavior of solutions to (1) with $f(u) = u^5$ and in one space dimension. This problem was most recently studied in [2]. We compare the asymptotic results 4, 5 with the numerical simulations of the full reformulated problem. The problem we solve is:

$$\begin{aligned} u_t &= u_{xx} + u^5; \qquad -a \leq x \leq a \\ u(\pm a, t) &= u_0(\pm a) \end{aligned} \tag{35}$$

These boundary conditions, were chosen for their simplicity. We found that the boundary conditions had little effect on the singularity structure provided that the boundary is far enough away from the maximum of u_0.

In this case, $p = 5$ and the appropriate transformation is

$$u(x,t) = [4v(x,t)]^{-\frac{1}{4}} \tag{36}$$

The reformulated problem is then

$$\begin{aligned} v_t &= -1 + v_{xx} - \tfrac{5}{4}v_x^2/v; \qquad -a \leq x \leq a \\ v(\pm a, t) &= T_0(\pm a) = u_0(\pm a)^{-4}/4 \end{aligned} \tag{37}$$

We now use the asymptotic result (4) with $f(u) = u^5$. When the second term is omitted, because it is asymptotically smaller than the first term, we obtain

$$v \sim (t_c - t) + \frac{(x - x_c)^2}{5|log(t_c - t)|} \tag{38}$$

To estimate t_c and x_c we use (5) with $\epsilon = 1$ and we get

$$\begin{aligned} t_c &= T_0(x_m) + T_0(x_m)T_0''(x_m) \\ x_c &= x_m - \frac{T_0(x_m)T_0'''(x_m)}{T_0''(x_m)} \end{aligned} \tag{39}$$

The 'higher order' corrections have been omitted. We cannot expect 39 to be very accurate since $\epsilon = 1$ is not small.

It is straightforward to see that the asymptotic form of v implies that the nonlinear term in 37 is bounded. Let

$$v_a(x,t) = (t_c - t) + \frac{x^2}{5|log(t_c - t)|} \tag{40}$$

Then

$$\frac{v_{ax}^2}{v_a} = \frac{4}{5}\frac{x^2}{[x^2 + 5(t_c - t)|log(t_c - t)|]} \tag{41}$$

This is clearly bounded, even as $t \to t_c$. Therefore, we shall use straightforward finite differences to solve the problem numerically.

We define w_j^n to be the numerical approximation to $v(n\Delta t, jh)$ where Δt and h are the temporal and spatial grid sizes respectively. The difference equation is

$$\frac{w_j^{n+1} - w_j^n}{\Delta t} = -1 + \frac{(w_{j-1}^n - 2w_j^n + w_{j+1}^n)}{h^2} - \frac{5}{16}\frac{(w_{j+1}^n - w_{j-1}^n)^2}{h^2}\frac{1}{w_j^n} \tag{42}$$

which is first order in time and second order in space. We also choose $\Delta t = \frac{1}{4}h^2$ to satisfy the stability requirement.

The numerical blow-up time t_c^* is chosen to be the first time that the numerical solution v becomes negative, and x_c^* is the position at which this occurs. A typical value of the numerical solution at t_c^* is -10^{-8}, so t_c^* is only a slight overestimate. Thus, the calculation actually continues the solution slightly beyond the blow-up time of the original solution u. The method becomes unstable a short time after the blow-up happens, however. The numerical blow-up structure will be compared to the predicted formula 38 of the parameters t_c and x_c.

The first calculation is the evolution of the parabola

$$v(x,o) = T_0(x) = 2^{-6} + \frac{x^2}{2}; \quad -1.5 \le x \le 1.5 \tag{43}$$

Here $x_m = 0$, $T_0''(0) = 1$ and $T_0'''(0) = 0$. The time evolution of the solution $v(x,t)$, with $h = .00586$, is shown in figure 1. The topmost curve is the initial condition ($t = 0$) and the lower curves are the solutions at successively later times. At the final time shown, the minimum value of v is $O(10^{-8})$. Note that the profile tends to flatten. This corresponds to a sharpening profile of the original untransformed solution u as is shown in figure 2. For figure 2, v was computed first and then u was determined from v via the transformation 36. Notice how sharply peaked the final profile is in figure 2. The final curve in figure 2 is shown at a slightly later time than that of figure 1. In figure 3, the values of u on the grid at the final time level are shown. It is clear from their distribution that a direct computation of u is very difficult as numerical resolution and stability become increasingly difficult to maintain. This is in contrast to figure 4 which shows the distribution of v on the grid. There is clearly no such problem here. This is the remarkable feature of the reformulation!

We now compare the asymptotic structure of the singularity (given by 38) with the numerical computation in figure 1. The result is shown in figure 5. The agreement is excellent and the curves lie on top of each other (at the singular point) to within the plotting resolution. The asymptotic predictions 39 for these data are $t_c = 3.125 \times 10^{-2}$ and $x_c = 0$. The computed values are found to be $t_c^* = 2.861 \times 10^{-2}$ and $x_c^* = 0$. The agreement for t_c is poor because $\epsilon = 1$ is not small.

The previous computation, as well as all the subsequent computations, takes less than 5 minutes to run on a Sun sparcstation 2. In the subsequent computations, we do not give the graphs of the original solution u as their form is the obvious modification of that from the parabolic computation.

We now consider initial data with more than one local minimum (maximum for the original solution u). Such data presents a challenge for rescaling algorithms as it is not always clear which maximum will blow-up first.

The second computation has as initial data the fifth order polynomial satisfying the conditions

$$\begin{aligned} T_0(\pm.5) &= .015625 = 2^{-6} \\ T_0'(\pm.5) &= 0 \\ T_0''(\pm.5) &= 1 \end{aligned} \tag{44}$$

which has local minima at $x = \pm.5$. The domain is $1.5 \leq x \leq 1.5$ and the third derivative is found to be $T_0'''(\pm.5) = \pm 6$. The time evolution of this solution is shown in figure 6 (h = .00586). Both minima tend to flatten and again there is no problem with resolution or stability. The minima pass through zero at the same time.

The asymptotic structure 38 is compared with the numerical results in figure 7 and again the agreement is excellent. The asymptotic results 39 predict that both minima will hit zero at the same time $t_c = 3.125 \times 10^{-2}$ and their positions will be $x_c = \pm.40625$. Numerically, both minima pass through zero at $t_c^* = 2.6485 \times 10^{-2}$ and at $x_c^* = \pm.428175$. Of course since $\epsilon = 1$ is not small, the lack of agreement is to be expected.

The third computation takes initial data with two local minima, but of different heights:

$$\begin{aligned} T_0(-.5) &= .015625 \\ T_0(+.5) &= .02 \\ T_0'(\pm.5) &= 0 \\ T_0''(\pm.5) &= 1.0 \end{aligned} \tag{45}$$

on the domain $-1.5 \leq x \leq 1.5$. The third derivatives are $T_0'''(-.5) = -5.7375$ and $T_0'''(+.5) = 6.2625$. The time evolution of this solution is shown in figure 8 (h = .00293). Both minima still tend to flatten, but the one corresponding to $x = -.5$, initially, passes through zero first.

The asymptotic results predict that the minimum located at $x = -.5$, initially, will hit zero first at $t_c = 3.125 \times 10^{-2}$ and at $x_c = .41305$. The numerical singularity time and position are found to be $t_c^* = 2.663 \times 10^{-2}$

and $x_c^* = -.4248$. The asymptotic structure 38 is compared to the numerical result in figure 9. The agreement is excellent.

The fourth computation takes initial data with two local minima with the same height, but different curvatures

$$
\begin{aligned}
T_0(\pm.5) &= .015625 \\
T_0'(\pm.5) &= 0 \\
T_0''(+.5) &= 1.0 \\
T_0''(-.5) &= 1.5
\end{aligned}
\tag{46}
$$

on the domain $-1.5 \leq x \leq 1.5$. Further, $T_0'''(-.5) = -10.5$ and $T_0'''(.5) = 4.5$. For this data, the asymptotic prediction is that the minimum located at $x = .5$ initially, will hit zero first at $t_c = 3.125 \times 10^{-2}$ and at $x_c = .440625$. This is indeed what happens numerically. The evolution is shown in figure 10. The numerical singularity time and position are found to be $t_c^* = 2.6901 \times 10^{-2}$ and $x_c^* = .43945$. The asymptotic structure 38 is compared to the numerical result in figure 11 where the agreement is excellent.

It is not surprising that the predicted singularity times and positions do not agree so well with their numerical counterparts. The asymptotic predictions were derived only for slowly varying initial data, an assumption t38s clearly violated by the initial data for the numerical computations. It is to be expected, however, that the predicted asymptotic form (38a) agrees so well with the full numerical simulations. The asymptotic form was derived under the assumption that $t_c - t$ is small. The singularity time, t_c, and position, x_c, are assumed to be given. In the comparison, the numerical values t_c^* and x_c^* are used to determine the asymptotic form. Since this form is then compared to the numerical solution at times near t_c^*, we have every reason to expect the agreement to be good.

4. Conclusions

We have presented asymptotic and numerical results for blowing-up solutions to a class of semilinear heat equations. We have shown how to reformulate the problem exactly to yield a new equation. This new equation has, as its essential feature, all of its terms bounded; a fact that can be seen from the asymptotics. Thus, straightforward finite difference techniques can be used to solve it numerically. This is a tremendous savings as previous numerical computations, for this type of problem, have relied on dynamical rescaling algorithms [2,4,12] and are therefore much more complicated. In future work, we will consider more general transformations for this problem. We hope that by deriving the transformation more generally, we will be able to apply our techniques to other problems with blowing-up solutions. Of course, as we have seen here, the success of such a transformation depends crucially on the asymptotic form of the blow-up up solution. Finally, we hope that the reformulation will also help to simplify the proofs of rigorous results. This is partially borne out in [11] where we give a simple proof of blow-up.

Acknowledgements

The authors are pleased to thank R. Kohn and M. Berger for some fruitful discussions. The authors are partially supported by NSF and ONR grants. The second author is also partially supported by an NSF postdoctoral fellowship. The current address of the second author is the School of Mathematics, Vincent Hall, 206 Church St. S.E., University of Minnesota, Minneapolis, MN 55455.

5. Appendix

In this appendix, we evaluate the integral

$$\int_0^t f'(u(t_c - \tau))d\tau \tag{47}$$

as follows. Let $v_0 = t_c - \tau$. Then $dv_0 = -d\tau$. Therefore the integral becomes

$$-\int_{v_0(0)}^{v_0(t)} f'(u(v_0))dv_0 \tag{48}$$

Now, recall that u and v_0 are related by the transformation

$$v_0 = \int_u^\infty \frac{d\tilde{u}}{f(\tilde{u})} \tag{49}$$

Change the variable of integration in (47) to $\tilde{u} = u(v_0)$. Thus $d\tilde{u} = -f(\tilde{u})dv_0$. This gives

$$\int_{u(v_0(0))}^{u(v_0(t))} \frac{f'(\tilde{u})}{f(\tilde{u})}d\tilde{u} = log\frac{f(u(t_c - t))}{f(u_0)} \tag{50}$$

since $u(v_0(0)) = u_0$.

References

1. Bebernes J. and Eberly D., **Mathematical Problems from Combustion Theory**, Springer-Verlag, 1989.
2. Berger M. and Kohn R., *A Rescaling Algorithm for the Numerical Calculation of Blowing-up Solutions*, Comm. Pure Appl. Math., v. 41, pp. 841-863, 1988.
3. Bressan A., *Stable Blow-up Patterns*, J. Diff. Eq., v. 98, pp. 75-75, 1992.
4. Chorin A., *Estimates of Intermittancy, Spectra and Blow-up in Developed Turbulence*, Comm. Pure Appl. Math., v. 34, pp. 853-866, 1981.

5. Filippas S. and Kohn R., *Refined Asymptotics for the Blowup of $u_t - \Delta u = u^p$*, Comm. Pure Appl. Math., v. 45, pp. 821-869, 1992.
6. Friedman A. and McLeod B., *Blowup of Positive Solutions of Semilinear Heat Equations*, Indiana Univ. Math. J., v. 34, pp. 425-447, 1985.
7. Galaktionov V.A. and Poshashkov S.A., *The Equation $u_t = u_{xx} + u^\beta$. Localization and Asymptotic Behavior of Unbounded Solutions*, preprint no. 97, 1985 (in Russian).
8. Galaktionov V.A. and Poshashkov S.A., *Application of a New Comparison Theorem to the Study of Unbounded Solutions of Nonlinear Parabolic Equations*, Differ. Uravnen., v. 22, pp. 1165-1173, 1986 (in Russian).
9. Herrero M. and Velasquez J., *Blowup Behavior of One Dimensional Semilinear Parabolic Equations*, Ann. Inst. Poincaré, Analyse Nonlinéaire, in press.
10. Herrero M. and Velasquez J., *Some Results on Blowup for Semilinear Parabolic Equations*, Proc. of I.M.A. Workshop on Degenerate Diffusions, May 1991.
11. Keller J.B. and Lowengrub J.S., in preparation.
12. LeMesurier B., Papanicolaou G., Sulem C., Sulem P., *The Focusing Singularity of the Cubic Schroedinger Equation*, Phys. Rev. A, v. 34, pp. 755-778.

6. Figure Captions

Figure 1. The time evolution of the parabolic initial data 43 is shown. The full computational domain is $-1.5 \leq x \leq 1.5$. ($h = .00586$, $t_c^* = 2.861 \times 10^{-2}$ and $x_c^* = 0$)

Figure 2. The time evolution of the original solution u (found by transforming v, of figure 1, by 36. ($h = .00586$)

Figure 3. The distribution of u on the grid, from figure 2, is shown.

Figure 4. The distribution of v on the grid is shown.

Figure 5. A comparison of the asymptotic form 38 with the full numerical simulation (from figure 1 at $t = 2.8607 \times 10^{-2}$) is shown. The parameters t_c and x_c, in formula 38 are replaced by t_c^* and x_c^* respectively.

Figure 6. The time evolution of the solution with data 44 is shown. The full computational domain is $-1.5 \leq x \leq 1.5$. ($h = .00586$, $t_c^* = 2.6485 \times 10^{-2}$ and $x_c^* = \pm.428175$)

Figure 7. A comparison of the asymptotic form with the full numerical simulation (from figure 6 at $t = 2.648 \times 10^{-2}$) is shown.

Figure 8. The time evolution of the solution with data 45 is shown. The full computational domain is $-1.5 \leq x \leq 1.5$. ($h = .00293$, $t_c^* = 2.6633 \times 10^{-2}$ and $x_c^* = -.4248$)

Figure 9. A comparison of the asymptotic form with the full numerical solution (from figure 8 at $t = 2.6631 \times 10^{-2}$) is shown.

Figure 10. The time evolution of the solution with data 46 is shown. The full computational domain is $-1.5 \leq x \leq 1.5$. ($h = .00293, t_c^* = 2.6901 \times 10^{-2}$ and $x_c^* = .43945$)

Figure 11. A comparison of the asymptotic form with the full numerical solution (from figure 10 at $t = 2.69 \times 10^{-2}$) is shown.

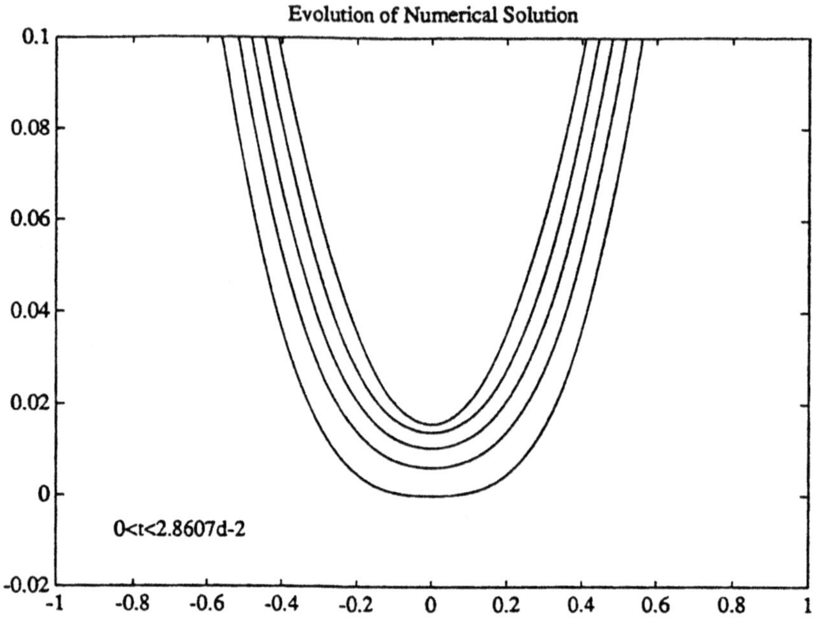

Figure 1.

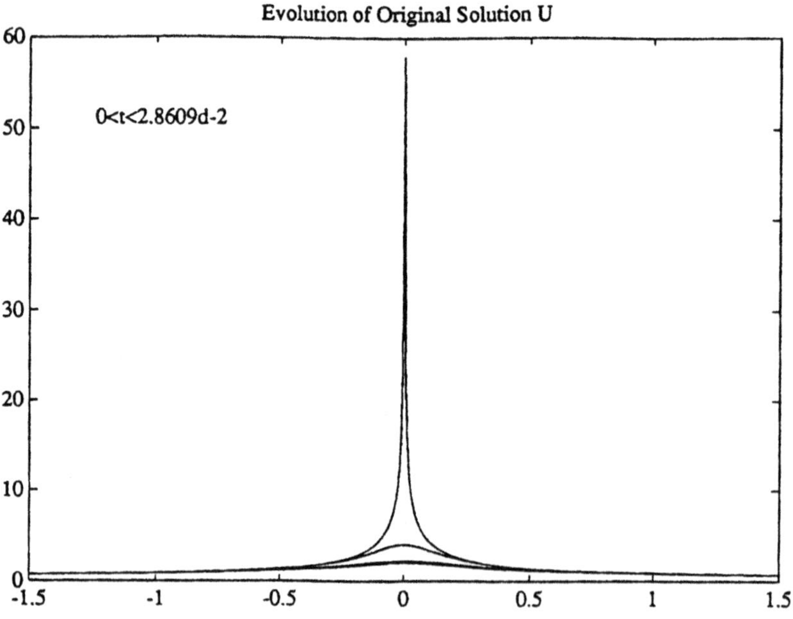

Figure 2.

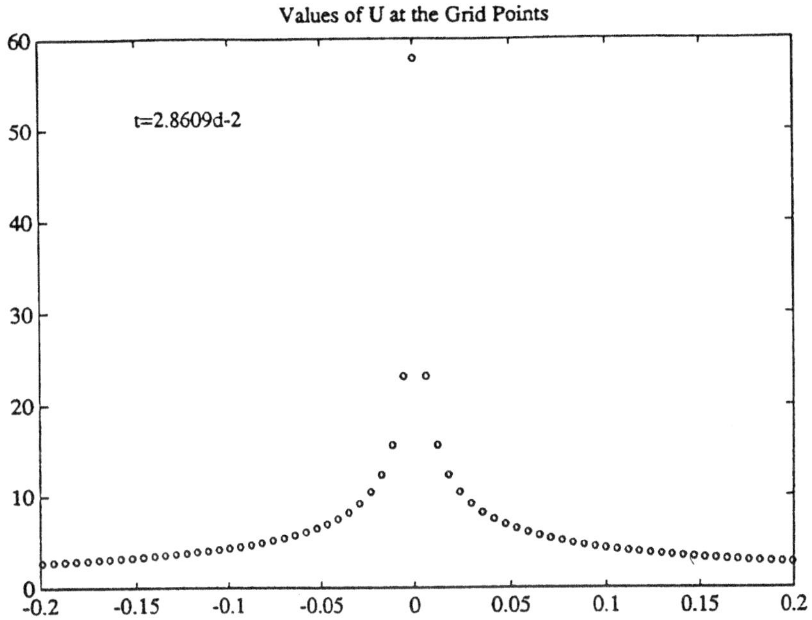

Figure 3.

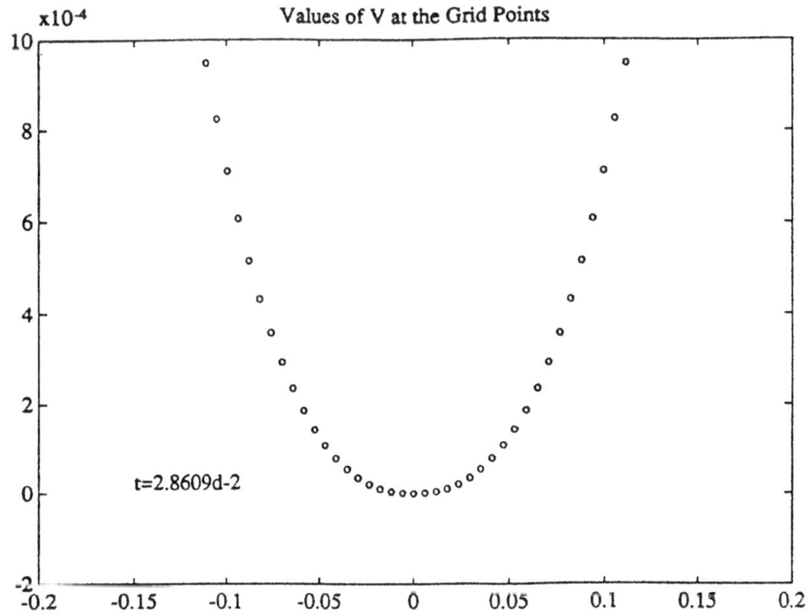

Figure 4.

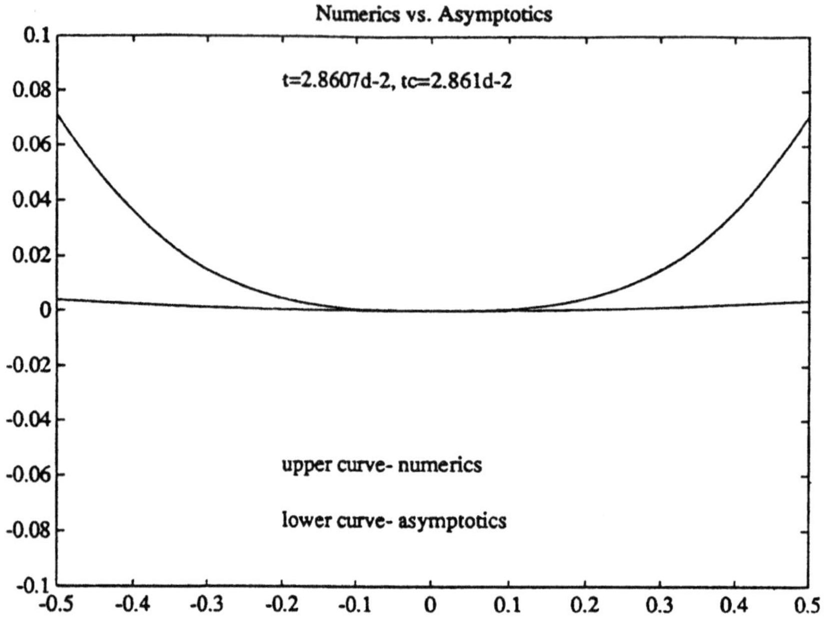
Numerics vs. Asymptotics
t=2.8607d-2, tc=2.861d-2
upper curve- numerics
lower curve- asymptotics
0.1
0.08
0.06
0.04
0.02
0
-0.02
-0.04
-0.06
-0.08
-0.1
-0.5
-0.4
-0.3
-0.2
-0.1
0
0.1
0.2
0.3
0.4
0.5

Figure 5.

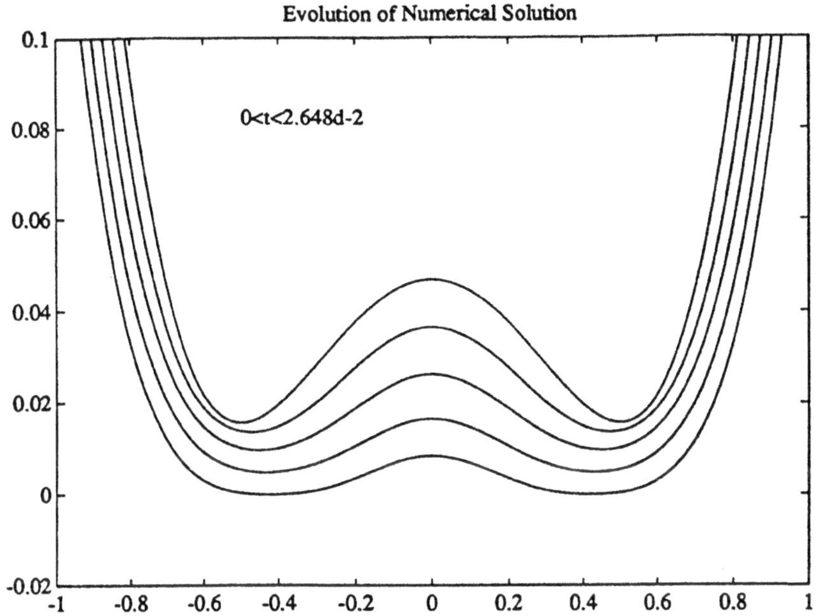

Figure 6.

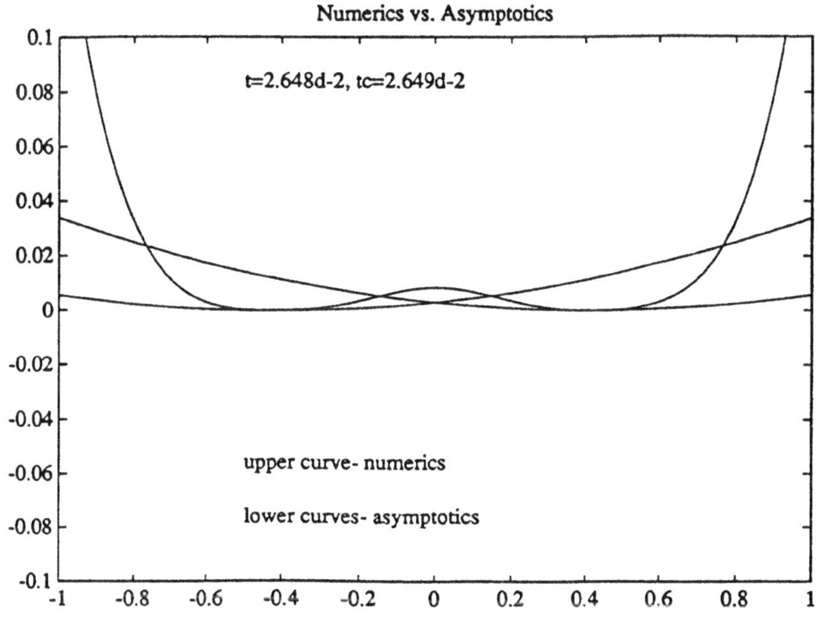

Figure 7.

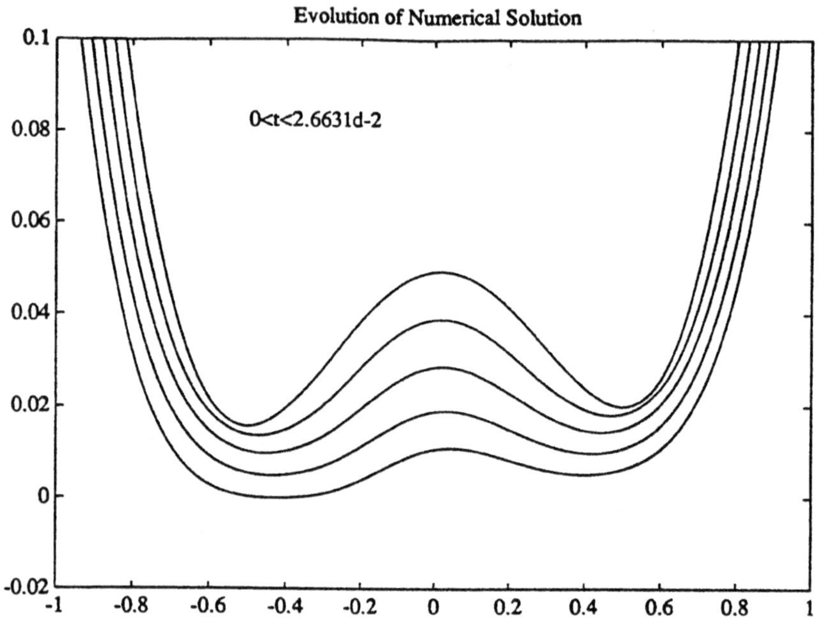

Figure 8.

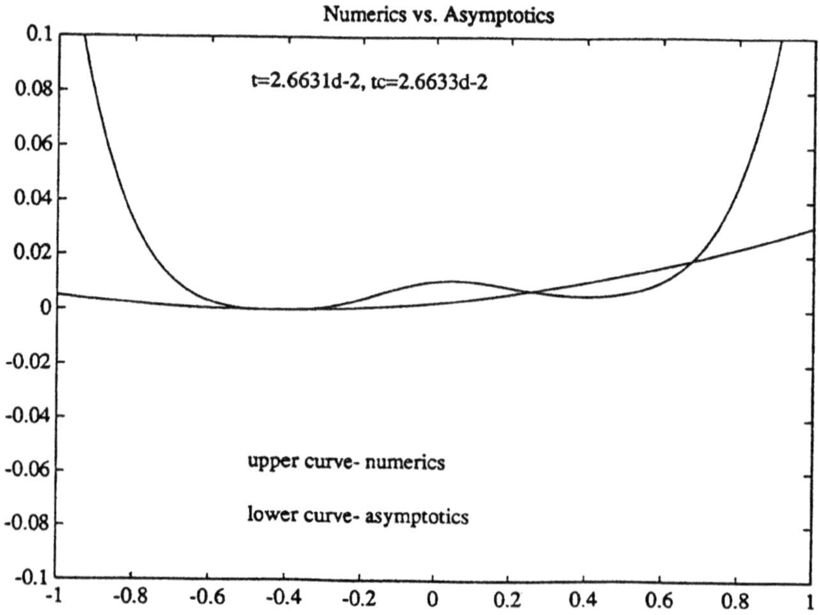

Figure 9.

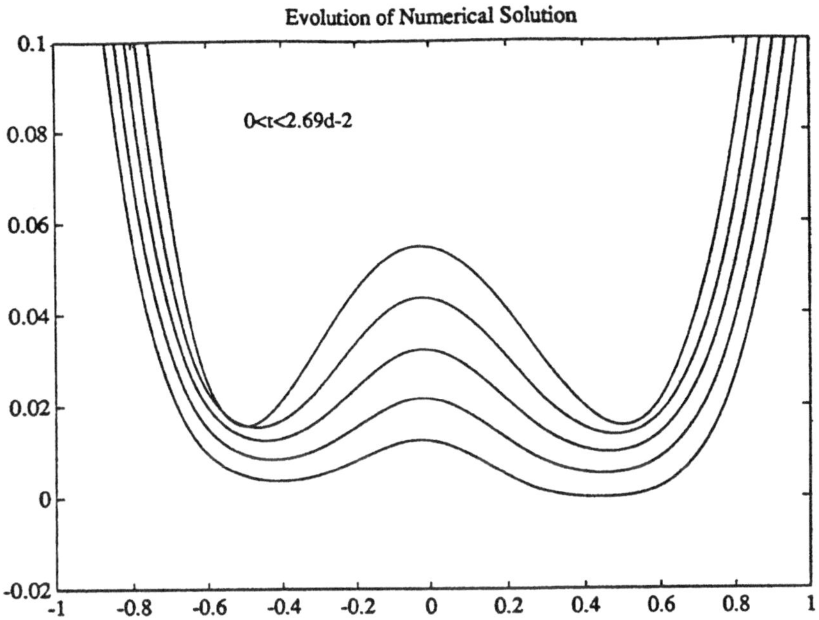

Figure 10.

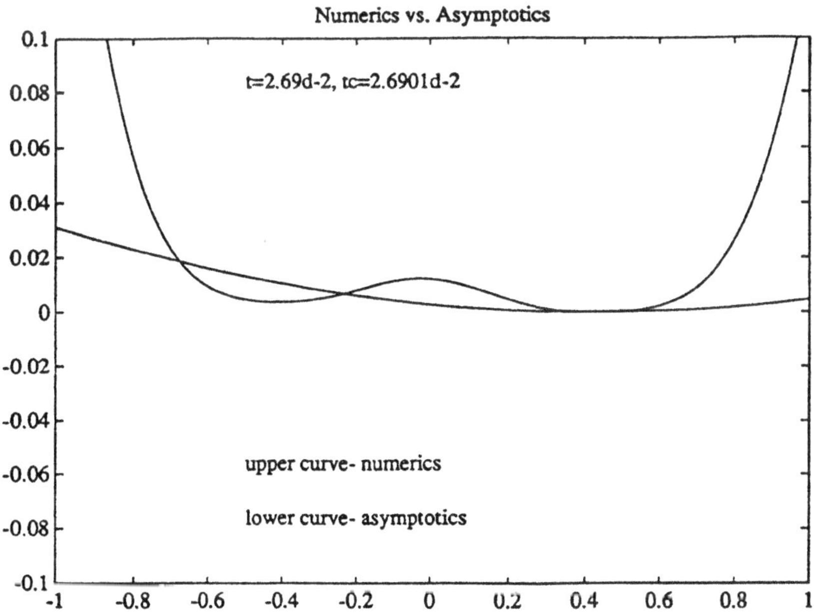

Figure 11.

SCALING OF A SINGULARITY OF EULER

R. M. Kerr
Geophysical Turbulence Program
National Center for Atmospheric Research
Boulder, CO 80307-3000, U.S.A.

ABSTRACT. Self-similar scaling in the collapse of perturbed anti-parallel vortex tubes towards a singularity of the three-dimensional, incompressible Euler equations is shown. Histograms of the normalized β-strain have a sharp peak at a positive value of about 0.4.

It has recently been shown[1] that the time development of the peak vorticity and axial strain in an Euler simulation of perturbed anti-parallel vortices is consistent with analytic constraints[2,3] for a singularity. In addition the evidence suggests that the exponent γ for the time dependence of peak vorticity ω_p, $1/(t_c - t)^\gamma$ with $\gamma \geq 1$, takes the limiting value of 1, consistent with the result of Biot-Savart calculations[4]. To maintain singular development all the components of the velocity stress $\partial u_i/\partial x_j$ in the vicinity of the peak vorticity ω_p would be expected to have the same $1/(t_c - t)$ time dependence. In addition the growth of the singular structure should be self-similar. To investigate these expectations figures 1 plots normalized values and locations of the maxima and minima of e_{xx}, e_{yy}, and e_{zz} for the singular calculation[1]. The normalizations are based upon the value and location of the peak vorticity ω_p. This analysis is presented only for the diagonal components of strain. For this flow $u_{z,x} << u_{x,z}$, so both $e_{xz} = \frac{1}{2}(u_{x,z} + u_{z,x})$ and $\omega_y = u_{z,x} - u_{x,z}$ are dominated by $u_{x,z}$, and in addition $u_{x,z}$ is also larger than e_{xx} and e_{zz}, so in the symmetry plane $u_{x,z}$ would dominate the determination of the two largest eigenvalues α and γ of the strain. This implies that analyzing e_{xz}, α and γ would not provide much information that has not already been provided by the vorticity analysis.

Figure 1a plots the maxima and minima of e_{xx}, e_{yy}, and e_{zz} divided by ω_p as functions of time. All of these ratios approach constant values toward the end of the calculation. Their y locations are all on the symmetry plane, $y = 0$, and x and z locations are given in figures 1b-d. Figure 1b plots normalized x positions for the maxima and minima of the strain components shown in figure 1a. The normalization is based upon distances from the estimated x

R. E. Caflisch and G. C. Papanicolaou (eds.), Singularities in Fluids, Plasmas and Optics, 131–135.
© 1993 *Kluwer Academic Publishers.*

position of the peak vorticity at the singular time x_c. That is, if $x(\omega_p)$ is the x position of the peak vorticity, then as $t \to t_c$, $x(\omega_p) \to x_c$. What figure 1b plots is $(x(t) - x_c)/(x(\omega_p) - x_c)$ for each maxima and minima. In this representation, those quantities whose relative x locations are less than 1 are in front of the peak vorticity and those whose relative x locations are greater than 1 are behind the peak vorticity. Both the maximum of the axial strain $e_{yy,p}$ and the minimum $e_{yy,m}$ are behind the peak, while the maxima and minima of the other strain components are in front of the peak vorticity.

Figure 1c plots to $t = 17$ the z locations divided by the z location of the peak vorticity for $e_{yy,p}$, $e_{xx,m}$, and $e_{zz,p}$. Again, relative location > 1 represents points above the peak vorticity and < 1 represents points below the peak vorticity. Only the z locations of these three quantities were determined with respect to the z location of ω_p because they are the only ones whose relative locations did not have large fluctuations. For each direction these were also the ones with the largest magnitude at the final time. That is, $|e_{yy,p}| > |e_{yy,m}|$, $|e_{xx,m}| > |e_{xx,p}|$ and $|e_{zz,p}| > |e_{z,m}|$. The z location of $e_{xx,m}$ is 0, that is it is on the dividing plane. Figure 1d plots the absolute z locations of $e_{yy,p}$, $e_{yy,m}$, $e_{xx,p}$, and $e_{zz,m}$, showing all of them converging linearly towards $z = 0$ at $t = 18.9$. A linear dependence of position on time to the singularity does not agree with the prediction made by a Biot-Savart analysis[4] which predicts square root dependence if the vorticity becomes increasingly isolated within a sphere of given radius. This is discussed further below.

These figures show that the ratios of the maxima and minima of the rates of strain to the peak vorticity and their locations with respect to the peak vorticity seem to be approaching fixed values as the singular time is approached. This indicates that the singular structure is becoming self-similar as the singular time is approached, which might be a condition for singular development. In the next section more graphics will be used to help visualize how these lows and highs are placed with respect to the vorticity and what the form of the self-similar structure is. One might also expect the development of self-similarity to appear in measures of the alignment of strains and vorticity. To study small scale alignment, histograms of the normalized correlation $\sqrt{6}\frac{\beta}{|e|}$, where β is the eigenvalue of the strain $e_{ij} = \frac{1}{2}(u_{i,j} + u_{j,i})$ with the smallest magnitude, have been found for homogeneous numerical flows[5]. In forced isotropic turbulence histograms of $\sqrt{6}\frac{\beta}{|e|}$ peak at a value of 0.5, with the peak becoming sharper if the flow was conditionally sampled for regions of strong dissipation[5]. At the point of peak vorticity in this flow, where $|e| \approx \frac{1}{\sqrt{2}}\omega_p$ and $e_{yy}|\omega_p/\omega_p \approx \frac{1}{20}$, $\sqrt{6}\frac{\beta}{|e|} \approx \frac{\sqrt{3}}{10} \approx 0.17$

Figure 2 show histograms of $\sqrt{6}\frac{\beta}{|e|}$ at a late time for four regions around the peak vorticity of this flow. The first region in figure 2a is fairly large, encompassing most of the region where there has been significant tilting of the vorticity out of y and into z. A clear peak at 0.4 appears. Figure 6 focus upon the immediate neighborhood of the peak vorticity and find that the peak is at a much smaller value, which given the smaller value of $\sqrt{6}\frac{\beta}{|e|}$ at the point of peak vorticity is not surprizing. To understand where the peak at 0.4 comes from, figures 2c and 2d use regions progressively further away from the symmetry plane in y. In figure 2c the histograms show extremely concentrated peaks at 0.4. As one moves out from the symmetry plane from figure 2c to 2d, the value of the peaks increase towards 0.4.

This trend suggests that $\sqrt{6}\frac{\beta}{|e|} = 0.4$ might gradually be filling the region near where the projected singularity will form. At the singular time only the symmetry plane itself would not have this alignment. That a single type of alignment might characterize the entire region when there is so much curvature in all three directions and clear distinctions between where a distorted vortex sheet is forming with high vorticity, and the surrounding regions with little vorticity, is surprizing. A broader distribution of alignments similar to what is found in studies of homogeneous numerical turbulence might have been expected.

These simulations[1] also leave open the question of a singularity of the velocity, where the peak velocity would obey $1/\sqrt{(t_c - t)}$. There is nothing in the literature that demands that if there is a singularity of the vorticity that the velocity blows up, although for Navier-Stokes there are some analytical results[6]. In addition, for vorticity growth all that is probably required is that there be alignment of the type just discussed. But given the assumption that if the analytic theory predicts a blowup for Navier-Stokes, the same should apply to Euler, one might expect the velocity to become singular with the same restrictions. A blowup of velocity is predicted by a Biot-Savart model[6], but only if the length scale goes as $1/\sqrt{(t_c - t)}$, not $1/(t_c - t)$ as the position of the peak vorticity does in these calculations. Therefore, as pointed out by A. Pumir at the workshop, velocity blowup can be consistent with how the position of the peak vorticity changes only if there are two length scales in the problem. Based upon the structure of the vorticity as the singularity is approached[1] this is possible. Two length scales might also be required for the blowup of global properties such as the enstrophy production and enstrophy, as is found for this problem[1]. What role the blowup of velocity and enstrophy play in the singular development still needs to be clarified.

Acknowledgements

NCAR is supported by the National Science Foundation.

References

R.M. Kerr, Evidence for a singularity of the three dimensional, incompressible Euler equations. submitted to Phys. Fluids (1992).

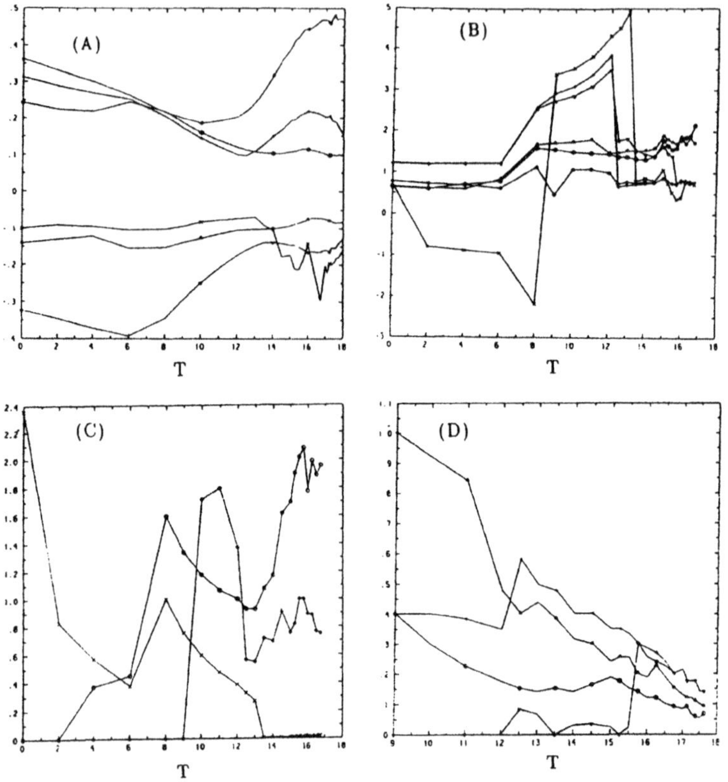

Fig. 1. Relative values and positions of maxima $e_{ii,p}$ and minima $e_{ii,m}$ of diagonal components of the strain $e_{ij} = \frac{1}{2}(u_{i,j} + u_{j,i})$ with respect to value and position of the peak vorticity ω_p. a: Values/peak vorticity, $e_{ii,(p,m)}/\omega_p$, b: Relative x location $(xloc(e_{ii,(p,m)}) - x_c)/(xloc(\omega_p) - x_c)$ where x_c is a projected x p c: Relative z location $zloc(e_{ii,(p,m)})/zloc(\omega_p)$ of $e_{22,p}$, $e_{11,m}$, and $e_{33,p}$. d: Absolute z location $zloc(e_{ii,(p,m)})$ of $e_{22,p}$, $e_{22,m}$, $e_{11,p}$ and $e_{33,m}$. $e_{22,p}$ (circle), $e_{22,m}$ (triangle), $e_{11,p}$ (plus), $e_{11,m}$ (cross), $e_{33,p}$ (diamond), and $e_{33,m}$ (inverted triangle).

E. D. Siggia and A. Pumir, Incipient singularities in the Navier-St okes equations. *Phys. Rev. Lett.* **55**, 1749–1752 (1985).

A. J. Majda, Vorticity, turbulence and acoustics in fluid flow. *SIAM Review* **33**, 349–388 (1991).

A. Pumir and E. D. Siggia, Vortex dynamics and the existence of solutions of the Navier-Stokes equations. *Phys. Fluids.* **30**, 1606–1626 (1987).

R. M. Kerr, Histograms of helicity and strain in numerical turbulence. *Phys. Rev. Lett.* **59**, 783–786 (1987).

J. Leray, Sur le mouvement d'un liquide visqueux emplissant l'espace. *Acta Math* **63**, 193 (1934).

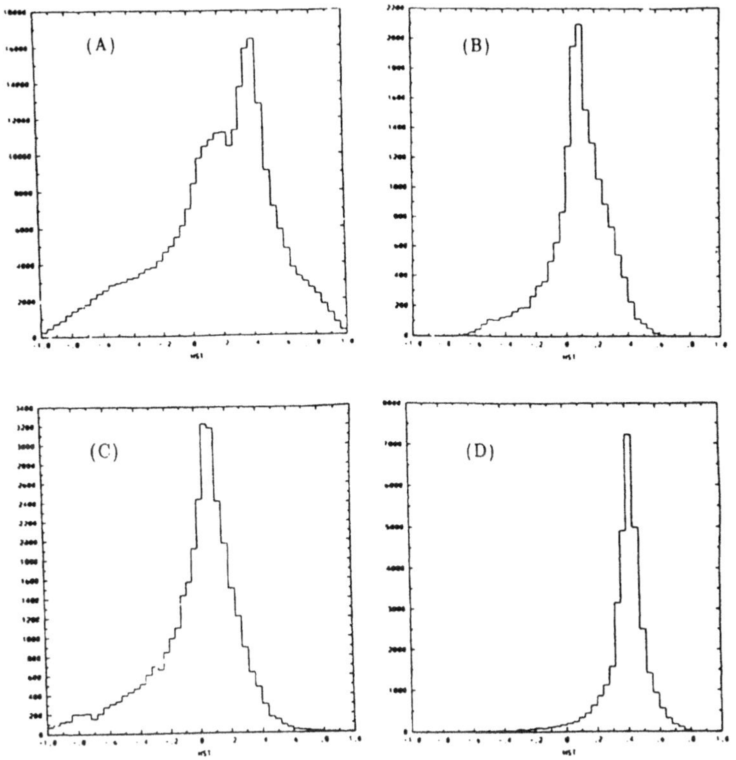

Fig. 2. Histograms of $\sqrt{6}\frac{\beta}{|e|}$ over 4 regions at two times. 17: t=12.5, 18: t=17. a: largest domain, $(x, y, z) = (3.04 : 9.32, 0 : 6.28, 0 : 3.14)$, b: inner domain $(x, y, z) = (3.24 : 6.28, 0 : 3.14, 0 : 1.57)$ c: middle domain $(x, y, z) = (3.24 : 6.28, 1.57 : 3.14, 0 : 1.57)$ d: outer domain $(x, y, z) = (3.24 : 6.28, 3.14 : 6.28, 0 : 1.57)$

THE RAYLEIGH CENTRIFUGAL INSTABILITY FOR VORTEX RINGS WITH SWIRL

ALEXANDER LIFSCHITZ
Department of Mathematics
University of Illinois
Chicago, Illinois 60680, U.S.A.

ABSTRACT. In a recent paper Lifschitz and Hameiri (1991) demonstrated how to analyze the stability of general three-dimensional flows of an ideal incompressible fluid via the geometrical optics method. In the present paper we solve the geometrical optics equations for vortex rings (without and with swirl) and prove that all of them are unstable with respect to short wavelength perturbations.

1. Introduction

Analyzing the stability of the fluid motion is necessary in order to understand its most fundamental features. The variety of conventional methods used for the stability analysis is limited to the normal mode expansion and the energy principle. The geometrical optics approach to the stability problem is an interesting alternative to the classical methods. It allows one to describe short wavelength instabilities of general flows of an inviscid incompressible fluid. They develop in thin tubes surrounding particular stream lines and in this sense can be called localized. These instabilities play an important role in many situations; in particular, it is very likely that they cause the transition from two-dimensional coherent structures (such as vortex rings) to three-dimensional chaotic ones (cf., for example, Bayly *et al.* 1988). During the last decade short wavelength instabilities have been considered by several researchers, particularly by Eckhoff (cf. Bayly 1988; Eckhoff 1981; Friedlander and Vishik 1991; Hameiri and Lifschitz 1990; Lifschitz 1992; Lifschitz and Hameiri 1991, 1992). In the present paper we use the geometrical optics approach and describe short wavelength instabilities of vortex rings (without and with swirl) in an inviscid incompressible fluid. We also demonstrate how the geometrical optics method can be used for analyzing the unstable part of the frequency spectrum for vortex rings without swirl thus narrowing the gap between conventional and nonconventional approaches to the stability

R. E. Caflisch and G. C. Papanicolaou (eds.), Singularities in Fluids, Plasmas and Optics, 137–149.
© 1993 *Kluwer Academic Publishers.*

problem (note that in a recent paper by Lebovitz and Lifschitz (1992) a similar problem has been solved for a differentially rotating fluid).

For over a century the vibrations of vortex rings attracted attention of numerous researchers. The original believe was that vortex rings are stable and most of the effort was concentrated on calculating their frequencies of vibrations (cf., for instance, W. Thomson (Lord Kelvin) 1867, J.J. Thomson 1883). Nevertheless, later experiments convincingly demonstrated that laminar vortex rings in ideal fluid are unstable with respect to bending waves (cf. Krutzsch 1939, Maxworthy 1972). The first asymptotic theory explaining this instability was proposed by Widnall and Tsai (1977) (cf. also Saffman 1978). Their theory is applicable to thin vortex rings without swirl and with constant vorticity in the circular core; it is based on the asymptotic expansion of the basic equations in powers of the small parameter being ratio of the core radius to the ring radius. We emphasize that our theory is based on a completely different asymptotic expansion. It allows one to study vortex rings without and with swirl which are not necessarily thin. Note that the case of a thin vortex ring without swirl with the circular core is the most difficult one from the geometrical optics view point because in general noncircular distorsions of the ring core and the presence of swirl are strong destabilizing factors. Using the results of our asymptotic analysis we demonstrate that all laminar vortex rings in an ideal fluid are unstable so that the development of turbulence in the vortex core is the only apparent mechanism for their stabilization.

2. Axisymmetric Vortex Rings

Steady motions of an incompressible, inviscid fluid are governed by the Euler equations of the form

$$V \cdot \nabla V + \nabla P = 0, \quad \nabla \cdot V = 0, \tag{2.1}$$

where V is the velocity and P is the kinematic pressure. In terms of the velocity V, the vorticty $W = \nabla \times V$ and the Bernoulli head function $H = \frac{1}{2}|V|^2 + P$ these equations can be rewritten as

$$W \times V + \nabla H = 0, \quad \nabla \cdot V = 0. \tag{2.2}$$

Below we restrict ourselves to flows asymptotically uniform at infinity and assume that

$$V(x) \rightarrow V_0 \quad as \quad |x| \rightarrow \infty. \tag{2.3}$$

Axisymmetric vortex rings represent an important class of steady solutions of the three-dimensional Euler equations which are invariant under rotations about the z-axis of the standard cylindrical coordinate system (the major axis). For each vortex ring one can distinguish the inside region where the the vorticity is nonzero (the core), and the outside region where the flow is irrotational. Stream surfaces inside the core form a family of nested tori (or, possibly, several families separated by the separatrix surfaces of complicated topology); the innermost torus of the family degenerates into a circle (the minor axis). Stream lines lying on a given toroidal surface either close after a few turns or cover this surface ergodically. In cylindrical coordinates (r, ϕ, z) with the z-axis being the axis of symmetry we can define the (Stokes) stream function Ψ and express the velocity field V and the vorticity field W of the vortex ring as

$$V = \nabla\phi \times \nabla\Psi + f\nabla\phi, \quad W = r^2 H'\nabla\phi - f'V, \tag{2.4}$$

where the prime denotes the derivative with respect to Ψ, $' = d/d\Psi$ (or $' = \partial/\partial\Psi$). Using the force balance condition (2.2) one can show that the swirling component of the velocity and the head function are functions of Ψ, $f = f(\Psi)$, $H = H(\Psi)$, while the stream function Ψ is governed by the following semilinear elliptic equation

$$\frac{\partial^2\Psi}{\partial r^2} - \frac{1}{r}\frac{\partial\Psi}{\partial r} + \frac{\partial^2\Psi}{\partial z^2} - r^2 H'(\Psi) + f(\Psi)f'(\Psi) = 0. \tag{2.5}$$

At present, it is not known how to solve this equation in the most general setting but particular families of its solutions have been studied in detail (cf. Moffatt 1988, Turkington 1989). Below we assume that certain solution of equation (2.5) is given and analyze the stability of the corresponding vortex ring. This solution represents a vortex ring without swirl if $f = 0$ and a vortex ring with swirl otherwise.

Stream surfaces of the vortex ring in question intersect the meridional plane $\phi = 0$ along level lines of the stream function Ψ. For simplicity we assume that this function has only one minimum and denote the point where it is attained by R, Z. Thus, locally stream surfaces form a family of nested tori with the minor axis being a circle $r = R, z = R, 0 \leq \phi < 2\pi$. Below we analyze the stability properties of the axis and all other tori separately.

Introduce so-called natural coordinates adequately describing the family of nested tori. We leave the ϕ coordinate intact and introduce appropriate coordinates in the r, z plane. We use Ψ as a label of stream surfaces and parametrize points on level lines of Ψ by an angle-like coordinate θ, $0 \leq \theta < 2\pi$, chosen in such a way that the coordinate system $x^1 = \Psi, x^2 = \theta, x^3 = \phi$ is orthogonal. Although the angle coordinate θ can be chosen in many ways our results are independent on its particular choice. For technical reasons it is more convenient to suppose that the angular variables θ, ϕ vary on the whole real axis, $-\infty < \theta, \phi < \infty$, imposing when necessary appropriate periodicity constraints.

Assuming that the functions $\Psi(r,z), \theta(r,z)$ are known we define the volume element J in coordinates Ψ, θ, ϕ, $J = r\partial(r,z)/\partial(\Psi,\theta)$, the physical basis vectors e_α, $e_\alpha = \nabla\alpha/|\nabla\alpha|$, $\alpha = \Psi, \theta, \phi$, and the Lame coefficients h_α, $h_\Psi = 1/|\nabla\Psi|$, $h_\theta = J|\nabla\Psi|/r$, $h_\phi = r$. Coordinates Ψ, θ, ϕ are particularly attractive for our purposes because in these coordinates the velocity field V (and the vorticity field W) assumes a very simple form: $V = V_\alpha e_\alpha$, $V_\Psi = 0$, $V_\theta = |\nabla\Psi|/r$, $V_\phi = f/r$. Note that in these coordinates equation (2.5) has a very simple form as well:

$$\frac{(JV_\theta^2)'}{J} - H'(\Psi) + \frac{f(\Psi)f'(\Psi)}{r^2} = 0, \tag{2.6}$$

Define the so-called poloidal turn-around time T and the velocity shear $\bar{q}$ being important internal characteristics of a given stream surface. The turn-around time, i. e. the time which is required for a fluid element to return to its original poloidal position, can be defined as follows

$$T = \oint \frac{dl}{V_\theta}, \tag{2.7}$$

where dl is the poloidal length element. During one full poloidal turn of a fluid element its toroidal coordinate ϕ increases by the amount $\Delta\phi$ which is given by the expression

$$\Delta\phi = \oint \frac{f}{r^2}\frac{dl}{V_\theta} = 2\pi\bar{q}, \tag{2.8}$$

where $\bar{q}$ is the so-called velocity shear. We emphasize that both T and $\bar{q}$ are defined in a way which is independent on the particular choice of the angle coordinate θ, so that they depend only on Ψ. When $\bar{q}(\Psi)$ is rational, $\bar{q}(\Psi) = N/M$, where M and N are mutually prime integers, all stream lines on the corresponding stream surface are closed curves turning M times around the minor axis of the torus and N times around its major axis; when $\bar{q}(\Psi)$ is irrational each stream line covers the corresponding surface ergodically.

3. Local Stability Conditions

In a recent paper Lifschitz and Hameiri (1991) demonstrated how to study the stability of general three-dimensional flows via the geometrical optics method. Without going into detail discussion we just present here the stream line, eikonal and transport equations governing the stability with respect to short wavelength perturbations. We write these equations in Cartesian coordinates. For the stream line passing through a point x_0 these equations have the form

$$\frac{dx}{d\tau} = V(x), \tag{3.1}$$

$$\frac{dk}{d\tau} = -L^{\dagger}k, \tag{3.2}$$

$$\frac{da}{d\tau} = -La + 2\frac{La \cdot k}{|k|^2}k \equiv Ma, \tag{3.3}$$

where L is the velocity gradient matrix, $L = (\partial V^i/\partial x^j)$, $L^{\dagger}$ is the transposed matrix, $L^{\dagger} = (\partial V^j/\partial x^i)$, and M is the matrix of the form

$$M = -L + 2\frac{kk}{|k|^2}L = -L - 2\frac{kdk/d\tau}{|k|^2}. \tag{3.4}$$

The corresponding initial conditions are

$$x(0) = x_0, \quad k(0) = k_0, \quad a(0) = a_0. \tag{3.5}$$

We emphasize that these equations have the same form in general curvilinear coordinates if the vector $da/d\tau$, the covector $dk/d\tau$, and the velocity gradient tensor L are defined according to the rules of differential geometry.

Equation (3.1) (the stream line equation) describes the stream line passing through the point x_0, equation (3.2) (the eikonal equation) describes the evolution of the wave vector with the original orientation k_0 along this stream line, finally equation (3.3) (the transport equation) governs the amplitude dynamics. We denote the corresponding solutions by $x(\tau; x_0)$, $k(\tau; x_0, k_0)$, $a(\tau; x_0, k_0, a_0)$ in order to emphasize their dependence on the initial data. Equations (3.1), (3.2) are Hamiltonian with the corresponding Hamiltonian $\Omega = k \cdot V$ which is an integral of motion. It can be shown that the transport equation has a linear integral of motion $I_k(a) = a \cdot k$. Below we restrict ourselves to incompressibe solutions a and assume that $I_k(a) = 0$.

Sufficient instability conditions with respect to short wavelength perturbations can be formulated as follows. The basic flow of an ideal incompressible fluid is linearly unstable in the velocity norm near the stream line passing through a point x_0 if for some k_0, a_0 the corresponding amplitude $a(\tau, x_0, k_0, a_0)$ has an unbounded growth, or, in other words, if

$$\sup_{\substack{x_0, k_0, a_0 \\ |k_0|=1, |a_0|=1, a_0 \cdot k_0 = 0}} \overline{\lim}_{\tau \to \infty} |a(\tau; x_0, k_0, a_0)| = \infty. \tag{3.6}$$

It is linearly and nonlinearly unstable in the vorticity (enstrophy) norm if the product of $|k(\tau; x_0, k_0)|$ and $|a(\tau; x_0, k_0, a_0)|$ grows without bound,

$$\sup_{\substack{x_0,k_0,a_0\\|k_0|=1,|a_0|=1,a_0\cdot k_0=0}} \overline{\lim}_{\tau\to\infty}|k(\tau;x_0,k_0)||a(\tau;x_0,k_0,a_0)| = \infty. \qquad (3.7)$$

Our objective in the rest of the paper is to analyze the stream line, eikonal and transport equations for vortex rings and to describe the corresponding instabilities.

4. Instabilities of the Minor Axis

Consider a vortex ring and assume that its minor axis intersects the meridional plane $\phi = 0$ at some point (R, Z). According to our assumptions Ψ has a minimum at (R, Z) and can be represented in the form

$$\Psi(r,z) = \Psi(R,Z) + \frac{1}{2}[a(r-R)^2 + 2b(r-R)(z-Z) + c(z-Z)^2] + \ldots, \qquad (4.1)$$

where $a = \partial^2\Psi/\partial r^2$, $b = \partial^2\Psi/\partial r\partial z$, $c = \partial^2\Psi/\partial z^2$, and dots stand for higher order terms. Due to the fact that (R, Z) is a point of minimum we have $a > 0$, $c > 0$, and $d = ac - b^2 > 0$. We also note that equation (2.5) is invariant under the transformations $\phi \to -\phi$, $z - Z \to -(z - Z)$, $\Psi \to -\Psi$, so that we may assume that $b \geq 0, f(R, Z) \geq 0$. If $f(R, Z) = 0$ we deal with a flow having points of stagnation; such a flow is unstable if level lines of Ψ near the point (R, Z) are not circular, i.e. if $(a - c)^2 + b^2 \neq 0$, (cf. Lifschitz and Hameiri 1991); thus without loss of generality we may assume that $f(R, Z) > 0$. Note that on the minor axis the vorticity W is parallel to the velocity V.

Write the geometrical optics equations for the minor axis. We use the physical components of the wave vector k and the amplitude a and introduce the following dimensionless variables and parameters $\bar{\tau} = (a+c)\tau/R$, $\bar{k} = Rk$, $\bar{a} = (1/R)a$, $\bar{A} = a/(a+c)$, $\bar{B} = b/(a+c)$, $\bar{C} = c/(a+c)$, $\bar{D} = d/(a+c)^2$, $\bar{F} = f(\Psi(R,Z))/R(a+c)$. According to our assumptions $0 < \bar{A} < 1$, $0 \leq \bar{B} < \sqrt{\bar{A}(1-\bar{A})}$, $0 < \bar{C} < 1$, $0 < \bar{F}$, and $\bar{A} + \bar{C} = 1$. Omitting the bars we write the geometrical optics equations in the form

$$\frac{dr}{d\tau} = 0, \quad \frac{d\phi}{d\tau} = F, \quad \frac{dz}{d\tau} = 0, \qquad (4.2)$$

$$\frac{d}{d\tau}\begin{pmatrix} k_r \\ k_\phi \\ k_z \end{pmatrix} = \begin{pmatrix} -B & 2F & A \\ 0 & 0 & 0 \\ -C & 0 & B \end{pmatrix}\begin{pmatrix} k_r \\ k_\phi \\ k_z \end{pmatrix}, \qquad (4.3)$$

$$\frac{d}{d\tau}\begin{pmatrix} a_r \\ a_\phi \\ a_z \end{pmatrix} = \begin{pmatrix} -B & 2F & -C \\ 0 & 0 & 0 \\ A & 0 & B \end{pmatrix}\begin{pmatrix} a_r \\ a_\phi \\ a_z \end{pmatrix} + \frac{2}{|k|^2}$$

$$\times \begin{pmatrix} Bk_r^2 - Fk_rk_\phi - Ak_rk_z & -Fk_r^2 & Ck_r^2 - Bk_rk_z \\ Bk_rk_\phi - Fk_\phi^2 - Ak_\phi k_z & -Fk_rk_\phi & Ck_rk_\phi - Bk_\phi k_z \\ Bk_rk_z - Fk_\phi k_z - Ak_z^2 & -Fk_rk_z & Ck_rk_z - Bk_z^2 \end{pmatrix} \begin{pmatrix} a_r \\ a_\phi \\ a_z \end{pmatrix}. \tag{4.4}$$

Without loss of generality we can write the solution of equation(4.2) in the form $r = R, \phi = F\tau, z = Z$. Up to an irrelevant multiplier and a time shift we can represent the general solution of equation (4.3) as

$$\begin{aligned} k(\tau) &= \cos\mu \left(\cos\sqrt{D}\tau, 0, \frac{B}{A}\cos\sqrt{D}\tau - \frac{\sqrt{D}}{A}\sin\sqrt{D}\tau \right) \\ &+ \sin\mu \left(-\frac{2FB}{D}, 1, -\frac{2FC}{D} \right), \end{aligned} \tag{4.5}$$

where μ is an angle-like parameter characterizing the toroidal component of the wave vector, $-\frac{\pi}{2} \leq \mu \leq \frac{\pi}{2}$. One can prove that the stability properties of the transport equation are independent on the sign of μ so that we can assume that $0 \leq \mu \leq \frac{\pi}{2}$. At first we consider $\mu = \frac{\pi}{2}$. For such a μ the corresponding wave vector k and the matrix M are time-independent. After some cumbersome but straightforward algebra one can show that the eigenvalues of the matrix M have the form

$$\lambda_2 = 0, \quad \lambda_\pm(A, B, F) = \pm \frac{D\sqrt{4F^2 - D}}{\sqrt{4F^2(B^2 + C^2) + D^2}}. \tag{4.6}$$

It turns out that the eigenvectors $e_\pm$ corresponding to $\lambda_\pm$ are orthogonal to k and represent incompressible amplitudes. If

$$2F > \sqrt{D}, \tag{4.7}$$

or, in other words, if the swirl component of the velocity is not too small (note that $D \leq 0.25$), then λ_+ is real and positive and the transport equation has an exponentially growing solution of the form

$$a(\tau) = \exp(\lambda_+\tau)e_+. \tag{4.8}$$

It is clear that in this case the flow is exponentially unstable both in the velocity and vorticity norms. Note that

$$\lambda_+(A, B, F) \to \frac{D}{\sqrt{B^2 + C^2}} \quad as \quad F \to \infty, \tag{4.9}$$

so that the limit value of the instability increment is finite.

In general, it is necessary to consider all possible values of μ, $0 \le \mu \le \frac{\pi}{2}$. When $\mu \ne \frac{\pi}{2}$ the coefficient matrix M is a periodic function of τ with the corresponding period equal to $2\pi/\sqrt{D}$. In order to analyze the behavior of solutions of the transport equation (4.4) one needs to construct the monodromy matrix and compute its eigenvalues. This equation has growing solutions if the corresponding monodromy matrix has an eigenvalue with the absolute value greater than unity. We denote this matrix by $Q(\mu, A, B, F)$ in order to emphasize its parametric dependence on μ, A, B, F (recall that $C = 1 - A$). According to the Liouville theorem $\det Q(\mu, A, B, F) = 1$. Besides, due to the fact that equation (4.4) has the integral $I_k(a) = a \cdot k$ one of the eigenvalues of the matrix Q, say ν_0, is equal to unity. It means that the characteristic equation for two other eigenvalues $\nu_\pm$ can be written in the form

$$\nu^2 - q(\mu, A, B, F)\nu + 1 = 0, \tag{4.10}$$

where $q(\mu, A, B, F) = trQ(\mu, A, B, F) - 1$. The eigenvalue with the absolute value greater than unity exists under the following condition

$$|q(\mu, A, B, F)| > 2. \tag{4.11}$$

If this condition is satisfied then the corresponding eigenvalue has the form

$$\nu_s(\mu, A, B, F) = \frac{1}{2}[(q(\mu, A, B, F) + s\sqrt{q^2(\mu, A, B, F) - 4}], \tag{4.12}$$

where $s = sign(q(\mu, A, B, F))$, and the transport equation has the exponentially growing solution of the form

$$a(\tau) = \exp(\lambda_s \tau)\tilde{a}(\tau), \tag{4.13}$$

where

$$\lambda_s(\mu, A, B, F) = \frac{\sqrt{D}}{2\pi} \log(\nu_s(\mu, A, B, F)), \quad \Re\lambda_s > 0, \tag{4.14}$$

and $\tilde{a}(\tau)$ is a periodic function of τ with period $2\pi/\sqrt{D}$, so that the basic flow is unstable.

In general, the monodromy matrix cannot be found analytically and one needs to evaluate it numerically. Fortunately, we already know that for sufficiently large F the flow is unstable so that we can restrict ourselves to the finite domain in the parameter space which coincides with the part of the ellipsoid $(A - \frac{1}{2})^2 + B^2 + 4F^2 \le \frac{1}{4}$ lying in the positive octant. Results of our computations show that for any triplet A, B, F in this octant we have

$$\bar{\lambda}(A,B,F) = \max_{0\le\mu\le\frac{\pi}{2}} \Re\lambda_s(\mu,A,B,F) > 0, \tag{4.15}$$

and we can conclude that the corresponding flow is unstable both in the velocity and vorticity norms even if the swirl component of the velocity is relatively small. The maximum growth rate of unstable short wavelength perturbations is equal to $\bar{\lambda}(A,B,F)$.

Our analysis shows that only vortex rings without swirl with circular stream lines can possibly be stable and that in general elliptic distorsions and swirl have a destabilizing impact on the flow. Thus, all generic vortex rings are unstable. This observation can be used in order to explain the results of recent computer experiments of Beale *et al.* (1991).

5. Instabilities of General Stream Lines

Consider a general stream line passing through some point with coordinates $(\Psi_0, \theta_0, 0)$. In natural coordinates we can write the stream line equation as follows

$$\frac{d\Psi}{d\tau} = 0, \quad \frac{d\theta}{d\tau} = \frac{1}{J}, \quad \frac{d\phi}{d\tau} = \frac{f}{r^2}, \tag{5.1}$$

so that the corresponding stream line has the form

$$\Psi = \Psi_0, \quad \phi = \int_{\theta_0}^{\theta} \frac{fJ}{r^2} d\theta. \tag{5.2}$$

The eikonal equation assumes the form

$$\frac{d}{d\tau}\begin{pmatrix} h_\Psi k_\Psi \\ h_\theta k_\theta \\ h_\phi k_\phi \end{pmatrix} = \begin{pmatrix} 0 & -(\frac{1}{J})' & -(\frac{f}{r^2})' \\ 0 & -\frac{\partial}{\partial\theta}(\frac{1}{J}) & -\frac{\partial}{\partial\theta}(\frac{f}{r^2}) \\ 0 & 0 & 0 \end{pmatrix} \begin{pmatrix} h_\Psi k_\Psi \\ h_\theta k_\theta \\ h_\phi k_\phi \end{pmatrix}, \tag{5.3}$$

its general solution can be written as

$$k_\Psi = h_\Psi\left[\int_0^\tau (\Omega J' - \Pi(\frac{fJ}{r^2})')\frac{d\tau}{J} + \Lambda\right], \quad k_\theta = h_\theta(\Omega J - \Pi\frac{fJ}{r^2}), \quad k_\phi = \frac{\Pi}{r}, \tag{5.4}$$

where Λ, Ω, Π are arbitrary constants. It can be shown that the wave vector $\mathbf{k}$ is periodic in τ if and only if $\Omega = (2\pi\bar{q}'/T')\Pi$. Below we restrict ourselves to this case because the opposite case is less dangerous from the stability view point. We also assume that the toroidal component of the wave vector k is not equal to zero, $\Pi \neq 0$.

In contrast to the stream line equation and the eikonal equation the transport equation has a complicated form in natural coordinates because its coefficients depend on the Lame coefficients and Christoffel symbols. In order to obviate this difficulty we use the fact that the amplitude a is orthogonal to the wave vector k and expand it as follows

$$a = \frac{d_\Psi}{h_\psi} E_\Psi + \frac{d_\theta}{V_\theta} E_\theta, \tag{5.5}$$

where E_Ψ, E_θ are linearly independent vectors of the form

$$E_\Psi = e_\Psi - k_\Psi \frac{k}{|k|^2}, \quad E_\theta = e_\theta - k_\theta \frac{k}{|k|^2}, \tag{5.6}$$

which are orthogonal to the wave vector k. After some very cumbersome algebra we obtain the following system of equations for d

$$\frac{d}{d\tau} d + Kd = 0, \tag{5.7}$$

where K is the traceless 2 by 2 matrix of the form

$$K = \begin{pmatrix} \delta r \frac{k_\Psi k_\theta}{|k|^2} & 2\frac{|V|^2}{V_\theta^2} \frac{\sigma_N}{|\nabla \Psi|} + v + \frac{\delta}{V_\theta^2} \frac{k_\theta^2}{|k|^2} \\ \delta r^2 V_\theta^2 - \delta r^2 V_\theta^2 \frac{k_\Psi^2}{|k|^2} & -\delta r \frac{k_\Psi k_\theta}{|k|^2} \end{pmatrix}. \tag{5.8}$$

Here $\delta = H' - 2\pi f'\bar{q}'/T'$, $v = [2\pi f'\bar{q}'/T' + ff'/r^2 - [f^2/r^2]']/V_\theta^2$, σ_N is the normal curvature of the stream line and k_α are given by expressions (5.4). Note that for vortex rings without swirl $v = 0$ because for such rings $f = 0$.

System (5.7) has periodic coefficients with period T parametrically depending on Ψ_0, θ_0, and $\hat{\Lambda} = \Lambda/\Pi$; its solutions can be analyzed via the Floquet method. In order to do that we introduce the monodromy matrix $Q(\Psi_0, \theta_0, \hat{\Lambda})$. It can easily be shown that without loss of generality we can put $\theta_0 = 0$ and consider the monodromy matrix $Q(\Psi_0, 0, \hat{\Lambda})$ depending on two parameters only. Thus, instead of considering all the monodromy matrices corresponding to different stream lines lying on the same stream surface we can restrict ourselves to the family of monodromy matrices corresponding to the stream line passing through the point with coordinates $(\Psi_0, 0, 0)$. By virtue of the Liouville theorem the determinant of the monodromy matrix is equal to one. Thus, the characteristic equation for its eigenvalues $\nu_\pm$ has the form

$$\nu^2 - q(\Psi_0, \hat{\Lambda})\nu + 1 = 0. \tag{5.9}$$

where $q = TrQ(\Psi_0, 0, \hat{\Lambda})$, it is similar to equation (4.10). The corresponding sufficient instability condition can be written as

$$|q(\Psi, \hat{\Lambda})| > 2, \tag{5.10}$$

for some $\Psi, \hat{\Lambda}$. When this condition is satisfied the monodromy matrix has the real eigenvalue of the form

$$\nu_s(\Psi, \hat{\Lambda}) = \frac{1}{2}[(q(\Psi, \hat{\Lambda}) + s\sqrt{q^2(\Psi, \hat{\Lambda}) - 4}], \tag{5.11}$$

where $s = sign(q(\Psi, \hat{\Lambda}))$, with the absolute value greater than unity. Accordingly, equation (5.7) has the exponentially growing solution of the form

$$d(\tau) = \exp(\lambda_s \tau)\tilde{d}(\tau), \tag{5.12}$$

where

$$\lambda_s(\Psi, \hat{\Lambda}) = \frac{1}{T}\log(\nu_s(\Psi, \hat{\Lambda})), \quad \Re\lambda_s > 0, \tag{5.13}$$

and $\tilde{d}(\tau)$ is a periodic function of τ with period T, so that the basic flow is unstable near the stream line in question. The maximum growth rate associated with a given stream surface can be represented as

$$\bar{\lambda}(\Psi) = \max_{\hat{\Lambda}} \Re\lambda_s(\Psi, \hat{\Lambda}), \tag{5.14}$$

it cannot be found without the detail analysis of the structure of the vortex ring. In principle, the maximum growth rate can be equal to zero which means that the flow is stable with respect to short wavelength perturbations localized near the stream surface in question. We see that for general stream lines as well as for the minor axis it is possible to reduce the stability analysis for the full three-dimensional problem to the stability analysis for a one-dimensional problem. Obviously, the latter problem is much simpler that the former.

In general, it is necessary to evaluate the trace of the monodromy matrix numerically, nevertheless, some simple sufficient conditions for the instability of vortex rings without swirl can be found analytically. It follows immediately from equations (5.4) that for such rings the physical components of the wave vector k have the form $k_\Psi = h_\Psi \Lambda$, $k_\theta = 0$, $k_\phi = \Pi/r$, so that equation (5.7) can be written as

$$\frac{d}{d\tau}\begin{pmatrix} d_\Psi \\ d_\theta \end{pmatrix} + \begin{pmatrix} 0 & 2\frac{\sigma_N}{|\nabla\Psi|} \\ H'\frac{|\nabla\Psi|^4}{r^2\hat{\Lambda}^2+|\nabla\Psi|^2} & 0 \end{pmatrix}\begin{pmatrix} d_\Psi \\ d_\theta \end{pmatrix} = 0. \tag{5.15}$$

Assuming for simplicity that all stream lines are convex, so that σ_N is negative (recall that Ψ increases outward) one can easily show that the general solution of equation (5.15) grows exponentially in time when $H' < 0$. Note that one cannot guarantee that the general solution of this equation is bounded in time when $H' > 0$ because of possible parametric resonances. Thus, a vortex ring without swirl having convex stream lines is unstable with respect to short wavelength perturbations if the corresponding Bernoulli function H is a decreasing function of Ψ on some interval.

For vortex rings without swirl there exists an interesting relation between the Bernoulli function H and the circulation $C = \oint V_\theta dl$ allowing one to reformulate the above condition in more familiar terms. Namely, it can be shown that $C' = TH'$. Using this observation we can prove that a vortex ring without swirl with convex stream lines is unstable if the circulation C is a decreasing function of Ψ on some interval, or, in other words, if the circulation decreases outward in some region. This condition is a natural generalization of the classical Rayleigh centrifugal instability condition for cylindrical flows.

For general vortex rings it is difficult to find the relation between the geometrical optics solutions and the standard normal modes. Nevertheless, for vortex rings without swirl such a relation can be established. Namely, it can be shown by means of the Weyl's sequence method (cf. Lifschitz 1989, Lebovitz and Lifschitz 1992) that all points $\sigma_\pm$ of the form

$$\sigma_\pm = \frac{1}{T} Log(\nu_\pm(\Psi, \hat{\Lambda})), \tag{5.16}$$

where $\nu_\pm$ are roots of equation (5.9) and *Log* is understood as a multi-valued complex function, belong to the so-called essential frequency spectrum. It is interesting to note that in the unstable case this part of the spectrum consists of *domains* in the complex plane rather than of isolated curves. This property of the spectrum seems to be very unusual.

6. Concluding Remarks

In the present paper we demonstrated how to use the geometrical optics method in order to study short wavelength instabilities of vortex rings in an ideal incompressible fluid. Our analysis indicates that all of them are unstable. Short wavelength instabilities always occur near the minor axis of a vortex ring; in general, they can occur near stream lines lying on toroidal stream surfaces such that $\bar{\lambda}(\Psi) > 0$ as well. Depending on the relative magnitude of the growth rate of unstable perturbations their development can either cause strong distorsions of the immediate neighborhood of the minor axis, or of some region lying between two toroidal stream surfaces. The direct numerical simulation of short wavelength instabilities of vortex rings is an important problem left beyond the scope of the present paper. We hope to be able to discuss results of such a simulation elsewhere.

Acknowledgements

The author thanks Professor Russel Caflisch for his kind invitation to participate in the NATO Advanced Research Workshop on Singularities in Fluids, Plasmas and Optics. This work was supported by the National Science Foundation under grant DMS-9100327.

References

Bayly, B.J., 1988 Three-dimensional centrifugal-type instabilities in inviscid two-dimensional flows. *Phys. Fluids*, **31**, pp. 56-64.

Bayly, B.J., Orszag, S.A. & Herbert, T., 1988 Instability mechanisms in shear-flow transition. *Ann. Rev. Fluid Mech.*, **20**, pp. 359-391.

Beale, J.T., Eydeland, A. & Turkington, B., 1991 Numerical tests of 3-D vortex methods using a vortex ring with swirl. *Lect. Appl. Math.*, **28**, pp. 1-9.

Eckhoff, K.S., 1981 On the stability for symmetric hyperbolic systems, I. *J. Differ. Equ.*, **40**, pp. 94-115.

Friedlander, S. & Vishik, M.M., 1991b Dynamo theory, vorticity generation, and exponential stretching. *Chaos*, **1**, pp. 198-205.

Hameiri, E. & Lifschitz, A., 1990 Ballooning modes in plasmas and in classical fluids. Proceedings of the Sherwood Theory Conference, Williamsburg, Virginia, USA.

Krutzsch, C.H., Uber eine Experimetell beobachtete Erscheining an Werbelringen bei ehrer translatorischen Beivegung in Werklechin. *Flussigheiter. Annln der Phys.*, **27**, pp. 497-523.

Lebovitz, N. & Lifschitz, A., 1992 Short wavelength instabilities of rotating, compressible fluid masses. *Proc. R. Soc. London Ser. A.*

Lifschitz, A., 1989 *Magnetohydrodynamics and spectral theory.* Kluwer.

Lifschitz, A., 1992 On the instability of three-dimensional flows of an ideal incompressible fluid. *Phys. Lett. A*, **167**, pp. 465-474.

Lifschitz, A. & Hameiri, E., 1991 Local stability conditions in fluid dynamics. *Phys. Fluids A*, **3**, pp. 2644-2651.

Lifschitz, A. & Hameiri, E., 1992 Localized instabilities of vortex rings with swirl. *CPAM.*

Maxworthy, T, 1972 The structure and stability of vortex rings. *J. Fluid Mech.*, **51**, pp. 15-32.

Moffatt, H.K., 1988 Generalised vortex rings with and without swirl. *Fluid Dynamics Res.*, **3**, pp. 22-30.

Saffman, P.G., 1978 The number of waves on unstable vortex rings. *J. Fluid Mech.*, **84**, pp. 625-639.

Thomson, J.J, 1883 *A treatise on the motion of vortex rings.* MacMillan.

Thomson, Sir W. (Kelvin, Lord), 1867 On vortex atoms. *Phil. Mag.*, **34**, pp. 15-42.

Turkington, B, 1989 Vortex rings with swirl: Axisymmetric solutions of the Euler equations with nonzero helicity. *SIAM J. Math. Anal.*, **20**, pp. 57-73.

Widnall, S.E. & Tsai, C.-Y., 1977 The stability of the thin vortex ring of constant vorticity. *Phil. Trans. R. Soc. Lond. A*, **287**, pp. 273-305.

DYNAMICS OF MAGNETIC VORTEX RINGS

N. PAPANICOLAOU
Department of Physics, University of Crete
and Research Center of Crete
Heraklion, Greece

ABSTRACT. The conservation laws of linear and angular momentum in the ferromagnetic continuum are expressed as moments of a suitable topological vorticity, thus establishing a direct link between the topological complexity of magnetic structures and their dynamics. In particular, we argue that finite-energy magnetic solitons with a nonvanishing Hopf index are stabilized by moving along the easy axis with constant velocity.

The study of the ferromagnetic continuum has been a subject of great interest since the pioneering work of Landau and Lifshitz [1] but reached maturity only in relatively recent years. The main structures encountered in practice are magnetic domain walls, which are essentially one-dimensional, and magnetic bubbles, whose essential features are two-dimensional. Both static and dynamical properties of domain walls are well understood, especially in view of an exact solution due to Walker [2]. The corresponding development in the case of magnetic bubbles proved to be more difficult but a reasonably complete theoretical understanding has been achieved [3,4], even though some of the more intriguing elements of bubble dynamics are still under investigation [5]. In contrast, progress with three-dimensional magnetic structures has been slow. To the best of our knowcledge there is no experimental evidence for the existence of genuinely 3-D magnetic solitons, whereas theoretical investigations have been scarce and somewhat unfocused [6]. It is the purpose of this talk to explore further the results of Ref. [5] and suggest that magnetic solitons with a nonvanishing Hopf index propagate along their symmetry axis and should be possible to establish at least numerically.

Some of the important features of the ferromagnetic continuum can be understood without explicit reference to the details of the dynamics. Hence we consider the Landau-Lifshitz equation in its general form

$$\dot{\mathbf{M}} = \mathbf{M} \times \mathbf{F} \qquad \mathbf{F} = -\,\delta W/\delta \mathbf{M} \qquad \mathbf{M}^2 = 1 \tag{1}$$

where $W = W(\mathbf{M})$ is the energy functional, and the constant magnitude of the magnetization $\mathbf{M} = \mathbf{M}(\mathbf{r},t)$ is normalized to unity by a suitable choice of rationalized units. Eq. (1) may be viewed as the Hamilton equation derived from the Hamiltonian W and the Poisson brackets

$$\{M_i(\mathbf{r}_1)\,,\, M_j(\mathbf{r}_2)\} = -\,\varepsilon_{ijk}\, M_k(\mathbf{r}_1)\, \delta(\mathbf{r}_1 - \mathbf{r}_2) \tag{2}$$

where Latin indices i,j,k,... take three distinct values and denote the usual Cartesian components. The explicit form of W may be found in Ref. [5]. We also neglect here the effect of dissipation.

The key quantity for the description of topological as well as dynamical properties of the magnetization is the vorticity γ whose components are given by

$$\gamma_i = 1/2\; \varepsilon_{ijk}\, (\partial_j \mathbf{M} \times \partial_k \mathbf{M}) \cdot \mathbf{M} \tag{3}$$

R. E. Caflisch and G. C. Papanicolaou (eds.), Singularities in Fluids, Plasmas and Optics, 151–158.
© 1993 *Kluwer Academic Publishers.*

Although this topological vorticity is not related in any real sense with rotational motion, its role in understanding the dynamics of the ferromagnetic continuum is fairly analogous to that of ordinary vorticity. For the moment, we are concerned with instantaneous properties of the magnetization $\mathbf{M}(\mathbf{r})$ at some instant t which is not displayed explicitly. An immediate consequence of the constraint $\mathbf{M}^2 = 1$ is that the vector field γ is solenoidal,

$$\nabla \cdot \gamma = 0 \tag{4}$$

and its sructural properties are similar to those of ordinary vorticity in fluid dynamics [7], or even those of magnetic induction in magnetohydrodynamics [8]. In particular, the vortex lines cannot terminate at any finite point in a three-dimensional ferromagnetic continuum without boundaries. A related fact is that the total vorticity

$$\Gamma = \int \gamma \, dV \tag{5}$$

vanishes for field configurations with reasonable behavior at infinity. This property follows from the relation $\partial_j(x_i\gamma_j) = \gamma_i$ which is a cosequence of eq. (4). Integrating both sides of this relation one finds that the total vorticity vanishes in the infinite-volume limit when γ decays at large distances faster than $1/r^3$. The above condition is always met for field configurations with finite energy.

A case of special practical interest is that of a magnetic film where vortex lines may terminate on the film boundaries and the total vorticity may be different from zero. Such a situation leads to some distinct effects in the dynamics of magnetic bubbles, as is discussed in Ref. [5]. Here we are concerned with the idealized case of a three-dimensional ferromagnetic continuum without boundaries.

In order to reveal a direct link that exists between the topology of vortex lines and the actual distribution of magnetization one may use the spherical parametrization

$$M_1 = \sin\Theta \cos\Phi \qquad M_2 = \sin\Theta \sin\Phi \qquad M_3 = \cos\Theta \tag{6}$$

in terms of which the vorticity takes the simple form

$$\gamma = -(\nabla\Pi \times \nabla\Phi) \qquad \Pi = \cos\Theta \tag{7}$$

This relation shows that the variables Π and Φ are analogous to the Clebsch potentials of fluid mechanics [9]. In particular, eq. (7) establishes that the intersection of any two surfaces $\Pi(\mathbf{r}) = c_1$ and $\Phi(\mathbf{r}) = c_2$, where c_1 and c_2 are constant, is a vortex line. In other words, the vortex lines coincide with the curves along which the magnetization remains constant.

To obtain a useful topological classification of the physically possible distributions of magnetization one may construct a quantity which measures the degree of knottedness, or helicity, of tangled vortex lines [7]. In the present context, such a quantity is more appropriately referred to as the Hopf index [9]. We consider the physically important class of configurations which approach the uniform magnetization $\mathbf{M}_o = (0,0,1)$ at spatial infinity, namely the ground state of the ferromagnet. Then the three-dimensional space is isomorphic to the sphere S^3 and a specific configuration $\mathbf{M} = \mathbf{M}(\mathbf{r})$ establishes a map from S^3 to S^2, where S^2 is the sphere defined from the constraint $\mathbf{M}^2 = 1$. Such a map is characterized by the integer-valued Hopf index defined as follows. Let $\mathbf{M}(\mathbf{r}) = \mathbf{M}_1$ and $\mathbf{M}(\mathbf{r}) = \mathbf{M}_2$ be any two vortex lines, according to the definition given in the preceding paragraph ($\mathbf{M}_1$ and $\mathbf{M}_2$ are constant unit vectors). The linking number of these two curves is independent of the specific choice of the pair of vortex lines and is called the Hopf index of the configuration $\mathbf{M}(\mathbf{r})$.

An analytical definition of the Hopf index is also possible by first recognizing that the solenoidal vorticity can be derived from a vector potential,

$$\gamma = \nabla \times \mathbf{a} \tag{8}$$

in terms of which the Hopf index is given by

$$N = 1/4\pi \int (\mathbf{a} \cdot \gamma)\, dV \tag{9}$$

Although the vector potential is unique only to within a gauge transformation, the integral in eq. (9) is gauge invariant for field configurations with sufficient decay at infinity. An explicitly gauge-invariant expression is obtained by inverting eq. (8) using, say, the familiar Coulomb gauge condition and by inserting the resulting nonlocal expression in eq. (9), thus expressing the Hopf index entirely in terms of the vorticity [5]. The remarkable fact is that the above integral is integer valued, and its explicit values coincide with those obtained through the linking-number definition [10].

It would appear that a bonafide local expression for the vector potential is possible in terms of the Clebsch potentials of eq. (7), namely

$$\mathbf{a} = -\Pi \nabla \Phi \tag{10}$$

provided that Π and Φ are differentiable functions of position. However, if such were the case, the Hopf index (9) would vanish identically. One may invert the argument to conclude that the functions $\Pi(\mathbf{r})$ and $\Phi(\mathbf{r})$ cannot be both globally differentiable, for field configurations with a nonvanishing Hopf index, even though the magnetization $\mathbf{M}(\mathbf{r})$ is always assumed to possess such a property. An explicit example worked out below demonstrates that the magnetization reaches the north as well as the south pole of the sphere $\mathbf{M}^2 = 1$ along certain (vortex) lines in the three-dimensional space, so the angle Φ becomes multivalued along those lines.

It should be clear that the preceding remarks do not invalidate the Clebsch representation of the vorticity given in eq. (7), as is evident from the original definition (3) and the assumed differentiability of the magnetization. One would thus be tempted to conclude that the difficulties associated with eq. (10) are largely irrelevant. Yet this equation reveals an important connection between topology and dynamics, for its right-hand side coincides with the standard definition of the linear momentum density. Recall that the variables $\Pi = \cos\Theta$ and Φ are canonically conjugate,

$$\{\Pi(\mathbf{r}_1)\,,\,\Phi(\mathbf{r}_2)\} = \delta(\mathbf{r}_1 - \mathbf{r}_2) \tag{11}$$

and that the Landau-Lifshitz equation (1) takes the standard hamiltonian form

$$\dot{\Phi} = \delta W/\delta\Pi \qquad\qquad \dot{\Pi} = -\,\delta W/\delta\Phi \tag{12}$$

which makes it apparent that the momentum density is indeed given by the right-hand side of eq. (10). Therefore the difficulties discussed in connection with eq. (10) also render the linear momentum ambiguous for field configurations with a nonvanishing Hopf index.

The ambiguities in the definition of the linear momentum were discussed previously on a number of occasions. Thus Slonczewski [11] was apparently the first to realize that a modification of the canonical definition is necessary in order to account for certain important aspects of the dynamics of magnetic bubbles with nonvanishing winding number; a brief survey of his arguments may be found on p. 227 of Ref. [3]. Haldane [12] and Volovik [13] also addressed this question from different perspectives. But we believe that the most direct resolution of the ambiguities appeared in our recent work on this subject [5] where both the linear and the angular momentum were expressed as moments of the topological vorticity.

The essential observation is that the time evolution of the vorticity is governed by the equation

$$\dot{\gamma}_i = \varepsilon_{ijk}\partial_j f_k \qquad f_k = -(\mathbf{F}\cdot\partial_k\mathbf{M}) = [(\delta W/\delta\mathbf{M})\cdot\partial_k\mathbf{M}] \tag{13}$$

which may be derived using the Landau-Lifshitz equation in its general form (1) and the explicit expression of the vorticity given in eq. (3). Furthermore, in the absence of external field gradients, one may invoke the more or less obvious identity

$$\int f_k\, dV = \partial_k W = 0 \tag{14}$$

to infer that the vector density f_k may be written as a divergence, $f_k = \partial_l \sigma_{kl}$, where the specific form of the tensor σ_{kl} depends on the details of the Hamiltonian and may be found in Ref. [5]. Thus we obtain the fundamental relation

$$\dot{\gamma}_i = \varepsilon_{ijk}\partial_j\partial_l\sigma_{kl} \tag{15}$$

which is a strong form of a local conservation law, in view of the double spatial derivative in its right-hand side, and provides the basis for the derivation of all conserved quantities in the Landau-Lifshitz equation. We should mention here that the density f_k played an important role also in the early work of Thiele [14].

A first consequence of eq. (15) is that the total (integrated) vorticity is conserved. In fact, the total vorticity vanishes for sufficiently localized configurations, as was already argued following eq. (5). Eq. (15) may also be used to establish that the Hopf index is time independent; the linking number of any two vortex lines cannot change under smooth evolution induced by the equation of motion.

But the main consequence of eq. (15) is that some of the low moments of the vorticity are also conserved, thus furnishing expressions for both the linear momentum **p** and the angular momentum **l** which are free of all ambiguities:

$$\mathbf{p} = 1/2\int(\mathbf{r}\times\gamma)\, dV \qquad \mathbf{l} = 1/3\int[\mathbf{r}\times(\mathbf{r}\times\gamma)]\, dV \tag{16}$$

Actually the orbital angular momentum given above is conserved only if the tensor σ_{kl} turns out to be symmetric, for a specific choice of the Hamiltonian, e.g. by neglecting the magnetostatic interaction. In general, the conserved quantity is the total angular momentum

$$\mathbf{J} = \mathbf{l} + \mathbf{m} \qquad \mathbf{m} = \int(\mathbf{M}-\mathbf{M}_o)\, dV \tag{17}$$

where **m** is the integrated magnetization, having subtracted the contribution of the ground state $\mathbf{M}_o = (0,0,1)$. Note finally that in the presence of a uniaxial anisotropy only the component of **J** along the symmetry axis is conserved. It is interesting that conservation laws completely analogous to (16) have long been known in the context of fluid dynamics, at least for incompressible fluids [15], a fact of which we were not aware during the writing of Ref. [5].

Independently of derivation, the correctness of the conservation laws (16) and (17) can be established by direct calculations. First, we note that the above expressions may be reduced to their standard canonical form if one freely performs partial integrations; but these are for the most part unjustified for field configurations with a nonvanishing Hopf index. Second, one may verify explicitly that the above quantities are conserved by virtue of the equation of motion (1). Third, one can work out the standard Poisson-bracket relations to demonstrate that **p** and **J** are indeed the generators of space translations and rotations, respectively. Nevertheless, some anomalous Poisson brackets emerge when the total vorticity (5) turns out to be nonvanishing.

Such may be the case in the quasi-two-dimensional continuum leading to some distinct effects in the dynamics of magnetic bubbles, analogous to the familiar Hall effect or even the Magnus "paradox" of fluid dynamics, which were the main focus of Ref. [5]. It was also stated in the above reference that nothing special happens in the dynamics of the three-dimensional continuum without boundaries, mainly because the total vorticity vanishes for field configurations with finite energy. In fact, this turns out to be a slight overstatement, as is shown below where the conservation laws (16) and (17) prove to be a valuable tool in analyzing some general aspects of the dynamics of 3-D magnetic solitons.

The main obstacle in establishing the existence of solitons in three dimensions is Derrick's scaling argument [16] which shows that nontrivial static solutions cannot exist. Thus, if magnetic solitons should be possible at all at D=3, they must be stabilized dynamically [6]. In this respect, Dzyaloshinskii and Ivanov [17] have suggested that topological magnetic solitons with Hopf index N=1 may be stabilized by precession of the magnetization. However, their proposal is unnatural in the context of the realistic ferromagnetic continuum for several reasons. First, the existence of a magnetostatic interaction makes it impossible to conserve independently the orbital angular momentum **l** and the total magnetization **m**, thus frustrating the ansatz of Ref. [17]. Second, as we shall see shortly, an axially symmetric soliton with N=1 generally exhibits a nonvanishing linear momentum along the easy axis, again frustrating the proposal of the above reference.

Therefore, it is more natural to suppose that once an initial configuration is somehow produced with a nonvanishing Hopf index it will eventually settle in a steady state in which the soliton propagates with constant speed, after some radiation is emitted in the form of spin waves. At any rate, a steady-state solution of the form $\mathbf{M} = \mathbf{M}(\mathbf{r}-\mathbf{v}t)$ is compatible with conservation of the linear momentum and satisfies the ordinary differential equation

$$-v_i\partial_i\mathbf{M} = \mathbf{M} \times \mathbf{F} \qquad \mathbf{F} = -\delta W/\delta\mathbf{M} \qquad \mathbf{M}^2 = 1 \tag{18}$$

which must be solved with the boundary condition $\mathbf{M}\rightarrow\mathbf{M}_0 = (0,0,1)$ at spatial infinity. Unfortunately, an analytical solution of (18) does not seem feasible for any realistic choice of the energy functional W. One must thus resort to a numerical solution, which is also difficult without a detailed qualitative understanding of the issues involved. The remainder of this paper is thus devoted to a qualitative analysis based on our preceding general discussion.

Some general properties of steady linear motion may be derived directly from eq. (18). We thus take the cross product of both sides of this equation with $\partial_j\mathbf{M}$ and subsequently the scalar product with **M** to obtain the relation

$$\varepsilon_{kij}v_i\gamma_j = f_k = \partial_l\sigma_{kl} \tag{19}$$

For field configurations with finite energy, integration of eq. (19) over all volume yields no information because the left-hand side becomes proportional to the vanishing total vorticity while the right-hand side also vanishes through Gauss' theorem. Such a seemingly trivial result is already in marked contrast with the case of magnetic bubbles where the analog of (19) leads to the important conclusion that a bubble with a nonvanishing winding number cannot be found in free translational motion unless an external magnetic field gradient is present [5]. No such restriction applies to 3-D solitons even when their Hopf index is different from zero.

Equation (19) may be iterated further to derive a useful virial relation.Thus we contract both sides of (19) with x_k, the Cartesian components of the position vector **r**, and then integrate over all space. Also employing a partial integration in the right-hand side one finds that

$$(\mathbf{v}\cdot\mathbf{p}) = 1/2\int \mathrm{tr}\sigma\, dV = [W_e + 3(W_a + W_m)]/2 \tag{20}$$

where **p** is the linear momentum defined in eq. (16) while W_e, W_a and W_m are the exchange, anisotropy and magnetostatic energy, respectively [5]. It is not difficult to show that the virial with finite energy because the right-hand side is positive definite. Nevertheless, eq. (20) does not contradict the existence of solutions with finite energy and momentum propagating with constant velocity.

A more definitive picture of propagating solitons is obtained by taking advantage of their possible rotational symmetry as well as topological complexity. In the presence of both the magnetostatic interaction and a uniaxial anisotropy, in addition to the exchange interaction, the symmetry group of the Landau-Lifshitz equation is the "diagonal" SO(2); that is, the group of transformations under which both the position vector **r** and the magnetization **M** are simultaneously rotated around the symmetry axis. An axially symmetric ansatz for the solution of eq. (18) is then given by $\Theta = \theta(\rho,z)$ and $\Phi = \phi + \chi(\rho,z)$, where (ρ,ϕ,z) are the usual cylindrical coordinates, provided that the velocity **v** points along the (symmetry) z-axis. Equation (18) then reduces to a coupled system of ordinary differential equations for the real functions θ and χ, which is still too difficult to solve analytically. Nevertheless we proceed with the description of the ansatz by expressing the vorticity, the linear momentum and the total angular momentum in terms of the above functions. Thus the three components of the vorticity are given by

$$\gamma_1 = c_1 \cos\phi - c_2 \sin\phi \qquad \gamma_2 = c_1 \sin\phi + c_2 \cos\phi \qquad \gamma_3 = c_3 \tag{21}$$

with

$$\begin{aligned} c_1 &= -\rho^{-1}\sin\theta\,(\partial\theta/\partial z) \\ c_2 &= \sin\theta\,[(\partial\chi/\partial\rho)(\partial\theta/\partial z) - (\partial\chi/\partial z)(\partial\theta/\partial\rho)] \\ c_3 &= \rho^{-1}\sin\theta\,(\partial\theta/\partial\rho) \end{aligned} \tag{22}$$

Employing these expressions in eqs. (16) and (17) we find

$$p_1 = 0 = p_2 \qquad p_3 = 1/2\int \rho c_2\, dV \qquad dV = 2\pi\rho\, d\rho\, dz \tag{23}$$

implying that only the third component of the linear momentum may be different from zero within the axially symmetric ansatz, whereas the conserved third component of the total angular momentum is found to vanish, $J_3=0$, as expected.

Hence an axially symmetric soliton propagating along the symmetry or easy axis is in principle compatible with the equation of motion and, in particular, with the virial relation (20). But do such solitons actually exist? Putting it differently, the question is whether an initial axially symmetric disturbance of the ferromagnetic medium will eventually settle into a coherent structure in rigid linear motion along the easy axis, after emitting some radiation to infinity in the form of spin waves, or it will completely disintegrate into spin waves. Although a rigorous answer to the above question is not available at this point, we would like to present some arguments to the effect that a coherent soliton does actually emerge at least when the initial configuration is endowed with a nonvanishing Hopf index.

As an example of such an initial condition one may take a configuration that is frequently used to illustrate the so-called Hopf fibration [10]. A convenient parametrization is obtained through the stereographic variable

$$\Omega = (M_1 + iM_2) / (1 + M_3) = \tan(\Theta/2)\exp(i\Phi) \tag{24}$$

in terms of which a configuration with N=1 is given by

$$\Omega = 2a\rho e^{i\phi} / [2az + i(\rho^2 + z^2 - a^2)] \tag{25}$$

where a is an arbitrary scale parameter. The most characteristic feature of this configuration is that the corresponding magnetization points toward the negative z-axis, $\mathbf{M} = (0,0,-1)$, around the ring ($\rho = a$, $z = 0$) which is thus a closed vortex line. Another vortex line that is easy to locate is the line along which the magnetization points toward the positive z-axis, $\mathbf{M} = (0,0,1)$. This is the z-axis itself which may be thought of as a closed vortex line by identifying all points at infinity where the magnetization also approaches the ground state configuration (0,0,1). The linking number of these two vortex lines, i.e. the Hopf index of configuration (25), is obviously equal to unity. The preceding remarks already demonstrate the important gross features of this configuration and are illustrated in Fig. 1a. More elaborate illustrations of the Hopf fibration may be found in the literature [6,17].

A detailed understanding of the structure of vorticity is obtained by recalling that the complex variable Ω remains constant along each vortex line. It is thus not difficult to see that the vortex lines are defined as the intersections of the two surfaces

$$(\rho - c)^2 + z^2 = c^2 - a^2 \qquad \tan(\phi - \phi_o) = (\rho^2 + z^2 - a^2) / 2az \tag{26}$$

where c and ϕ_o are arbitrary constants in the intervals $c \geq a$ and $0 \leq \phi_o \leq 2\pi$. The first equation in (26) defines a torus symmetrically disposed around the z-axis. The torus contains the fundamental ring described in the preceding paragraph, to which it reduces in the limit $c \to a$. Thus, for any fixed c, the vortex lines are confined on the surface of the torus which they cover completely as the constant ϕ_o in the second equation (26) varies through its respective interval, as is demonstrated in Fig. 1b. All vortex lines are closed and the linking number of any two of them is equal to unity. The observed ring-like structure of the topological vorticity explains the title adopted for this talk, at the risk of implying a possibly misleading correspondence with vortex rings encountered in fluid dynamics [15].

The picture of vortex lines described above makes it apparent by simple inspection that the linear momentum of eq. (16) develops a nonvanishing component along the symmetry axis. In any case, an explicit computation using as input the configuration (25) yields

$$p_1 = 0 = p_2 \qquad p_3 = 8\pi^2 a^2 \tag{27}$$

while the total angular momentum is, of course, equal to zero because of the invariance of (25) under the diagonal SO(2). For completeness, we also quote here the exchange energy of this

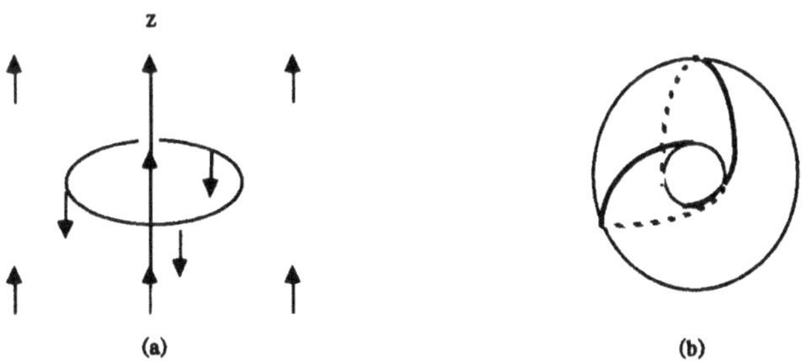

(a) (b)

FIGURE 1. Schematic illustration of some important features of an axially symmetric configuration with N=1 (a) The two limiting vortex lines along which the magnetization denoted by an arrow points toward the positive and negative z-direction, respectively. (b) Sketch of two generic vortex lines viewed from the positive z-axis.

configuration, namely

$$W_e = 1/2 \int (\partial_i \mathbf{M} \cdot \partial_i \mathbf{M})\, dV = 16\pi^2 a \tag{28}$$

Needless to say, the field (25) does not solve the equation of motion but is viewed here as a more or less typical axially-symmetric configuration with a nonvanishing Hopf index (N=1). An important lesson from our analysis is that a nonvanishing linear momentum along the symmetry axis appears inevitable when $N \neq 0$, thus explaining our reservations about the ansatz of Ref. [17] expressed earlier in the text. There is a sense in which the above heuristic arguments could be made rigorous by deriving, for instance, a lower bound for the magnitude of linear momentum that would entail some combination of the Hopf index and, say, the exchange energy to account for the correct scaling dimensions. But the most immediate task is to verify numerically that the configuration (25), used as an initial condition, will evolve according to the Landau-Lifshitz equation (1) into a coherent magnetic vortex ring in rigid linear motion along the z-axis [18].

I am grateful to Theodore Tomaras and Michael Marder for numerous discussions, and to George Papanicolaou for giving me the opportunity to present this material at the NATO workshop. The work was supported in part by a grant from the EEC (SC1-CT-91-0705) and a grant from the Greek Secretariat for Research and Technology (ΠΕΝΕΔ 9013502).

REFERENCES

[1] L. Landau and E. Lifshitz, Physik A (Soviet Union) 8 (1935) 153
[2] N.L. Schryer and L.R. Walker, J. Appl. Phys. 45 (1974) 5406
[3] A.P. Malozemoff and J.C. Slonczewski, *Magnetic Domain Walls in Bubble Materials* (Academic Press, New York, 1979)
[4] T.H. O'Dell, *Ferromagnetodynamics, the Dynamics of Magnetic Bubbles, Domains and Domain Walls* (Wiley, New York, 1981)
[5] N. Papanicolaou and T.N. Tomaras, Nucl. Phys. B360 (1991) 425
[6] A.M. Kosevich, in *Solitons*, ed. by S.E. Trullinger, V.E. Zakharov and V.L. Pokrovsky, p. 555 (Elsevier Science Publishers B.V., 1986)
[7] H.K. Moffatt, J. Fluid Mech. 35 (1969) 117
[8] H.K. Moffatt, *Magnetic Field Generation in Electrically Conducting Fluids* (Cambridge University Press, 1978)
[9] E.A. Kuznetsov and A.V. Mikhailov, Phys. Lett. 77A (1980) 37
[10] R. Bott and L.W. Tu, *Differential Forms in Algebraic Topology* (Springer, New York, 1982)
[11] J.C. Slonczewski, J. Magn. Magnetic Materials 12 (1979) 108
[12] F.D.M. Haldane, Phys. Rev. Lett. 57 (1986) 1488
[13] G.E. Volovik, J. Phys. C: Solid State Phys. 20 (1987) L83
[14] A.A. Thiele, Phys. Rev. Lett. 30 (1973) 230; J. Appl. Phys. 45 (1974) 377
[15] G.K. Batchelor, *An Introduction to Fluid Dynamics* (Cambridge University Press, 1967)
[16] G.H. Derrick, J. Math. Phys. 5 (1964) 1252
[17] J.E. Dzyaloshinskii and B.A. Ivanov, JETP Lett. 29 (1979) 540
[18] M. Marder and N. Papanicolaou, work in progress.

MORPHOLOGY AND SINGULARITIES OF VORTEX FLOWS

ALAIN PUMIR* , and ERIC D. SIGGIA
LASSP, Cornell University, Ithaca
NY14853
USA

ABSTRACT. Recent results on the 3d incompressible, axisymmetric Euler equations are reviewed. The connection with the 2d Boussinesq equations is explained, and used to simulate the formation of point like singularities. One of the upshots of this study is the spontaneous appearance of shocks. This phenomenon can be intuitively understood with the help of the Clebsch formulation of the Euler equations, and of a simplified model (equations of porous media convection).

1. Introduction

Understanding the precise mechanisms responsible for the physics of 3 d, turbulent, incompressible flows remains a major challenge. The equations describing the fluid motion, the Navier Stokes equations :

$$\partial \mathbf{u} + (\mathbf{u} \cdot \nabla)\mathbf{u} = -\nabla p + \nu \nabla^2 \mathbf{u} \tag{1}$$

$$\nabla \cdot \mathbf{u} = 0 \tag{2}$$

where $\mathbf{u}(\mathbf{x}, t)$ is the velocity field and ν is the viscosity have been known for over a century. It is surprising to realize that many of their properties are still unknown. In particular, the low viscosity limit ($\nu \to 0$), relevant to the problem of turbulence still resists an analytic treatment, and remains a numerical challenge.

The 3d Euler equations :

$$\partial \mathbf{u} + (\mathbf{u} \cdot \nabla)\mathbf{u} = -\nabla p \tag{3}$$

$$\nabla \cdot \mathbf{u} = 0 \tag{4}$$

obtained by setting the viscosity ν to 0 in the 3d Navier-Stokes equations, have been thought to be a good starting point to analyze the fluid mechanical

* Permanent address: LPS, Ecole Normale Supérieure, 75231 Paris,France

R. E. Caflisch and G. C. Papanicolaou (eds.), Singularities in Fluids, Plasmas and Optics, 159–166.
© 1993 *Kluwer Academic Publishers.*

processes involved in 3d fluid turbulence. It has to be emphasized however that the connection between the solutions of the Navier-Stokes equations in the limit $\nu \to 0$ and the solutions of the Euler equations is not completely obvious.

It is useful to write the Euler equations in term of the vorticity field : $\omega \equiv \nabla \wedge \mathbf{u}$:

$$\partial \omega_i + (\nabla \cdot \mathbf{u})\omega_i = s_{ij}\omega_{ij} \tag{5}$$

$$s_{ij} = \frac{1}{2}(\frac{\partial u_i}{\partial x_j} + \frac{\partial u_j}{\partial x_i}) \tag{6}$$

The matrix s_{ij} is symmetric ($s_{ij} = s_{ji}$) and traceless; hence s_{ij} has three real eigenvalues ($\lambda_1 \geq \lambda_2 \geq \lambda_3$), with $\lambda_1 + \lambda_2 + \lambda_3 = 0$, corresponding to three orthogonal eigenvectors, $\hat{e}_i$. The component in the directions $\hat{e}_i$, corresponding to a positive eigenvalue grows. This important effect is called vortex stretching, and is believed to play a key role in turbulent flows. Because s_{ij} can be expressed as a linear functional of the vorticity field, the growth of ω comes from a nonlinear term, which can possibly induce a blow up (think of $dx/dt = x^2$). The singularity issue turns out to be rather subtle and not completely understood. The way the vorticity couples with the rate of strain it generates is non trivial. This leads to ask how does the vorticity field organizes itself, and how does it couple to the strain it generates. The same question is also asked in the context of 'coherent structures', which are thought to be important in many turbulent flows.

Numerical simulations of the 3d Euler equations show the emergence of vortex sheets, *i.e.*, surfaces of velocity discontinuity[1,2,3,4]. The dynamics of these objects is essentially 2 dimensional, so the self stretching almost turns itself off. Note that simulations of the full 3 dimensional Navier-Stokes equations show rather the formation of vortex tubes[5,6,7]. This prediction has been recently confirmed experimentally[8]. A detailed theoretical understanding is still lacking.

In this paper, we summarize our recent work on a restricted class of 3 dimensional Euler flows (axisymmetric flows), and on a attempt to understand the origin of vortex sheets in the solutions to the 3d Euler equations. For more details, one may consult references 9, 10 and 11.

2. Axisymmetric Euler flows

An axisymmetric flow is defined by a velocity field independent of the azimuthal angle, ϕ around the symmetry axis, Oz. The velocity field, $\mathbf{u}(r, z, t)$ satisfies the 3d Euler equations :

$$\partial_t(ru_\phi) + (\mathbf{u}_{||} \cdot \nabla_{||})(ru_\phi) = 0 \tag{7}$$

$$\partial_t(\omega_\phi/r) + (\mathbf{u}_{||} \cdot \nabla_{||})(\omega_\phi/r) = -\frac{1}{r^4}\partial_z(ru_\phi)^2 \tag{8}$$

$$\frac{1}{r}\partial_r(ru_r) + \partial_z u_z = 0 \tag{9}$$

$$\omega_\phi = \partial_z u_r - \partial_r u_z \tag{10}$$

where $\mathbf{u}_{||} = (u_r, u_z)$ and $\nabla_{||} = (\partial_r, \partial_z)$.

Equation (7) expresses the conservation of circulation around a loop ($r = cst, z = cst.$), and equation (9) is the usual incompressibility condition. The mechanism of generation of azimuthal vorticity, as described from Eq. (8) is a consequence of the centripetal acceleration. It is very instructive to compare the system (7-10) with the 2-d Boussinesq equations with a constant gravity, $\mathbf{g} = -g\hat{e}_y$:

$$\partial_t\theta + (\mathbf{u}\cdot\nabla)\theta = 0 \tag{11}$$

$$\partial_t\omega + (\mathbf{u}\cdot\nabla)\omega = \partial_x\theta \tag{12}$$

$$\partial_x u_x + \partial_y u_y = 0 \tag{13}$$

$$\omega = \partial_x u_y - \partial_y u_x \tag{14}$$

A comparison of the two systems of equations shows that the quantity $(ru_\phi)^2$, which is advected by the flow, (Eq(7)), plays the same role as the temperature as the source of vorticity (see Eq. (8) and (12)). The 'gravity' in the axisymmetric problem is directed radially outwards, and varies with the distance to the axis, r, as $\mathbf{g} = \hat{e}_r/r^4$. The velocity field (u_x, u_y) corresponds to (u_z, u_r). Rigorously, this analogy is valid only when the typical length scale of the solution is much smaller than the distance to the axis, $\bar{r}$. In this case, one has the correspondence :

$$(\bar{r}\omega_r, \omega_\phi, \bar{r}\omega_z) \leftrightarrow \left(-\partial_x\theta^{1/2}, \omega, \partial_y\theta^{1/2}\right) \tag{15}$$

The analogy between 2d Boussinesq and the 3d axisymmetric problem is of no use to understand the evolution of the solutions right at $r = 0$. In the following, we will restrict ourselves to solutions well away from the symmetry axis.

Grauer and Sideris [12] first noticed that solutions to the axisymmetric Euler equations may very well develop singular solutions in a finite time, yet the problem is technically 2-dimensional. The analogy we have noted with the Boussinesq problem suggests that the occurence of singularities in axisymmetric flows at finite distance and well away from the symmetry axis is equivalent to the occurence of singularities in the Boussinesq problem.

In order to study numerically a blow up problem, it is crucial to resolve smaller and smaller scales. For this purpose, we have used both adaptive mesh techniques, and a continuous rescaling of the equations. The Boussinesq equations can be rescaled by the following transformation : $X = x/a$,

$\Omega = \omega a^{1/2}$, $\Psi = \psi/a^{3/2}$, $\theta = \Theta$ and $\partial_T t = a^{1/2}$. The equations then read :

$$\partial_T \Theta + \alpha \mathbf{X} \cdot \nabla_X \Theta + \frac{\partial(\Psi, \Theta)}{\partial(X, Y} = 0 \tag{16}$$

$$\partial_T \Omega + \alpha \mathbf{X} \cdot \nabla_X \Omega + \frac{1}{2} \alpha \Omega + \frac{\partial(\Psi, \Omega)}{\partial(X, Y)} = \partial_X \Theta \tag{17}$$

with the definitions :

$$\nabla_X^2 \Psi = \Omega, \alpha = \partial_T lna \tag{18}$$

If the resulting system can be integrated from $T = 0$ out to ∞ with $0 < \alpha 1 \leq \alpha \leq \alpha 2 < \infty$, then, it is easy to check that the solutions to the 2d Boussinesq equations blow up in a finite time, with $||\nabla \theta||_\infty \sim 1/(t^* - t)^2$ and $t^* = \int_0^\infty dt' exp(-1/2 \int_0^{t'} dt'' \alpha(t''))$. This possibility of continuously rescaling the equations makes it much easier to study numerically the Boussinesq problem. In the following, we will focus on this problem.

We use finite differences. The interval $(0, 1)$ is mapped onto $(-\infty, +\infty)$ by an analytic change of variables, and then discretized. Far away from the origin of the coordinates, the resolution is low. When the solution is continuously rescaled, parts of the solution are pushed away, and then become poorly resolved. We had then to truncate the solution at ' ∞', to avoid having large gradients on a couple of lattice points. We checked extensively that this does not affect the evolution of our solution. Physically the contribution to the strain from the truncated region is indeed relatively small.

Details of our numerical procedures can be found in ref. 11 (see also ref. 2).

Our numerical results can be understood from relatively simple considerations. Consider as an initial condition a bubble, with a smooth distribution of temperature (we have chosen $\theta = (1 + 0.2y)/(1 + x^2 + y^2)^2)$. The fluid at infinity is heavier (colder) than the fluid at the center. Because of buoyancy, the light (hot) fluid tends to rise in the heavier (colder) fluid. Intuitively, one expects that the regions of high temperature will tend to catch up with regions of lower temperature. This mechanism suggests the formation of shocks, as it is the case in the Burgers equations. Obviously, the analogy with Burgers is only qualitative, since there is no non trivial velocity field of the form $\mathbf{u} = (0, f(y))$.

Numerical results confirm this picture, and show that the cap of the bubble develops very strong gradients. One is therefore lead to a sharp interface, separating cold fluid on top of warmer fluid. The emergence of very sharp fronts in the Boussinesq problem means that vortex sheets form in the axisymmetric Euler problem. Note that the emergence of sheets in this problem has a simple origin, thanks to the analogy with Boussinesq flows.

As it is well known, a flat interface separating light fluid, underneath heavy fluid is Rayleigh-Taylor unstable. One would naively expect that the sharp interface should undergo rapid instabilities. Because the bubble is rising, a stagnation point flow develops near the tip, and tends to restabilize the interface [13], thus delaying the development of the instability. This leads to very long transients, where the cap of the bubble has a radius of curvature very large compared to its width, and where the growth of the gradients is slower than exponential.

As the fronts become sharper, the interface eventually becomes unstable. The destabilisation induces a rather complicated shape with much smaller radii of curvature, and rolling up vortex sheets. The growth of the gradients also becomes much faster. Over 5 orders of magnitude, we find that the gradient of θ grows like $1/(t^* - t)^2$. Simultaneously, the maximum vorticity (in the axisymmetric Euler problem, the azimuthal component of the vorticity vector) grows like $1/(t^* - t)$. Our numerical data could be fit with the following functional dependences :

$$\theta = \theta_0 + (t^* - t)^\eta f(\mathbf{x}/(t^* - t)^{2+\eta}, ln(t^* - t)) \tag{19}$$

$$\omega = 1/(t^* - t) g(\mathbf{x}/(t^* - t)^{2+\eta}, ln(t^* - t)) \tag{20}$$

with $\eta = 0.2 \pm 0.1$. Note that the Boussinesq equations are consistent with these scalings for any value of η. The dependence on $ln(t^* - t)$ in these equations is non trivial, and presumably very hard to determine analytically.

Useful insight can be obtained by ignoring the actual structure of the interface, and using a contour dynamics limit. In this approximation, one describes the front by a temperature discontinuity, located on a curve $\mathbf{x}(\lambda, t)$, where λ is a Lagrangian parameter. The following equation can be derived (see ref. 11 for details) :

$$\partial_t |\nabla\theta|^2 \sim c \int_0^t |\nabla\theta|^2(t')\kappa(t')dt' \tag{21}$$

where κ is the radius of curvature of the interface, $1/|\nabla\theta|$ is the stretching and c is a constant.

Equation (21) is useful to understand qualitatively the dynamics in the two regimes we found numerically. Before the cap of the bubble becomes unstable, $\kappa \sim cst.$, and Eq.(21) predicts that $|\nabla\theta|$ grows exponentially. When the instability starts, our numerical results suggest that $\kappa(t) \sim |\nabla\theta|$. This has to be understood in a statistical sense, since we never reached a truly self similar state. In this case, it is easy to show that Eq. (21) leads to a growth asymptotically like : $|\nabla\theta| \sim 1/(t^* - t)^2$.

Several salient conclusions for axisymmetric Euler flows should be noted. The singularity is pointlike, although its shape is unsteady. The r, z components of the 3d vorticity diverges like $(t^* - t)^{-2}$, while $\omega_\phi \sim (t^* - t)^{-1}$. Simple estimates, using the scalings in Eq. (19-20), show that $\int |\omega|^{2+a} d^3x$ diverges for $a > \eta$. Last, the vorticity in these flows is nearly parallel to the second eigenvector of the rate of strain matrix, like in all the other flows that have been studied numerically.

3. Vortex sheet formation in Euler flows

One of the upshots of the Boussinesq analogy is that it provides a natural and intuitive way to understand why vorticity tends to concentrate on sheets. Obviously it is by no means obvious that the physics of a thermal plume will be applicable to the most general Euler flow. We suggest here a more general explanation for the origin of vortex sheets formation in 3d Euler flows.

As noted by Clebsch [14] (see also reference 15 for a more recent review), it is possible to write the velocity field $\mathbf{u}$ in terms of the scalar functions :

$$\mathbf{u}(\mathbf{x}, t) = \lambda \nabla \mu + \nabla \phi \tag{22}$$

Vorticity is simply the cross product of the gradients of λ and μ :

$$\omega = \nabla \lambda \wedge \nabla \mu \tag{23}$$

Note that a canonical transformation of (λ, μ) leaves the vorticity unchanged. The function λ and μ are simply advected by the flow :

$$\partial_t \lambda + \mathbf{u} \cdot \nabla \lambda = 0 \tag{24a}$$

$$\partial_t \mu + \mathbf{u} \cdot \nabla \mu = 0 \tag{24b}$$

These equations correspond to a particular choice of 'gauge'. The most general form of the equations can be obtained by replacing the zero right hand side of Eq. (24a,b) by the infinitesimal canonical transformation : $(\partial_\mu S(\lambda, \mu, t), -\partial_\lambda S(\lambda, \mu, t))$. We will ignore these subtleties here, and stick to Eqns. (24a,b). The incompressibility condition imposes that :

$$\nabla^2 \phi + \nabla \cdot (\lambda \nabla \mu) = 0 \tag{25}$$

Because of the gauge inveriance, it is conceivable, although not obvious that one can impose that the gradient of λ, say, be much smaller than the gradient of μ. Imagine now that the gradient of λ is approximately frozen. Then, we have for μ :

$$\partial_t \mu + (\mu \hat{y} - \nabla \phi) \cdot \nabla \mu = 0 \tag{26}$$

$$\nabla^2 \phi + \partial_y \mu = 0 \tag{27}$$

Equation (26,27) is nothing but the equation of convection in porous media. The velocity is related to the temperature, μ in the same way as the acceleration is related to θ in the Boussinesq problem. Note that Eq.(26,27) is the incompressible version of the Burgers equation :

$$\partial_t u + u\partial_x u = 0 \tag{28}$$

The Burgers equation is known to develop shocks :

$$u(x,t) \sim (t^* - t)^{1/2} f(x/(t^* - t)^{3/2}) \tag{29}$$

Because of the incompressibility condition, the above quasi 1 dimensional solution of Eq.(28) is not a solution of Eq. (26,27).

As in the case of Boussinesq (axisymmetric) flows,it is intuitively very clear why sharp gradients in μ will develop (hot fluid tends to catch up with cold fluid, thus leading to steepening fronts ahead of rising bubble). An extended numerical study of the equations (26,27) confirmed this picture, thus suggesting a mechanism of sheet formation in Euler flows.

A very different question consists in understanding whether the porous media equations develop finite time singularities. Using a pseudo spectral code, and starting from random initial conditions in μ in a periodic, 2 dimensional domain, we have studied numerically the evolution of the field. A few bubbles (or plumes) spontaneously developed. In several cases it was possible to observe a destabilisation of the caps of these thermals, thus leading to a rapid deterioration of the numerical resolution. Using adaptive mesh techniques, we have been able to push the calculations up to an amplification of 10^5 of the gradient, involving several destabilisation of the cap of the bubble, and without any obvious impediment to further integration. The physics of the growth of infinite gradients in a finite time in Equation (26,27) is very reminiscent of the one we found for the Boussinesq problem. The singularity we found numerically suggests the following asymptotic form :

$$\mu = (t^* - t)^{\eta} M(\mathbf{x}/(t^* - t)^{1+\eta}, ln(t^* - t)) \tag{30}$$

Note the similarity with Eq. (19,20). The value of the small exponent η is : $\eta \sim 0.1$. Note also that we have developed a 'contour dynamics' formulation of this problem, which largely confirms the numerical results we found in our study of the partial differential equation (26,27)[16].

The structure of the collapse found numerically for all these problems is also very reminiscent of the collapse we found for the Biot-Savart model of vortex filaments[17]. The analytic description of the development of singularities in these problems will probably involve a mechanism of proliferation and propagation of singularities in the 'unphysical domain' (see ref. 18, and also the contributions by T. Dombre and S. Tanveer in these proceedings).

Of course, the connection of this model with the 3d Euler equations has to be understood better. The singular solutions to the porous media convection equations we found are presumably very loosely related to the Euler problem. Nevertheless, the mechanism of sheet generation looks very robust. Preliminary numerical results suggest that sheet formation in the 3d Euler equations is somehow related to such a mechanism.

Acknowledgements

The authors have benefited from many discussions with Boris Shraiman and Thierry Dombre. The financial support of the AFOSR (Grant No. 91-0111) is gratefully acknowledged.

References

1. M. Brachet, D. Meiron, B. Nickel, S. Orszag and U. Frisch (1983) *J. Fluid Mech.* **130**, 411.
2. A. Pumir and E. Siggia (1990) *Phys. Fluids* **A2**, 220.
3. M. Brachet, M. Meneguzzi, H. Politano, P.L. Sulem and A. Vincent (1992), Private communication
4. R. Kerr (1992), Private communication
5. E. Siggia *J. Fluid Mech.* (1982) **107**, 357.
6. A. Vincent and M. Meneguzzi (1991) *J. Fluid Mech.* **225**, 1.
7. Z.S. She, E. Jackson and S. Orszag (1990) *Nature (London)* **344**, 226.
8. S. Douady and Y. Couder (1991), *Phys. Rev. Lett.* **67**, 983.
9. A. Pumir and E. Siggia (1992), *Phys. Rev. Lett.* **68**, 1511.
10. A. Pumir, B. Shraiman and E. Siggia (1992), *Phys. Rev A* **45** R5351.
11. A. Pumir and E. Siggia (1992) *Phys. Fluids* **A4**, 1472.
12. R. Grauer and T. Sideris (1991), *Phys. Rev. Lett.* **67**, 3511.
13. P. Pelcé (1988), *Dynamics of curved fronts* Perspective in Physics, Academic Press.
14. A. Clebsch (1859), *J. Angew. Math.* **56**, 1.
15. R. Salmon (1988), *Ann. Rev. Fluid Mech.* **20**, 225.
16. A. Pumir and T. Dombre (1992), in preparation.
17. A. Pumir and E. Siggia (1987), *Phys. Fluids* **30**, 1606.
18. T. Dombre, A. Pumir and E. Siggia (1992), *Physica* **D**, to appear.

Topological Transitions in Hele-Shaw Flow

Michael J. Shelley
Courant Institute of Mathematical Sciences
New York University
251 Mercer St.
New York City NY 10012
Raymond E. Goldstein
Department of Physics
Joseph Henry Laboratories
Princeton University
Princeton NJ 08544
and
Adriana I. Pesci
Department of Chemical Engineering
Princeton University
Princeton NJ 08544

ABSTRACT. The processes which lead to topological reconfiguration of fluid interfaces are studied in Hele-Shaw flow. This is motivated by recent experiments of droplet formation through the osculation of two interfaces separating immiscible fluids. In the Hele-Shaw approximation, such a configuration reduces to interface dynamics through their representation as vortex sheets. To begin, we focus on thin fluid layers and develop an asymptotic theory which yields simplified, nonlinear equations for the local thickness. As particular examples, we consider the dynamics of gravity driven fluid jets, as well as the motion of unstably stratified fluid layers. In both cases, the bounding interfaces collide at a finite time, with an associated singularity in the fluid velocity. Some comparison is made with simulations of the full Hele-Shaw equations, and connections with experiments are discussed.

1. Introduction

Despite their frequent occurence in nature, topological transitions in fluid systems are poorly understood. Especially difficult are those transitions involving material surfaces in the fluid, as their reconfiguration must entail intrinsically singular behavior. An additional obstacle has been the lack of systems in which topological transitions can be controlled to produce satisfactory experiments. However, a successful set of observations has recently

R. E. Caflisch and G. C. Papanicolaou (eds.), Singularities in Fluids, Plasmas and Optics, 167–188.
© 1993 *Kluwer Academic Publishers.*

been performed (Goldstein, Mason, & Shyamsunder, 1991; Dance, *et al.* 1992) on a surprisingly simple example of such a transition: the splitting of a drop of fluid in Hele-Shaw flow. This fission was achieved either by applying pressure across a thin layer of fluid, or by the action of gravity on a two-dimensional pendant drop.

The first case was considered previously (Constantin, *et al.* 1992) wherein heuristic arguments were used to obtain a nonlinear PDE for the layer thickness $h(x,t)$. This PDE is a nonlinearly degenerate, fourth-order diffusion equation. Within this description, a topological transition is signalled by the vanishing of $h(x,t)$ at some point, in either a finite or infinite time, from strictly positive initial data. We shall refer to this process as "pinching." The equation of interest is posed on a finite interval, and the system is closed by giving boundary data for h and h_{xx}, the latter by relating the curvature to the pressure jump through the Gibbs-Thompson relation.

In this paper, we focus on gravity-driven interface pinching in two related physical situations. In the first, the surfaces bounding a thin layer of fluid in the Hele-Shaw cell are being driven together by a Rayleigh-Taylor instability. Using the aspect ratio of the layer as a small parameter, we derive a governing PDE for the layer thickness from a boundary integral formulation of the problem. This PDE is again a nonlinearly degenerate diffusion equation, but now the stabilizing fourth-order diffusion competes with a destabilizing second-order anti-diffusion. In numerical simulations of the model equation, the layer thickness becomes zero at a finite time, a phenomenon accompanied by a singularity of the PDE. These results are compared with numerical simulations of the full Hele-Shaw flow. In a situation perhaps more relevant to the experiments, we also consider the behavior of gravity driven jets in the Hele-Shaw cell. Employing the same asymptotic methods, we find again for the thickness a degenerate fourth-order diffusion equation. However, the destabilizing anti-diffusion is replaced by a nonlocal and dispersive term involving a Hilbert transform. Numerical simulations of an approximation to this equation reveals that pinching and finite-time singularities do occur readily, particularly when the system is subject to certain kinds of boundary conditions. In both of these instances, corrections to the model equations can be found directly by further expansion in the aspect ratio.

In section 2, we review the vortex sheet formulation of Hele-Shaw dynamics. The asymptotic analysis appropriate to thin fluid layers is given in section 3, along with derivations of the governing PDEs for each case. We also present a variational formulation of these equations. Section 4 presents the methods and results of numerical studies. We conclude in section 5 with a summary of open questions, directions for future work, and some comments on experimental realizations of these phenomena.

2. The Equations of Motion

We are interested here in the motion of the two interfaces which bound a thin layer of fluid trapped between two other fluids, as shown in Figure 1. To summarize the essential aspects of the vortex sheet formalism, consider first a single interface Γ that separates two Hele-Shaw fluids of differing viscosities and densities. We adopt the convention used in Fig.1 that the lower fluid is labeled 1, the upper labeled 2, and similarly for their respective densities ρ_j and viscosities μ_j. In each fluid we have Darcy's law:

$$v_j = -\frac{b^2}{12\mu_j}\left(\nabla P_j - \rho_j F\right) \ , \tag{1}$$

where b is the gap width of the cell, v and P are the velocity and pressure in each fluid, and $F = -\nabla\phi$ is a divergence free body force (i.e. gravitational force). By the incompressiblity constraint,

$$\nabla \cdot v_j = 0 \ , \tag{2}$$

the pressure is harmonic and plays the role of a velocity potential, while the velocity field is irrotational.

At the interface Γ, we impose the Gibbs-Thompson boundary condition which relates the pressure jump to the curvature κ by

$$P_1 - P_2 = -\tau\kappa. \tag{3}$$

where τ is the surface tension. From the requirement that the interface move with the fluid, there is also the kinematic boundary condition,

$$n \cdot (v_1 - v_2)\,|_\Gamma = 0 \ , \tag{4}$$

where n is the normal to Γ.

It follows from the above assumptions that the velocity field can be represented as a vortex sheet on Γ (Tryggvason & Aref, 1983; Dai & Shelley, 1992). This is done most conveniently in complex variables. Let $Q^* = u - iv$ be the complex conjugate velocity field evaluated at $\eta = x + iy$, and $z(p) = x(p) + iy(p)$ describe the location of Γ as a function of a real parameter p. Then Q^* is given by

$$Q^*|_\eta = \frac{1}{2\pi i}\int_{-\infty}^{+\infty} \frac{\gamma(p')}{\eta - z(p')} dp' \ . \tag{5}$$

For a real vortex sheet strength γ, the kinematic boundary condition (4) is automatically satisfied. The requirement that the interface move with the fluid is enforced through the evolution equation

$$\begin{aligned} \frac{\partial z^*}{\partial t}(p,t) &= \frac{1}{2\pi i} P\int_{-\infty}^{+\infty} \frac{\gamma(p')}{z(p) - z(p')} dp' \\ &\equiv Q^*(z(p)) \ , \end{aligned} \tag{6}$$

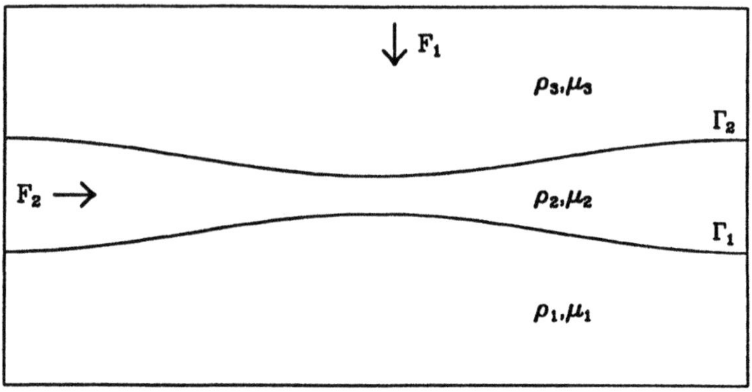

Fig. 1. A schematic of a thin layer of fluid, trapped between two others, in the Hele-Shaw cell.

Here P denotes the principal value of the Birkhoff-Rott integral in Eq. (6). The vortex sheet strength γ is itself determined by the Gibbs-Thompson condition, and can be shown to satisfy the integral equation

$$\gamma(p) + 2A^{\mu}\text{Re}\left\{z_p(p)Q^*(z(p))\right\} = \frac{1}{\theta}\partial_p\left(-\tau\kappa(p) - 2\Delta\rho\phi(z(p))\right) . \qquad (7)$$

Here $A^{\mu} = (\mu_1 - \mu_2)/(\mu_1 + \mu_2)$ is the Atwood ratio for the viscosities, $\Delta\rho = \rho_1 - \rho_2$, and $\theta = 12\bar{\mu}/b^2$, with $\bar{\mu}$ being the average of the viscosities, and the subscripts indicate partial derivatives.

Eqs. (6) and (7) constitute a complete formulation of the interface evolution problem. Eq. (7) is a Fredholm integral equation of the second kind for $\gamma(p)$. It can be shown to be contractive, and thus has a unique solution which can be found through fixed point iteration (see, for example, (Kellog, 1929)).

By choosing a length L for the system (say a periodicity length of the initial data), and a scale G for the body force, Eqs. (6) and (7) can be nondimensionalized. The integral equation can then be given as

$$\gamma(p) + 2A^{\mu}\ \text{Re}\left\{z_p(p)Q^*(z(p))\right\} = \partial_p\left(S\kappa(p) - R\phi(z(p))\right) , \qquad (8)$$

where $S = \tau/\theta L^2\sqrt{LG}$ is a positive, nondimensional surface tension parameter, and $R = 2\Delta\rho G/\theta\sqrt{LG}$ is a nondimensional quantity whose sign depends upon the sign of $\Delta\rho$.

The vortex sheet formulation is straightforward to extend to the case of two interfaces, Γ_1 and Γ_2. Their motion obeys the equation

$$\begin{aligned}\frac{\partial z_j^*}{\partial t}(p,t) &= \frac{1}{2\pi i} P \int_{-\infty}^{+\infty} \frac{\gamma_1(p')}{z_j(p) - z_1(p')} dp' \\ &+ \frac{1}{2\pi i} P \int_{-\infty}^{+\infty} \frac{\gamma_2(p')}{z_j(p) - z_2(p')} dp' \\ &\equiv Q^*(z_j(p)) \, . \end{aligned} \tag{9}$$

for $j = 1, 2$. When $j = 1$ the first integral must be interpreted as being of principal value type, and similarly when $j = 2$ for the second integral. The two vortex sheet strengths γ_1 and γ_2 now are determined through the coupled Fredholm integral equations

$$\gamma_j(p) + 2A_j^\mu \; Re\left\{z_{j,p}(p) Q^*(z_j(p))\right\} = \partial_p \left(S_j \kappa_j(p) - R_j \phi(z_j(p))\right) \, , \tag{10}$$

for $j = 1, 2$. Again, Eqs. (9) and (10) constitute a complete formulation of the evolution problem.

3. The Lubrication Approximation

In this section we derive an equation of motion for the thickness $2h$ of a thin layer of trapped fluid in two special cases. In the first case, gravity acts perpendicular to the layers ($F = -g\hat{y}$) and with $\rho_1 < \rho_2 < \rho_3$ the trapped fluid is unstably stratified relative to both the upper and lower fluids. In the second case, gravity acts parallel to the layers ($F = g\hat{x}$) and with $\rho_1 = \rho_3$ this models the flux of fluid in a gravity driven jet in a Hele-Shaw flow. In the first case, we find a purely local PDE for h, while in the second the PDE is nonlocal, as it involves a Hilbert transform. In either case, if $g = 0$, the lubrication approximation studied earlier (Constantin, *et al.* 1992) is recovered.

These results pertain to thin fluid layers, where the notion of "thin" is made concrete as follows. Consider a layer of length L and area A, whose mean thickness is $w = A/L$. By defining the parameter ϵ through $w = \epsilon L$, then a thin layer means $\epsilon << 1$. Moreover, ϵ is a conserved quantity since the area, $A = \epsilon L^2$, is conserved due to the fluid's incompressibility. The dynamics which result from retaining only the leading-order terms in the expansion in powers of the small parameter ϵ is the *lubrication approximation*.

The present analysis is also based on the assumption that the initial data (and subsequent evolution) are *antisymmetric*, as in Fig. 1. Thus, in the vortex sheet formulation of Eq. (9), the two interface locations are related by $z_1(p,t) = z_2^*(p,t)$, and their vortex sheet strengths by $\gamma_2 = -\gamma_1 \equiv \gamma$.

Before summarizing the expansion technique and results, a short technical digression is necessary. We have given a *Lagrangian* description of the interface evolution. For our purposes, it is more convenient to recast it in an *Eulerian* frame. Fixing the variable parametrizing Γ_1 and Γ_2 to be x, the interfaces are given by $z_1(x,t) = x - ih(x,t)$ and $z_2(x,t) = x + ih(x,t)$. Then so long as the interfaces do not fold over (i.e. become multiply defined in x) the thickness $2h(x,t)$ is governed by the equation

$$h_t(x,t) = -v + uh_x(x,t) , \tag{11}$$

where $u - iv = Q^*(x + ih(x,t))$ is the conjugate velocity on Γ_2.

Returning to the lubrication approximation, we define H by

$$h(x,t) = \epsilon H(x,t) , \tag{12}$$

and thus

$$z_1(x,t) = x - i\epsilon H(x,t) \ \ and \ \ z_2(x,t) = x + i\epsilon H(x,t) . \tag{13}$$

In general, the vortex sheet strength depends upon local properties of the interface, such as its position ϵH and curvature

$$\kappa = \frac{\epsilon H_{xx}}{(1+\epsilon^2 H_x^2)^{3/2}} , \tag{14}$$

and so γ itself will have an expansion in ϵ:

$$\gamma = \gamma^{(0)} + \epsilon\gamma^{(1)} + \epsilon^2\gamma^{(2)} + \cdots \tag{15}$$

Thus, in the context of the vortex sheet formulation, we seek expansion in ϵ of expressions of the form

$$\begin{aligned} I[\gamma]\,|_{z_1=x+i\epsilon H} &= \frac{1}{2\pi i}\int_{-\infty}^{+\infty} \frac{\gamma(x')dx'}{(x-x') + i\epsilon(H(x)+H(x'))} \\ &- \frac{1}{2\pi i} P\int_{-\infty}^{+\infty} \frac{\gamma(x')dx'}{(x-x') + i\epsilon(H(x)-H(x'))} \\ &\equiv I_1[\gamma] - I_2[\gamma] . \end{aligned} \tag{16}$$

The method of expansion of such integrals is discussed in detail elsewhere (Baker & Shelley, 1991). The principal value integral I_2 is actually the more straightforward, and involves geometric expansions of the integrand, together with integrations by parts. This is possible due to the uniform boundedness of $(H(p) - H(q))/(p - q)$ for smooth H. It is the 'smooth' integral I_1 that is the more difficult. There the expansions are accomplished through Taylor expansion in ϵ. Local terms arise now due to the presence of a δ-function in the integrands as $\epsilon \to 0$. This amounts to the application of the Plemelj formulae to take these limits. We give only the results of this expansion procedure here; details will be presented elsewhere (Goldstein, Pesci, & Shelley, 1992). Keeping terms up to $O(\epsilon^2)$ in $-v + \epsilon uH_x$, we find the equation of motion

$$\epsilon H_t = -\partial_x \left\{ \epsilon H \left(\gamma^{(0)} + \epsilon\gamma^{(1)} + \epsilon\mathcal{H}\left[(\gamma^{(0)}H)_{x'} \right] \right) \right\} , \tag{17}$$

where $\mathcal{H}$ is the Hilbert transform,

$$\mathcal{H}[f] = \frac{1}{\pi} P \int_{-\infty}^{+\infty} \frac{f(x')dx'}{x - x'} . \tag{18}$$

Eq. (17) is a central result upon which all subsequent analysis depends.

The general equation of motion in Eq. (17) has several important features. First, it is in the form of a conservation law for H (or equivalently for $h = \epsilon H$),

$$h_t + j_x = 0, \quad j = hU \tag{19}$$

where j is a current, and U is some mean velocity across the layer. A second fundamental consequence of this form is that vanishing of h also implies a singularity in the velocity U (Constantin, *et al.* 1992). Let $X(t)$ locate a minimum of h that reaches zero at a finite time t^*. Then

$$\frac{d}{dt} h(X(t), t) = -h(X(t), t) U_x(X(t), t) . \tag{20}$$

It follows then that if $\lim_{t \to t^*} h(X(t), t) = 0$, then

$$\lim_{t \to t^*} U_x(X(t), t) = +\infty, \tag{21}$$

and moreover

$$\lim_{t \to t^*} \int^{t^*} U_x(X(s), s) ds = +\infty . \tag{22}$$

For the two cases considered here, we will see that Eqs. (21) and (22) imply that at the least h_{xxxx} will diverge at the pinch point.

In applying these general results to the two cases outlined earlier, we shall assume for simplicity that there is no viscosity contrast between the two fluids: $A_j^\mu = 0$. Aside from being a good approximation to many experimental situations, this has the technical virtue of rendering the integral equations (10) explicit expressions for the vortex sheet strengths.

3.1. THE UNSTABLE LAYER

For the Rayleigh-Taylor instability, consider the case in which the interfacial tensions τ_1 and τ_2 are equal, and the densities are such that

$$\rho_2 = \frac{1}{2}(\rho_1 + \rho_3) . \tag{23}$$

By this choice of physical parameters, we have $S_1 = S_2 \equiv S$ and $R_1 = R_2 \equiv R$. The vortex sheet strengths are then

$$\gamma_j(p) = \partial_p \left(S\kappa_j(p) - R y_j(p) \right) . \tag{24}$$

Note, as is generally the case, that each γ_j is a perfect derivative. This has the consequence that the velocity integrals in Eq. (9) are independent of the choice of parameterization of the curve.

Expanding the vortex sheet strengths in Eq. (24) in powers of ϵ, we find

$$\gamma^{(0)} = 0, \quad \gamma^{(1)} = S H_{xxx} - R H_x . \tag{25}$$

Then, reconstituting h from ϵH, we obtain

$$h_t(x,t) = -\partial_x \left\{ h \left(S h_{xxx} - R h_x \right) \right\} . \tag{26}$$

Absorbing the dimensionless coefficient S by defining a rescaled time $t' = St$ (and immediately dropping the $'$), we have finally the purely local PDE

$$h_t(x,t) = -\partial_x \left\{ h \left(h_{xxx} - \alpha h_x \right) \right\} , \tag{27}$$

where

$$\alpha = \frac{2g\Delta\rho L^2}{\tau} \tag{28}$$

is a Bond number for the system. It measures the relative importance of buoyancy to the restoring force of surface tension. When $\alpha = 0$, we recover the equation studied in earlier work (Constantin, *et al.* 1992). If we rewrite Eq. (27) in the form

$$h_t(x,t) = -h \left(h_{xxxx} + \alpha h_{xx} \right) - \alpha h_x^2 - h_x h_{xxx} , \tag{29}$$

we see a qualtitative resemblance to the Kuramoto-Sivashinsky (KS) equation (Kuramoto & Tsuzuki, 1976; Sivashinsky, 1977),

$$h_t(x,t) = -h_{xxxx} - \nu h_{xx} - \lambda h_x^2 . \tag{30}$$

While the KS equation does not satisfy a variational principle, we shall see below that Eq. (29) does.

At this point, it is instructive to consider the linear stability of a flat interface $h = \bar{h}$ both in the context of the lubrication theory result Eq. (27) and the exact vortex sheet calculation. Now set $L = 2\pi$. Then in the lubrication approximation, the growth rate of a disturbance of wavevector k is

$$\sigma_k^a = -\bar{h} \left(k^4 + \alpha k^2 \right) , \quad (\textit{lubrication theory}) \tag{31}$$

whereas the exact result is

$$\sigma_k^a = -\frac{1}{2} \left(|k|^3 + \alpha |k| \right) \left(1 - e^{-2|k|\bar{h}} \right) . \quad (\textit{exact}) \tag{32}$$

The superscript "a" reminds us that these quantities pertain to antisymmetric perturbations of the interface. Note that Eq. (31) follows from Eq. (32) by expansion in small $k\bar{h}$. Thus, at least as far as linear analysis is concerned, *the lubrication approximation is valid for fluid layers whose modulations are on length scales large compared to their thickness.*

There are of course two independent eigenfunctions found in the linearization of the motion about a flat layer, symmetric and antisymmetric. The symmetric eigenfunction which we have so far ignored corresponds to motion in which the lower interface is the vertical translate of the upper interface. A straightforward calculation of the exact growth rate for symmetric perturbations yields

$$\sigma_k^s = -\frac{1}{2}\left(|k|^3 + \alpha|k|\right)\left(1 + e^{-2|k|\bar{h}}\right) . \quad (exact) \tag{33}$$

This linear analysis indicates that symmetric data is actually more unstable than antisymmetric (i.e. $\sigma_k^s > \sigma_k^a$). A partial justification for ignoring these potentially important modes is that if one were to look for pinching solely on the basis of the linear analysis, for geometrical reasons this would occur most naturally for the antisymmetric linear motion, in which opposing segments of the interfaces are driven toward each other. At present, the lubrication theory is much more easily accomplished for antisymmetric data, but a more general theory is possible using the techniques discussed in (Baker & Shelley, 1991).

The dispersion relation Eq. (31) embodies the familiar result that the fluid is unstably stratified if $\alpha < 0$, i.e. the heavier fluid on top. This is the Rayleigh-Taylor instability of two Hele-Shaw fluids and σ_k shows the competition between this instability and the restoring force of surface tension. The most unstable mode has a wavevector

$$k^* = \left[\frac{-\alpha}{2}\right]^{1/2} , \tag{34}$$

If a perturbation $\zeta(x,t)$ to the flat layer has a sinusoidal form with wavevector k corresponding to a positive growth rate, we may interpret the positive (i.e. upward) lobes of ζ at the upper interface as lighter fluid ascending, and the negative lobes as heavier fluid falling under the influence of gravity. The assumption of antisymmetry of the two interfaces then implies that at the lower interface the upward- and downward-moving domains of fluid are half a period out of phase with those from above. Thus, the falling fluid 3 will collide with the rising fluid 1.

3.2. THE GRAVITY-DRIVEN JET

In this case, we again assume the equality of surface tensions, and set

$$\rho_1 = \rho_3 \ , \quad with \quad \rho_2 \quad arbitrary. \tag{35}$$

The vortex sheet strengths are

$$\gamma_j(p) = \partial_p \left(S_j \kappa_j(p) + R_j p\right) \ , \tag{36}$$

with $S_1 = S_2$ and $R_1 = -R_2$, and the ϵ expansion yields

$$\gamma^{(0)} = R, \quad \gamma^{(1)} = SH_{xxx} \ . \tag{37}$$

Again absorbing the dimensionless coefficient S by defining a rescaled time $t' = St$ and dropping the $'$, the equation of motion for the thickness is

$$h_t(x,t) = -\partial_x \left\{h\left[h_{xxx} + \alpha\left(1 + \mathcal{H}\left[h_x\right]\right)\right]\right\} \ . \tag{38}$$

Now, because $\gamma^{(0)}$ was nonzero, the equation of motion is nonlocal through the presence of the Hilbert transform. This nonlocality is essentially a dispersive contribution to the interface motion, and would be expected to be small if the gradient $h_x << 1$, as is implicitly assumed in the lubrication approximation. In the context of discussions below concerning variational principles, as well as in numerical studies, we shall find it useful to consider the version of the jet dynamics with this term omitted,

$$h_t(x,t) = -\partial_x \left\{h\left[h_{xxx} + \alpha\right]\right\} \ . \tag{39}$$

Once again, it is instructive to examine the linearization of the full lubrication theory result (38) about a jet of mean thickness $\bar{h}$, comparing with the exact analysis from the vortex sheet formulation. We obtain, respectively,

$$\sigma_k^a = -\bar{h}k^4 - i\alpha k\left(1 - \bar{h}|k|\right) , \quad (lubrication\, theory) \tag{40}$$

and

$$\sigma_k = -\frac{1}{2}|k|^3\left(1 - e^{-2|k|\bar{h}}\right) - i\frac{\alpha k}{2}\left(1 + e^{-2|k|\bar{h}}\right) \ . \quad (exact) \tag{41}$$

Thus the density stratification, regardless of the sign of α, leads only to linearly dispersive waves, in addition to the damping given by surface tension. The lubrication approximation is again valid in the limit $|k|\bar{h} << 1$.

3.3. A VARIATIONAL PRINCIPLE

Here we remark on the existence of a variational principle governing the interface evolution equation for the density-stratified problem (27), and for the simplified fully local dynamics (39) of the gravity-driven jet. Both of these dynamics have the variational form

$$h_t(x,t) = \partial_x \left\{h\partial_x \left(\frac{\delta \mathcal{F}}{\delta h}\right)\right\} \ , \tag{42}$$

for some functional $\mathcal{F}[h]$. The form of this gradient flow was observed previously (Constantin, *et al.* 1992) for motion in the absence of gravity. Its presence here suggests that it is much more general. We interpret the quantity

$$U \equiv \partial_x \left(\frac{\delta \mathcal{F}}{\delta h} \right) \tag{43}$$

as the mean velocity in Eq. (19). Its form, the derivative of a functional derivative, is reminiscent of that which pertains to Euler's equation, as seen for instance in the motion of the boundary of a vortex patch (Goldstein & Petrich, 1992), and derives here from Darcy's law relating velocities to gradients of pressure.

A consequence of this variational structure is that the function $\mathcal{F}$ evolves in time as

$$\mathcal{F}_t = -\int_{-\infty}^{\infty} dx h U^2 . \tag{44}$$

$\mathcal{F}$ is monotonically decreasing in time provided the half-width h is a positive quantity. As we have seen, U is a characteristic velocity of the flow, so the form of (44) is in accord with the dissipation expected from Darcy's Law. That is, the balance between the pressure gradient and the viscous dissipation implies that the local viscous force is proportional to the velocity (unlike in the full Navier-Stokes equation in which it is associated with *gradients* of the velocity), so the rate of energy dissipation per unit area of the flow should scale as the square of the velocity. The total energy dissipation rate is thus proportional to an integral of U^2 over the two dimensional region containing the Hele-Shaw fluid; in the lubrication approximation, the velocity field is constant across the layer, so the area integral becomes a one-dimensional integral over the extent of the thin layer, the integrand being hU^2.

The two functionals which generate Eqs. (27) and (39) from Eq. (42) are, respectively,

$$\mathcal{F} = \int_{-\infty}^{\infty} dx \left\{ \frac{1}{2} S h_x^2 + \frac{1}{2} R h^2 \right\} \tag{45}$$

and

$$\mathcal{F} = \int_{-\infty}^{\infty} dx \left\{ \frac{1}{2} S h_x^2 - R h x \right\} \tag{46}$$

We interpret the term $(1/2)h_x^2$ as the small-gradient expansion of the excess arclength over that associated with a flat interface

$$\int_{-\infty}^{\infty} dx \left\{ \sqrt{1 + h_x^2} - 1 \right\} \simeq \int_{-\infty}^{\infty} dx \frac{1}{2} h_x^2 \, . \tag{47}$$

so these terms in the energy functional, appearing multiplied by the surface tension parameter S, have the interpretation as the excess energy associated with a curved interface. The terms proportional to R (and hence to $\Delta\rho$) in the Lyapunov functions (45) and (46) can be interpreted as the gravitational potential energy associated with the presence of the intermediate fluid layer. In the Rayleigh-Taylor problem, the energy of interest is that of a column of fluid of height h, which scales in the familiar way as h^2. However, the energy density of $\mathcal{F}$ is positive-definite only for $R > 0$, the stable situation. For the jet, the interpretation follows from recognizing the product $h(x)dx$ as the differential area of fluid at height x.

At present, it is not clear whether the nonlocal contributions to the equation of motion of the jet can be incorporated into this variational form (say as nonlocal contributions to the density of functional $\mathcal{F}$). If this is possible, a direct physical interpretation of such terms presumably centers around the kinetic energy from the interacting vortex sheets.

4. Numerical Results

In this section, we present some preliminary numerical results on the two lubrication approximations (27) and (39), as well as some simulations of the full Hele-Shaw equations for the first case. The PDE simulations for the Rayleigh-Taylor geometry were performed with pseudo-spectral techniques, while the full Hele-Shaw simulations utilized methods developed in (Baker & Shelley, 1986) and (Shelley, 1992). The jet dynamics were studied with an implicit finite-difference technique adapted from earlier work (Constantin, *et al.* 1992). Details of the numerical methods will appear elsewhere (Goldstein, Pesci, & Shelley, 1992).

4.1. THE UNSTABLE LAYER

We consider first the case of the unstable layer. The upper box of figure 2 shows the evolution of h for Eq. (27) at several times. The initial condition is the simple sinusoidal perturbation on its mean

$$h(x, t = 0) = \bar{h} + a\cos x \quad 0 \le x \le 2\pi \ , \tag{48}$$

where $\bar{h} = 0.1$, $a = 0.05$, and $\alpha = -10$. This initial condition is an unstable eigenfunction with $k = 1$ for the linearized problem. Moreover, we see from Eqs. (31) and (34) that all linear modes with $0 < k < \sqrt{10}$ are unstable, with the fastest growing mode have $k^* = \sqrt{5} \approx 2.2$. The "small" parameter ϵ (the aspect ratio) used in our expansion procedure has the value $\epsilon = \frac{2\bar{h}}{L} = \frac{1}{10\pi} \approx 0.03$. The calculation begins with 64 computational points. When the number of active Fourier modes approaches the Nyquist frequency, the number of points are doubled to 128, and then again to 256.

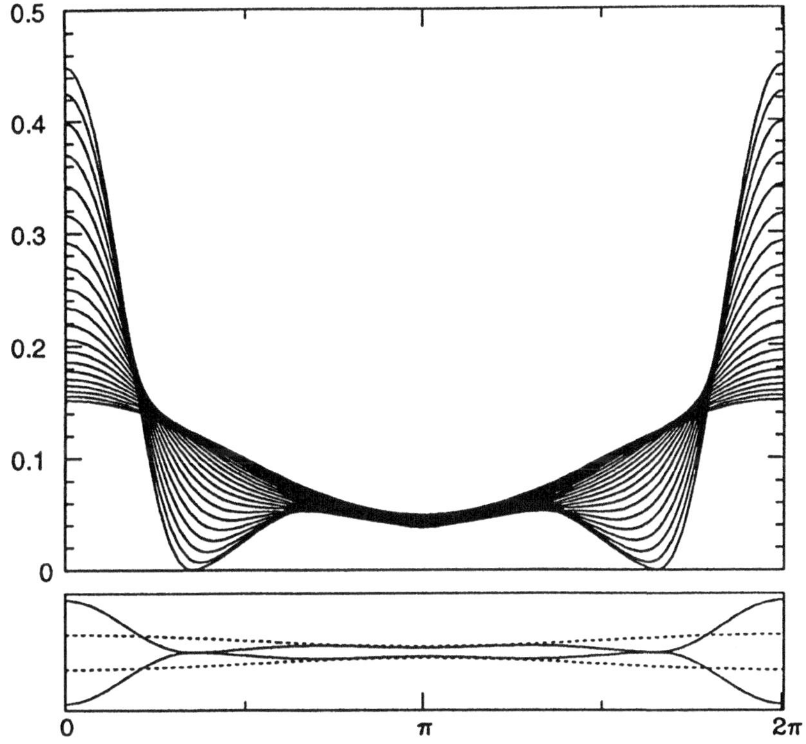

Fig. 2. Pinching of the fluid layer, as modeled by the lubrication approximation, through the Rayleigh-Taylor instability.

where $\bar{h} = 0.1$, $a = 0.05$, and $\alpha = -10$. This initial condition is an unstable eigenfunction with $k = 1$ for the linearized problem. Moreover, we see from Eqs. (31) and (34) that all linear modes with $0 < k < \sqrt{10}$ are unstable, with the fastest growing mode have $k^* = \sqrt{5} \approx 2.2$. The "small" parameter ϵ (the aspect ratio) used in our expansion procedure has the value $\epsilon = \frac{2\bar{h}}{L} = \frac{1}{10\pi} \approx 0.03$. The calculation begins with 64 computational points. When the number of active Fourier modes approaches the Nyquist frequency, the number of points are doubled to 128, and then again to 256.

These simulations indicate that $h(x,t)$ reaches zero in a finite time. The lower box shows to scale the fluid layer initially (dashed) and at the last time shown above (solid). In contrast to experiment (Dance, *et al.* 1992), the two osculating interfaces meet to form a cusp at the pinch point, rather than as a corner. Additionally, the pinch point does not occur at the point of initial minimum thickness ($x = \pi$), as would be indicated naively by the linear analysis, but rather at two symmetric secondary minima which develop dynamically. This is not surprising. For initial conditions supported at a single mode, the generic short-time effect of nonlinearity is to produce an exponential cascade of energy to smaller scales, which includes that of the fastest growing linear mode. This is the scale of the secondary structures seen here. Simulations for which k^* is close to 1 show pinching, but not the obvious development of such secondary structures. Concomitant with the pinching, and perhaps even driving it, is the rapid rise of light fluid spikes into the fluid above (and thus also below).

The upper box of Fig. 3 shows the minimum value of $h(x,t)$ (solid) as a function of time, as well as $h(\pi,t)$ (dashed), the layer half-thickness at the center. Initially, the minimum of h is at $x = \pi$ so that the two curves coincide. However, the local minimum at $x = \pi$ is eventually overtaken by the newly developed minima in the secondary lobes, giving the break observed in the graph. ¿From this figure, it appears that the critical time for pinching is $t_c \approx 0.98$. As implied by Eq. (21), pinching requires a divergence in $U_x = h_{xxxx} + \alpha h_{xx}$ at the point of pinching. The lower box in Fig. 3 shows h_{xx} (solid) and $-h_{xxx}$ (dashed) at point of minimum h as functions of time. It appears from this figure that h_{xx} remains bounded at the critical time (and thus a cusp is formed between the two osculating interfaces), while h_{xxx} diverges. And while not shown, h_{xxxx} also displays a divergence as the critical time is approached. Thus, the singularity is stronger than is required by Eq. (21), and while U_x diverges, is is only through one of its constituents, not both.

Our numerical solution of the lubrication approximation (27) suggests that the bounding interfaces of such a fluid layer can touch in Hele-Shaw flow. But is it a truly accurate diagnostic of what can happen in the un-approximated flow? While there are checks of self-consistency within the lubrication approximation, the advantage of this case is that the full flow can be computed directly, and thus compared.

Figure 4 shows some preliminary simulations of the full Hele-Shaw flow for the same physical parameters and initial condition as above. While these calculations are not (yet) as accurate as those for Eq. (27), the results are very encouraging. To the eye, figure 2 and figure 4 are nearly identical. As before, the upper box of Fig. 5 shows the minimum value of $h(x,t)$ (solid) as a function of time, as well as $h(\pi,t)$ (dashed). Again, the behavior is remarkably similar to that predicted by the lubrication approximation, the only notable difference being a delayed critical time, now about $t_c \approx 1.21$. However, one need only compare Eqs. (31) and (32) to see that the growth rate of a perturbation within the lubrication theory is always larger than that for the full system .

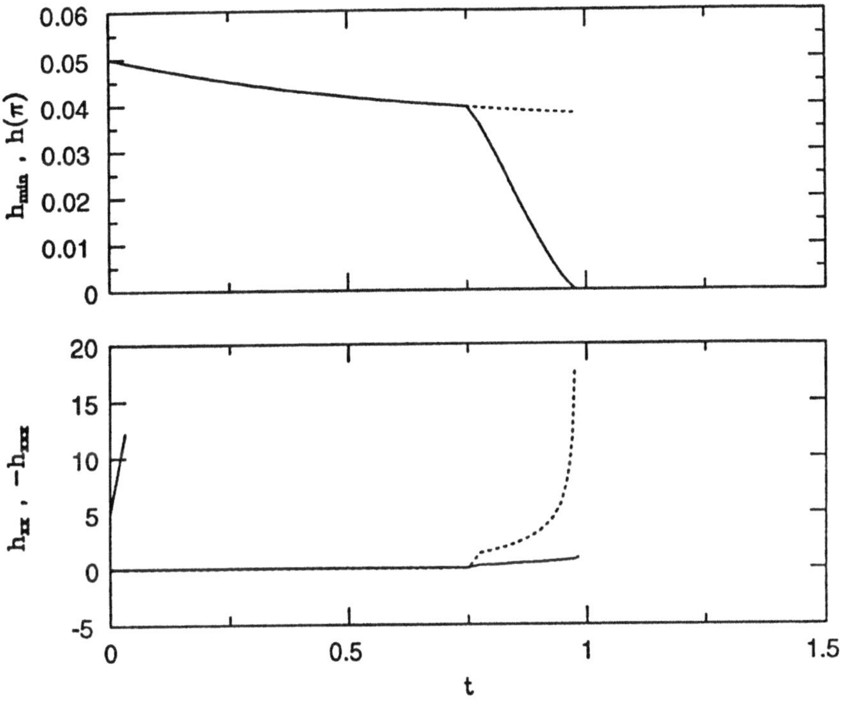

Fig. 3. The upper box shows $h_{min}(t)$ and $h(\pi,t)$ from the lubrication approximation. The lower box shows h_{xx} and h_{xxx} at the point of least thickness.

These simulations of the full system use only 129 computational points, and it is clear to us that this is insufficient. One of the obstacles to increasing resolution is the stiffness of the system due to surface tension. We are currently implementing a new numerical method (Hou, Lowengrub & Shelley, 1992) which will allow us to use much higher resolutions without a strong stability constraint.

4.2. THE GRAVITY-DRIVEN JET

Here we focus on the local approximation to the jet dynamics, Eq. (39). While it is possible to absorb the advective term $-\alpha h_x$ into a redefinition of the parameter x (a Galilean transformation), it is important to recognize the role played by boundary conditions in such a transformation. Consider for instance the situation in which the jet emanates from an aperture with fixed width $2\bar{h}$ at $x = 0$, and thus forms the neck of a growing drop. In this case a transformation to a moving frame results in boundary conditions applied at a moving point, so there is no overall simplification to the dynamics.

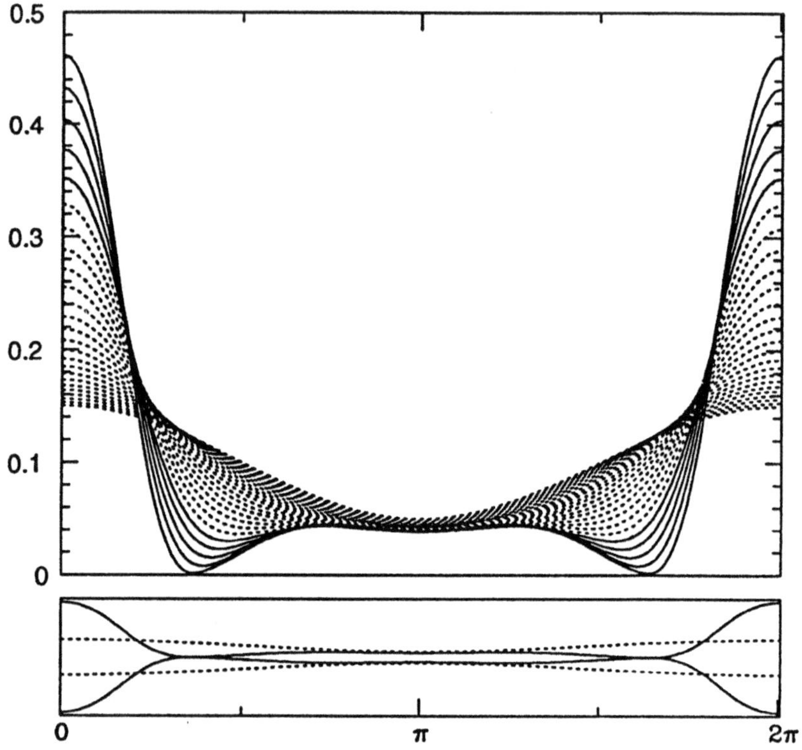

Fig. 4. Pinching of the fluid layer, from full simulations of the Hele-Shaw flow, through the Rayleigh-Taylor instability.

On the other hand, were there no boundary conditions except at infinity (or periodic boundary conditions on a fixed interval), such a frame change would simplify the dynamics.

Following previous work, we are led to consider boundary conditions which fix the width h at the two ends of the physical interval, and which satisfy the Gibbs-Thompson condition consistent with an externally applied pressure P:

$$h(0,t) = \bar{h}, \quad h_{xx}(0,t) = P \ , \tag{49}$$

and

$$h(L,t) = \bar{h}, \quad h_{xx}(L,t) = P + \alpha L \ . \tag{50}$$

where the additional term αL accounts for the increase in pressure due to the greater depth in the fluid. The as yet unspecified location L of the bottom

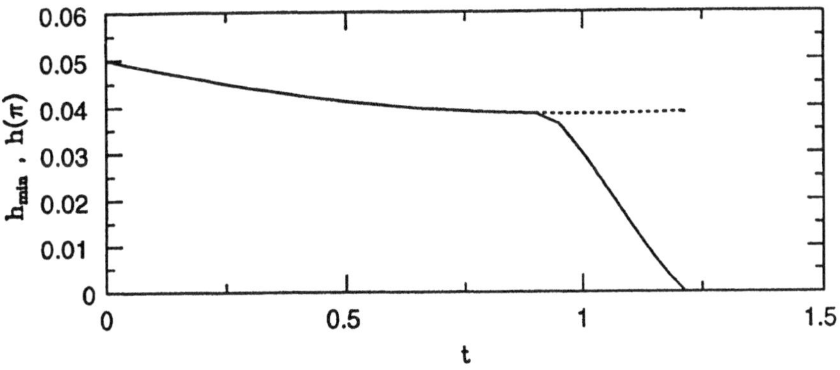

Fig. 5. $h_{min}(t)$ and $h(\pi, t)$ from the simulations of Hele-Shaw flow.

boundary is now determined by a simple physical argument. Consider a layer which exists on the interval $x \in [0, L(t)]$, and let $\mathcal{M}$ be the area of fluid bounded by the function $h(x,t)$,

$$M(L,t) = \int_0^{L(t)} dx h(x,t). \tag{51}$$

Using the equation of motion for h, Eq. (39), the rate of change of this enclosed area is

$$\frac{dM}{dt} = h(0,t)\,(h_{xxx}(0,t) + \alpha) + h(L,t)\,[L_t - (h_{xxx}(L,t) + \alpha)]\,. \tag{52}$$

The first terms, proportional to $h(0,t)$, represent the contribution from flow into the jet at the top, the second and third, proportional to $h(L,t)$ from flow out the bottom. We choose to move the endpoint L such that the entire contribution to dM/dt is from inward flow at the top;

$$L_t = h_{xxx}(L,t) + \alpha. \tag{53}$$

It is convenient for numerical work to introduce a rescaled coordinate $\xi = x/L(t)$. The dynamics of h then transforms as $h_t \to h_t + h_\xi \xi_t$, with $\xi_t = -(\xi/L)L_t$. Combining this with the evolution equation for L yields the rescaled equation of motion,

$$h_t(\xi,t) = -\frac{1}{L^4}\partial_\xi\,(h h_{\xi\xi\xi}) - h_\xi\left(\alpha\frac{(1-\xi)}{L} - \frac{\xi}{L^4}h_{\xi\xi\xi}(1,t)\right)\,, \tag{54}$$

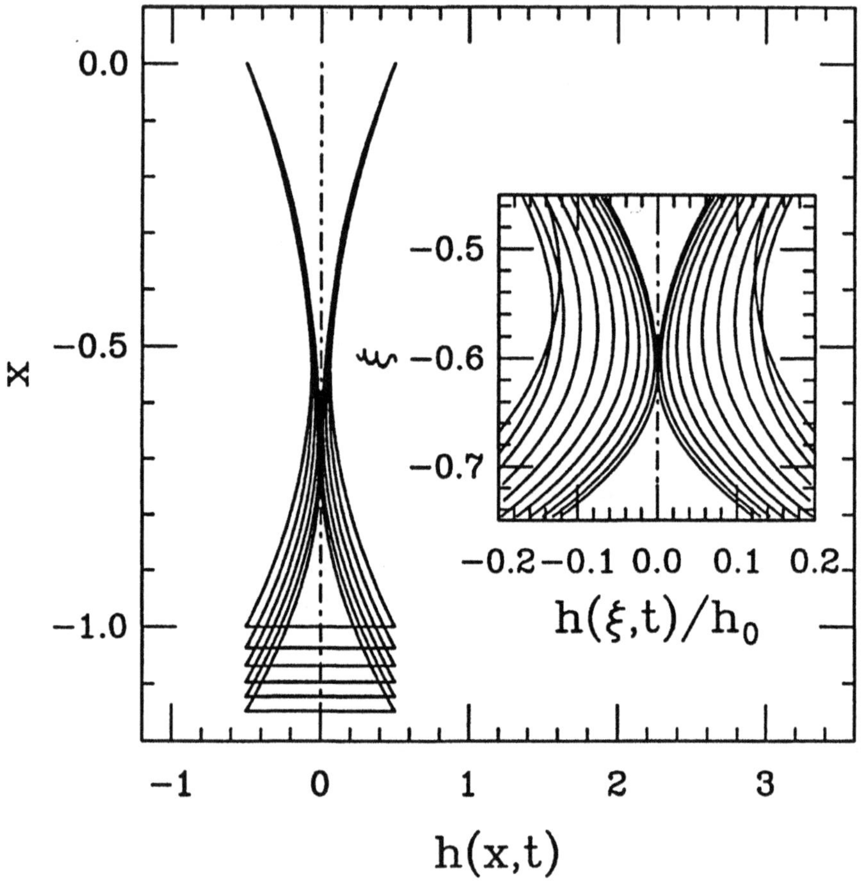

Fig. 6. The pinching of a gravity-driven jet. Inset shows magnified view of the approach to osculation.

for $\xi \in [0,1]$, with the length parameter evolving as

$$L_t = \frac{1}{L^3} h_{\xi\xi\xi}(1,t) + \alpha \; . \tag{55}$$

The rescaled boundary conditions are then $h(0,t) = h(1,t) = \bar{h}$ and

$$h_{\xi\xi}(0,t) = PL(t)^2, \quad h_{\xi\xi}(1,t) = PL(t)^2 + \alpha L(t)^3 \; . \tag{56}$$

Figure 6 shows the shape of the jet, both in its entirety with the moving bottom boundary evident, and also in an enlarged version of the region near the osculation of the two interfaces, in the rescaled variable ξ. The parameters are $P = 2.5$, $h_0 = 0.5$, $\alpha = 2.0$, with 200 grid points in the

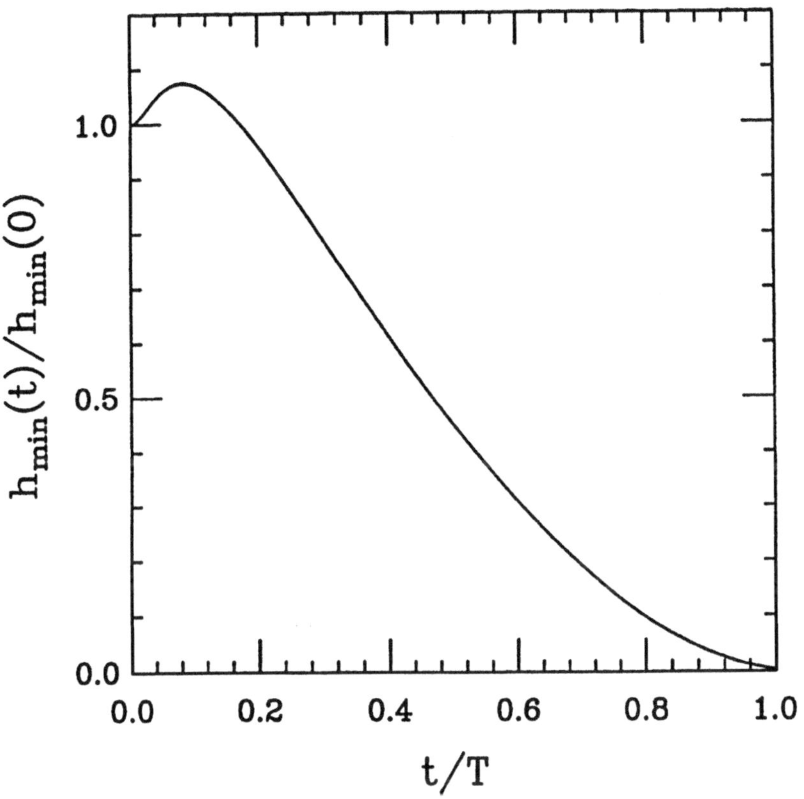

Fig. 7. The minimum width of the fluid layer as a function of the dimensionless time variable t/T, where T is the time to the singularity.

discretization. The finite-time vanishing of the minimum value of $h(\xi, t)$ is shown in Figure 7. Figure 8 shows the third derivative at equally-spaced time intervals, and, in the inset, the growth of the maximum of h_{xxx}. The case for the existence of a finite-time singularity is very strong.

Preliminary numerical simulations of the nonlocal PDE for the jet using periodic boundary conditions reveal no pinching, even in infinite time. Indeed, the solution ultimately relaxes to the flat state. Conversely, in a previous study of gravity-free pinching (Constantin, *et al.* 1992), the boundary data is chosen there to strongly bow the interface, and for the class of *initial data* studied, pinching is observed in infinite time. This suggests that for this problem the boundary conditions are crucial in producing the finite-time singularity. This makes comparison of the model PDE with full numerical simulations of the Hele-Shaw equations difficult.

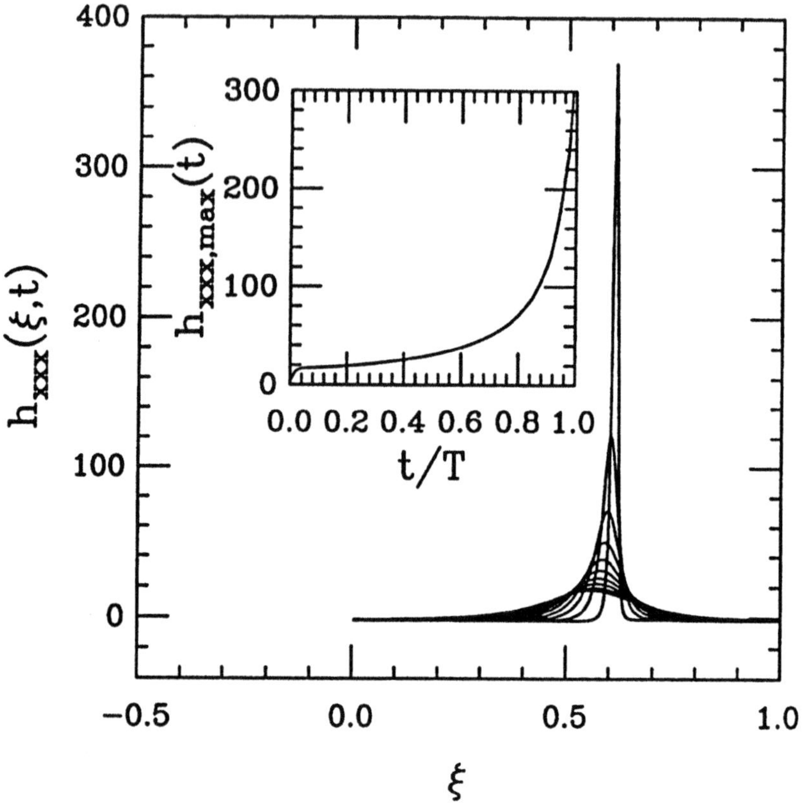

Fig. 8. The quantity h_{xxx} as a function of the rescaled coordinate ξ, at various times. Inset shows the growth of the maximum value of h_{xxx}.

5. Conclusions

Building upon an earlier study (Constantin, *et al.* 1992), we have illustrated through interface dynamics in the Hele-Shaw cell, one theoretical approach to the understanding of topological transitions in fluid systems. In two distinct contexts, unstably-stratified fluid layers and gravity-driven fluid jets, the finite-time osculation of bounding interfaces emerges readily from the PDEs obtained by the lubrication approximation. Naturally associated with this event are singularities at the osculation point in both the shape of the interfaces and in the velocity field. Preliminary simulations of the full Hele-Shaw flow for the unstably stratified layer show good agreement with predictions of the lubrication theory.

As we remarked earlier, the appearance of cusps at the point of contact differs from the experimental observations on gravity-driven droplet fission

(Dance, *et al.* 1992). The experiments indicate that the form of the interface pinch, immediately before the the reconnection of the interfaces, is in the shape of an $\times$. Thus, the experimental evidence appears to imply a diverging interfacial curvature at the topological transition, whereas from our lubrication theory it is only the *derivative* of the curvature which diverges for the cusp. The experiments also show that the collapse accelerates significantly as the interfaces come together. This suggests that inertial effects have become important. We speculate that osculation point shape and and the lack or presence of inertial effects are closely connected; this is an area of current interest. In the absence of gravity, experiments (Goldstein, Mason, & Shyamsunder, 1991) on topological transitions driven by applied pressure across a thin neck of fluid *do* show shapes at the pinch point much like those predicted by the lubrication approximation in the absence of gravity. We are not aware of any experimental studies of topological transitions resulting from the Rayleigh-Taylor instability considered here. Experiments on this phenomenon should prove most illuminating.

The pinching of the unstably-stratified layer may be related to singular behavior postulated to occur in porous media convection (Pumir, Shraiman, & Siggia, 1992). Numerical studies of unstable, continuously stratified porous medium suggest that gradients of the density field diverge at a finite time. This divergence pinches together contours of the density field, which are also material surfaces. The unstably stratified layer is the analog to porous media convection where the density is taken to be piece-wise constant.

Finally, the development of some understanding of the analytic form of the finite-time singularities found here remains a largely open problem, although there has been some recent progress in this area (Zhou, 1992).

Acknowledgements

We are indebted both to our experimental colleagues M. Dance, T. Mason and E. Shyamsunder and our theoretical colleagues P. Constantin, T. Dupont, L.P. Kadanoff, and S. Zhou for ongoing collaborations. REG was supported in part by the National Science Foundation, through Grant CHE-9106240. MJS was partially supported by the NSF through grants DMS-915792 and DMS-9100383, and by the Department of the Air Force through grant F-49620-92-J-0023F. Finally, it is a pleasure to thank the organizers of this workshop for a most interesting conference!

References

Baker, G. and Shelley, M. 1986 Boundary integral techniques for multi-connected domains. *J. Comp. Phys.* **64**, 112.

Baker, G., and Shelley, M. 1991 On the relation between thin vortex layers and vortex sheets. *J. Fluid Mech.* **215**, 161.

Carrier, G., Krook, M. & Pearson, C. 1966 *Functions of a Complex Variable*, New York: Mcgraw-Hill.

Constantin, P., Dupont, T., Goldstein, R., Kadanoff, L., Shelley, M. & Zhou, S. 1992 Droplet Separation in the Hele-Shaw Cell? submitted to *Physical Review A*.

Dai, W., and Shelley, M. 1992 A numerical study of the effect of surface tension and noise on an expanding Hele-Shaw bubble. submitted to *Physics of Fluids A*.

Dance, M., Goldstein, R.E., Kachru, S. & Shyamsunder, E. 1992. unpublished.

Goldstein, R.E., Mason, T. & Shyamsunder, E. 1991. unpublished.

Goldstein, R.E., Pesci, A.I. & Shelley, M.J. 1992 Interface motion and topological transitions in Hele-Shaw flow. preprint.

Goldstein, R.E. and Petrich, D. 1992 Solitons, Euler's equation, and vortex patch dynamics. *Phys. Rev. Lett.* **69**, 555.

Hou, T, Lowengrub, J. & Shelley, M.J. 1992. In preparation.

Kellogg, D. *Foundations of Potential Theory*, New York: Dover, 1929.

Kuramoto, Y. and Tsuzuki, T. 1976 Persistent propagation of concentration waves in dissipative media far from thermal equilibrium. *Prog. Theor. Phys.* **55**, 356.

Shelley, M. 1992 A study of singularity formation in vortex sheet motion by a spectrally accurate vortex method. *J. Fluid Mech.*, in press.

Pumir, A., Shraiman, B. & Siggia, E. 1992 Vortex morphology and Kelvin's theorem. *Phys. Rev. A* **45**, 5351.

Sivashinsky, G.I. 1977 Non-linear analysis of hydrodynamic instability in laminar flames. 1. Derivation of basic equations. *Acta Astronaut* **4**, 1177.

Tryggvason, G. and Aref, H. 1983 Numerical experiments on Hele-Shaw flow with a sharp interface. *J. Fluid Mech.* **136**, 1.

Zhou, S. 1992 . Ph.D. thesis, University of Chicago, unpublished.

SEPARATRICES AND SINGULARITIES

A Bhattacharjee
Department of Applied Physics Columbia University New York New York 10027 USA

ABSTRACT. Singularities tend to occur near separatrices. Two examples are described in support of this assertion. The first example describes current sheets in coronal plasmas that grow algebraically in time. The second example describes a vortex singularity which develops in finite time in three-dimensional Euler flows.

1. Introduction

The formation of singularities from smooth initial conditions is a phenomenon of great interest for plasmas and fluids. The main theme of this paper is that singularities tend to form near separatrices. I shall illustrate this theme by two examples from plasma physics and fluid dynamics.

My own understanding of separatrices comes mostly from studies of magnetic reconnection in toroidal plasmas with nested magnetic surfaces. (A magnetic surface is a surface on which field-lines lie.) In such systems it is quite easy to identify the separatrices. These are the so-called rational surfaces on which field lines close on themselves. The possible formation of singularities at these separtrices has been the subject of considerable research in plasma physics. Since it is known where singularities should appear, the main focus of this research has been on the structure and topology of the singularities [*Grad*, 1967, *Rosenbluth et al.*, 1973, *Moffatt*, 1985, *Cary and Kotschenreuther*, 1985, *Hegna and Bhattacharjee*, 1991].

If we consider three-dimensional magnetic fields which are not necessarily confined to a toroidal domain, it is not obvious where the separatrices are. In a pioneering paper which emphasizes geometry over dynamics, Greene [1988] has elucidated the role of magnetic nulls in three-dimensional fields. (Magnetic nulls are points were the magnetic field vanishes identically.) He has shown that in the presence of nulls, the geometrical structure of magnetic fields allows us to identify separatrices where magnetic reconnection and singular currents will tend to occur. Greene's ideas have been further devel-

R. E. Caflisch and G. C. Papanicolaou (eds.), Singularities in Fluids, Plasmas and Optics, 189–195.
© 1993 *Kluwer Academic Publishers.*

oped by Lau and Finn [1990] who have explored the relationship between two-dimensional reconnection and three dimensional reconnection, both with and without nulls.

The framework devoloped by Greene and Lau and Finn provides a very useful point of departure for the study of singularity formation in plasmas as well as fluids. What is valuable about this framework is that it addresses the question of *where* singularities should form if the dynamics permit. These singularities may occur in finite-time or infinite time. Whereas geometrical considerations can suggest the possible location of singular structures, only dynamics can determine how rapidly in time such structures grow.

In this paper I shall discuss two recent developments in the subject of singularity formation, one in plasmas and the other in fluids. The first example deals with current singularities in coronal magnetic fields, the second deals with the formation of vortex singularities in three-dimensional Euler flows. In both cases we show that singularities tend to develop at separatrices. However, the time of singularity formation is quite different in the two cases. In the coronal plasma model, the current singularity tends to develop algebraically in time. In the model of three-dimensional Euler flows, the vortex singularity occurs in finite time.

2. Current Singularities in the Solar Corona

The study of current singularities in the solar corona is motivated by the search for a mechanism to heat the coronal plasma. The resistivity of this plasma is so very low that is is not possible to generate much Ohmic dissipation out of a smooth current distribution. In 1972, Parker proposed a model for the solar corona in which current singularities (also called current sheets) are formed as a result of the random motion of photospheric footpoints. In Parker's 1972 model, the initial state is assumed to be a straight uniform magnetic field

$$\mathbf{B} = B_0 \hat{\mathbf{z}}, \tag{1}$$

confined between two horizonal planes at $z = \pm L$. This is a simplified model for a coronal loop in which the effects of geometrical curvature have been neglected. The footpoints of the magnetic field are anchored at $z = \pm L$ where they move about deforming the vacuum field (1) to a state prescribed by

$$\mathbf{B} = B_0 \hat{\mathbf{z}} + \mathbf{b} = B_0 \hat{\mathbf{z}} + \nabla\chi \times \hat{\mathbf{z}}, \tag{2}$$

where χ is the flux function. Unlike (1), (2) represents a magnetic field which carries a current $\mathbf{J} = c\nabla \times \mathbf{B}/4\pi \neq 0$. The main question now is whether there are final states, accessible from the initial state, in which the current resides in sheets.

For long-thin coronal loops, Wang and I have recently given conditions under which current sheets form [*Bhattacharjee and Wang*, 1991]. The long-thin approximation is appropriate for coronal loops which typically have a transverse dimension $a \sim 10^8\,cm$ and a length $L \sim 10^{10}\,cm$. This yields an aspect ratio of $L/a \sim 100$. The point of view we have adopted is that if the z-dependance of **B** is weak, then the dynamics of such a long-thin tube is not significantly different from an infinitely long plasma with $B_z \neq 0$. The reduction of the effective dimensionality of the problem from three to two allows us to track the separatrices easily. It is clear that near the edges $z = \pm L$, there are boundary-layers. However, these boundary-layers do not interfere in an essential way with the formation of internal separatrices.

As discussed in a paper by Wang and myself [1992], the timescale of the photospheric footpoint motion determines the qualitative nature of the dynamics. The twisting motion of the footpoints occurs on a characteristic timescale $\tau_0 \sim a/v_0$, where v_0 is a typical twisting velocity. The magnetic field between the two planes respond to the twisting motion on the boundaries on the Alfvén timescale $\tau_A = v_A/L$ where $v_A = B_0/\sqrt{4\pi\rho}$ is the Alfvén speed and ρ is the mass density of the plasma. Compared with τ_A, the timescale of resistive diffusion, $\tau_R = 4\pi a^2/\eta c^2$, where η is the plasma resistivity, is very slow in the solar corona. We assume that $\tau_A \ll \tau_0 \ll \tau_R$. The inequality $\tau_A \ll \tau_0$ introduces a major simplification in that the plasma can then be assumed to relax through a sequence of quasi-equilibrium states. This assumption is reasonably good for most regions of the plasma except for a narrow region near the separatrix where the effects of inertia and resistivity are extremely important. The demonstration of current sheet formation then involves two levels of calculation. The first level involves the exterior region (far away from the separatrix) where we calculate neighboring equilibrium states. The second involves the inner region near the separatrix (where the effects of inertia and resistivity are retained) and the asymptotic matching of the inner and outer region solutions. A solution with a current sheet is one in which the magnetic field in the inner region exhibits a discontinuity when extrapolated to the separatrix. However, this discontinuity will be resolved in reality by even a small but finite resistivity which makes its presence felt in the inner region.

The separatrix in the configuration (2) lies where $\nabla\chi = 0$. Even though the initial state is smooth it may, under certain circumstances, evolve into a neighbouring equilibrium with a current sheet. This may occur either as the result of an instability ("free reconnection") or by means of boundary perturbations in a stable equilibrium ("forced reconnection"). It turns out that this current sheet solution may be described by an integral equation first discovered by Rosenbluth *et al.* [1973] and subsequently discussed by others [*Waelbroeck*, 1989, *Bhattacharjee and Wang*, 1991].

We now discuss how rapidly in time the current sheet grows. If there is an instability, then the sheet amplitude grows exponentially for a short time. However, nonlinear effects slow down the instability from an exponential to an algebraic growth in time [*Waelbroeck*, 1989]. In the absence of an instability, the perturbation induced at the boundary can also cause a current sheet to grow at the the separatrix. However, in this case the sheet grows out of the stable magnetohydrodynamic continuum at an algebraic rate even in the linear regime. An intial value calculation described elsewhere [*Hahm and Kulsurd*, 1985, *Wang and Bhattacharjee*, 1992] gives the amplitude of the current at the seperatrix

$$J \simeq \frac{c}{2\pi^2} \frac{a}{L} \frac{k_0^2 B_0 \xi_0}{\sinh(k_0 a)} \left(\frac{t}{\tau_A}\right), \tag{3}$$

which is bounded in a region of width

$$\Delta \simeq \frac{\pi}{2k_0} \frac{\tau_A}{t}. \tag{4}$$

Here ξ_0 is the amplitude of the boundary perturbation and $2\pi/k_0$ is its wavelength. This, then, is no finite time singularity, for the amplitude increases to infinity as the width shrinks to zero in infinite time. Resistivity intervenes before the singularity can be realized, and the ideal phase is followed by a reconnection phase in which the magnetic field lines reconnect to form islands at the seperatrix. It can be shown [*Wang and Bhattacharjee*, 1992] that in the nonlinear regime the amplitude of the current sheet reaches a maximum value of

$$J \simeq \frac{\sqrt{2}c}{4\pi} \frac{B_0}{L} \left(\frac{\tau_R}{\tau_A}\right)^{1/2} \left[\frac{k_0 \xi_0}{\sinh(k_0 a)}\right], \tag{5}$$

after which is diffuses away on the time scale of resistive diffusion. Estimates show that such a current sheet is singular enough to cause intense heating and the numbers look reasonable for "quiet" loops, but not quite enough for "active" loops.

3. Vortex Singularities in Three-Dimensional Euler Flows

The example of current sheet formation considered in Section 2 occurs in a quasi-toroidal domain. We have taken the point of view that a long-thin cylindrical loop of extremely large aspect ratio is essentially the same as an infinite cylinder which, in turn, is equivalent to a periodic cylinder with identified ends. The identification of separatrices in general three-dimensional configurations which are not neccessarily toroidal is a problem of greater subtlety. It seems that this is one of the difficulties posed by three-dimensional Euler flows; it is not obvious where to look for singularities. One way out of this difficulty is to follow Greene's ideas on three-dimensional separatrices involving nulls [*Greene*, 1988]. I now report a recent result by Wang and myself on the development of a finite-time singularity in a highly symmetric Euler flow involving nulls [*Bhattacharjee and Wang*, 1992].

In our model, the singularity occurs at a velocity field null (stagnation point) which lies on the line joining two vorticity field nulls. The singularity is realized in finite time as the two vorticity nulls approach each other and the velocity field null. (Dr. M. Brachet has asked me about the physical significance of a velocity null which can be transformed away by a Galilean transformation. My answer to this question is that a Galilean transformation can, of course, eliminate a velocity null but not a vorticity null. The vortex singularity in our model is seen in a reference frame in which the two vortex nulls approach each other.)

I now describe our initial conditions. We take an initial background flow of the form $v_x(\mathbf{x}, t=0) = F(y+z)$, $v_y(\mathbf{x}, t=0) = F(z+x)$, and $v_z(\mathbf{x}, t=0) = F(x+y)$, where F is a smooth function. Hence the condition $\nabla \cdot \mathbf{v} = 0$ is satisfied identically. We assume that $\mathbf{v}$ has a null (stagnation point) at the origin $x = y = z = 0$. Then near the null, $\mathbf{v} = \mathbf{x} \cdot (\nabla \mathbf{v})_0$. The tensor $(\nabla \mathbf{v})_0$, which has zero trace for incompressible flows, has two degenerative eigenvalues $\lambda_- = -F'(0)$, and a third eigenvalue $\lambda_+ = 2F'(0)$. We require that $F'(0) \equiv v_0/L$ be positive; here v_0 and L are positive constants. Then two of the eigenvalues λ_- are negative and the third eigenvalue λ_+ is positive. This null is type A. The two streamlines originating from the null along the eigenvector for λ_+ define the so-called γ_A-line which is a one dimensional unstable manifold. In the vicinity of the null, the eigenvectors for the the two negative eigenvalues λ_- lie on a two-dimensional plane, called the Σ_A-surface, which is a two-dimensional stable manifold. Near the null, the background flow $\mathbf{v}$ is irrotational.

Superimposed on the initial background flow, we have a smooth incompressible localized flow of the form $u_x(\mathbf{x}, t=0) = f(y)$, $u_y(\mathbf{x}, t=0) = f(z)$, $u_z(\mathbf{x}, t=0) = f(x)$ with $f(0) = 0$. The flow $\mathbf{u}$ is also has a null at the origin where $\mathbf{u} = \mathbf{x} \cdot (\nabla \mathbf{u})_0$. The tensor $(\nabla \mathbf{u})_0$ has the eigenvalues $\lambda_0 = f'(0)$ and $\lambda_\pm = f'(0)\exp(\pm i2\pi/3)$. We choose $f(x) = u_0 \exp(-x^2/a^2)$, where u_0 and a are positive parameters and $a \ll L$.This choice is sufficient to localize the flow $\mathbf{u}$ near the origin. Since $f'(0) = u_0/a > 0$, the null is of type A_s. The subscript s denotes the spiraling trajectories of the streamlines into the null in the two-dimensional surface Σ_{A_s} containing eigenvectors for the complex eigenvalues $\lambda_\pm$. This two-dimensional surface is the stable manifold for the null of $\mathbf{u}$, whereas the γ_{A_s}-line is the unstable manifold associated with the eigenvector for the positive eigenvalue λ_0. The global geometry of a type A_s null is similar to that of a type A null. The only difference between the two types of nulls is the spiraling structure near the null in A_s.

The localized flow $\mathbf{u}$ (as well as the background flow $\mathbf{v}$) obey the symmetry relations $u_x(x,y,z) = u_y(z,x,y) = u_z(y,z,x)$. An advantage of this high degree of symmetry is that the knowledge of a single component of the velocity is sufficient to describe the two other components.

Unlike the background flow **v** which is irrotational near the origin, the localized flow **u** produces a vorticity field $\omega_x = -f'(z)$, $\omega_y = -f'(x)$, $\omega_z = -f'(y)$. Since $f'(x) = (u_0/a)(1 - 2x^2/a^2)\exp(-x^2/a^2)$, it follows that the vorticity has two nulls at $x = y = z = \pm a/\sqrt{2} \equiv a_\pm$. It is easy to see that the null $x = y = z = a_+$ is of type A_s, whereas the null $x = y = z = a_-$ is of type B_s. A null of type B_s is characterized by one real, negative eigenvalue and two complex eigenvalues. The two vortex lines originating from the B_s-null along the eigenvector for the real, negative eigenvalue form the γ_{B_s} curve which is a stable manifold. The two-dimensional surface Σ_{B_s} which contains the eigenvectors for the two complex eigenvalues in the vicinity of the B_s-null is an unstable manifold. Vortex lines for a B_s-null can be obtained by simply by reversing the directions of the vortex lines for an A_s-null. The straight line connecting the two vorticity nulls in our intial conditions intersects the origin which is a null for the total velocity field.

The time-evolution of the intial state described above can be obtained by asymptotic analysis of the time-dependent Euler equation. The asymptotic analysis is based on the scale-separation between the background and local flows in the intial state which provide a small parameter a/L. Details are presented elsewhere [*Bhattacharjee and Wang*, 1992]. The result of this analysis is the prediction of a finite-time vortex singularity, with $\omega_x \sim \omega_y \sim \omega_z \sim (t_c - t)^{-1}$. The singularity in vorticity occurs at a three-dimensional X-point. Those who are familiar with the history of magnetic reconnection will be reminded here of the Petscheck model which predicts very fast reconnection in the presence of an X-point [see, for instance, *Parker*, 1979].

Acknowledgements

The results reported in this paper are the product of my collaboration with Dr. Xiaogang Wang, formerly my student and presently a postdoctoral colleague, from whom I have learned much. I am grateful to Professor R. Caflisch for inviting me to this workshop and for providing finanical support through the auspices of NATO.

References

Bhattacharjee, A., and Wang, X., Current sheet formation and rapid reconnection in the solar corona, *Ap. J*, **372**, 321-328, 1990.

Bhattacharjee, A., and Wang X., Finite time vortex singularity in a model of three-dimensional Euler flows, *Phys. Rev. Lett.*, **69**, 2196-2199, 1992.

Grad, H., Toroidal containment of a plasma, *Phys. Fluids*, **10**, 137-154, 1967.

Greene, J. M., Geometrical properties of three-dimensional reconnecting magnetic fields with nulls, *J. Geophys. Res.*, **93**, 8583-8590, 1988.

Hegna, C. C., and Bhattacharjee, A., Islands in three-dimensional steady flows, *J. Fluid Mech.*, **227**, 527-542, 1991.

Lau, Y. T., and Finn, J. M., Three-dimensional kinematic reconnection in the presence of field nulls and closed field lines, *Ap. J.*, **350**, 672-691, 1990.

Moffatt, H. K., Magnetostatic equillibrium and analogous Euler flows of arbitrarily complex topology: Part 1, Fundementals, *J. Fluid Mech.*, **159**, 359-378, 1985.

Parker, E. N.,Topological dissipation and the small-scale fields in turbulent gases, *Ap. J.*, **174**, 499-510, 1972.

Parker, E. N., Cosmic magnetic fields, Clarendon Press, Oxford, 392-439, 1989.

Rosenbluth, M. N., Dagazian, R. Y., and Rutherford, P. H., Nonlinear properties of the internal $m = 1$ instability in the cylindrical tokamak, *Phys. Fluids*, **16**, 1894-1902, 1973.

Waelbroeck, F., Current sheet formation and nonlinear growth of the $m = 1$ kink-tearing mode, *Phys. Fluids B*, 1, 2372-2380, 1989.

Wang, X., and Bhattacharjee, A., Forced reconnection current sheets and coronal heating, *Ap. J*, to appear.

NONLINEAR DYNAMICS OF ULTRA-COLD GAS

V.Flambaum* and E.Kuznetsov**
* - Institute for Nuclear Physics, 630090 Novosibirsk, Russia.
** - Institute of Automation and Electrometry, 630090 Novosibirsk, Russia.

January 8, 1993

1. In recent years it has become possible to cool and to trap neutral atoms using the resonant laser radiation (see,for example, [1] and references therein). The main idea of cooling and trapping is connected with the dependence of radiation pressure taking into account the resonant absorption and photon radiation by atoms (see for example, [2]). The temperature of the cooling of neutral atoms (use H and hydrogen-type atoms Na, Cs) has been reached up to the fraction of mK for relatively large samples of atoms and for minutes in a time that provided by engaging of atoms by magnetic trap with zero field at the center of the system. The typical density n was about $10^{11} - -10^{12}\mathrm{cm}^{-3}$. For such densities the quantum effects become essential close to the Bose condensation temperature

$$T_B = \frac{3.31\hbar^2 n^{2/3}}{g^{2/3}m}$$

and below. For example, this temperature for H has the order of 1 mK for such density that is more or close to the achieved cooling temperature. For heavy atoms the reached temperature until yet is far from the corresponding T_B but there exist some ideas how to get lower temperature [3]. Therefore one can hope that in the nearest years it will be get at the experiment the unique quantum gas.

The purpose of this paper is to treat the behavior of such gas for the temperature less than the condensation temperature when almost all particles are

R. E. Caflisch and G. C. Papanicolaou (eds.), Singularities in Fluids, Plasmas and Optics, 197–203.
© 1993 *Kluwer Academic Publishers.*

in the condensate state (for simplicity we restrict ourselves by $T = 0$) when the radiation is absent. It should be noted that resonant electromagnetic wave absorption and forthcoming reemission lead to the repulsing between atoms of the Coulomb type [2]. Without radiation the interaction between atoms in gas phase for long distance more than Bohr radius a is defined by the Van der Waals attraction forces, $U = -\gamma/r^6$. For the distance comparable with a the main interaction is the exchange one depending on the total spin of electrons of scattering atoms. We, nevertheless, shall not take into account the spin dependence in the scattering. Then in the gas approximation following to [4] for oscillations of the condensate one can get the standard Hamiltonian for the nonlinear Schrödinger equation (NLSE)

$$H = \int \left[\frac{\hbar^2}{2m} |\nabla\psi|^2 + \frac{G}{2}|\psi|^4 + \frac{\beta}{3}|\psi|^6 \right] dr. \tag{1}$$

Here ψ is the wave function of the condensate, so that $n = |\psi|^2$ is the gas density , the constant G is defined over pair potential U : $G = \int U\, dr$. This constant is connected with the scattering length $\alpha = (m/2\pi\hbar)G$. The character of the interaction depends significantly on the α sign. For example, for hydrogen α =1 (in Bohr radiuses), for Cs α=100, but signs of α are unknown. Therefore we consider both cases. In the repulsion case we neglect the next term of expansion over n proportional to β. The main attention will be paid to the attraction because in opposite case there exists a lot of papers about, starting from the Bogolyubov famous paper.

2. Instability and Collapse.

We start our consideration from the attraction case. The equation corresponding to H is well-known NLSE, which we write in dimensionless variables for $G < 0$:

$$i\psi_t = -1/2\Delta\psi - f(|\psi|^2)\psi \tag{2}$$

where $f(|\psi|^2) = |\psi|^2 - \beta|\psi|^4$. We shall assume that initial state of the gas is homogeneous with constant density n so that $\beta|\psi|^2 << 1$. Because the theory of the NLSE is well known for this case [5]-[10] we present here only the main facts from this theory paying the attention to the application to this concrete physical situation.

First of all remind that such state will be unstable with growth rate

$$\Gamma = k(nf' - k^2/4)^{1/2}. \tag{3}$$

The threshold of this instability is defined from $f' = 0$. It gives $n = 1/2\beta$, that can be considered as condensed phase density. If we start from gas we have the instability. The nonlinear stage of this instability will be the collapse , i.e., the formation of the singularity for finite time.

It should be said a few words about 3-D solitons as one possible variant of the instability development.The simplest solutions of this type $\psi = \phi \exp(i\lambda^2 t)$ represent the stationary point of H for fixed number of particles $N = \int |\psi|^2 \, dr$:

$$-\lambda^2\phi + \frac{1}{2}\Delta\phi + |\phi|^2\phi - \beta|\phi|^4\phi = 0. \tag{4}$$

These solutions are unstable in gas phase according to the criterion : $dN/d\lambda^2 < 0$ [10]. The qualitative arguments of this instability can be get from the scaling transformations, $\psi \to a^{-3/2}\psi(r/a)$ remaining N. As a result of H becomes the function of parameter a :

$$H = I_1/a^2 - I_2/2a^3 + \beta I_3/3a^6 \tag{5}$$

where $I_1 = \frac{1}{2}\int |\nabla\psi|^2 \, dr$, $I_2 = \int |\psi|^4 \, dr$, $I_3 = \int |\psi|^6 \, dr$. Maximum of this function corresponds to the unstable soliton as to the saddle point. Besides this function for $\beta = 0$ occurs to be unbounded from below. It is the main reason of collapse in this system . From this point of view the collapse can be considered as process of the falling of some "particle" in unbounded potential. If β is finite then this process can be stopped for very large amplitude that corresponds to the minimum of this function, i.e.,to the soliton of the condensed phase.It should be noted that for $\beta = 0$ there are no stabilization or saturation of the instability. The compression will continue until the formation of singularity. It follows strictly from the virial theorem:

$$\frac{d^2}{dt^2}\int r^2 \mid \psi \mid^2 \, dr < 4H. \tag{6}$$

After twice integration we have that positive value $\int r^2 \mid \psi \mid^2 \, dr$ vanishes for finite time. Because of for the initial state H are negative after the formation of clusters with characteristic size of $l \propto k_0^{-1}$ the collapse becomes inevitable, where $k_0 = \sqrt{2n_0}$ corresponds to the maximum of Γ (3). Next questions appearing here are about possible regimes of collapse. The first variant of the collapse is so called strong regime of compression when all particles in the cavity occur in the singularity [6]. It is easy to understand that this process is semi-classical one because from the usual quantum mechanics it is known that in the potentials $V \propto -r^{-\sigma}$ for $\sigma > 2$ the falling becomes while approach the center more classical. But such collapse is unstable with respect to short-wave perturbations resulting in the formation of weak singularities. For this type of collapse the captured number of particles is formally equal to zero, in practice is quite small. The most simple argument why instead of adiabatic compression the weak singularity appears is the following. Let us consider some region Ω with negative Hamiltonian H (we call this region as a cavity). Then with the help of the mean value theorem follows the inequality

$$\max \mid \psi \mid^2 \geq \mid H_\Omega \mid /2N_\Omega. \tag{7}$$

Here N_Ω is the number of particles in Ω. Let us imagine that from Ω we have radiation of waves (or emission of particles). Because of these emitted particles carry out the positive portion of H (strictly speaking it is valid for the separate cavity) that results in the reduction of H for Ω, i.e., H for this region occurs to be more negative. It is evident that N_Ω vanishes . So the ratio $\mid H_\Omega \mid /2N_\Omega$ tends to the infinity together with $\max \mid \psi \mid^2$. It means that the radiation of waves promotes the collapse, leads to its acceleration. The corresponding solution to this regime is the self-similar one of (2) for $\beta = 0$:

$$\psi = (t_0 - t)^{-1/2-i\epsilon} g\left(r/(t_0 - t)^{1/2}\right) \tag{8}$$

where $g(\xi)$ obeys the equation

$$-(i - 2\epsilon)g - (\xi\nabla)g + \Delta g + 2 \mid g \mid^2 g = 0. \tag{9}$$

Hence the characteristic radius of collapsing solution is proportional to $(t_0 - t)^{1/2}$. From this fact follows that particles reached the critical density $n = 1/\beta$ will form the drop with the radius $r_0 = \beta^{1/2}/ \mid g(1) \mid$. The consequent number of particles in the drop can be estimated as

$$N \approx \mid g(1) \mid^{-3} \beta^{1/2} 4\pi/3. \tag{10}$$

Because of $\beta \ll 1$ this number is less than one. It means that at the collapse time t_0 the formation of condensed phase does not perform, secondly, the energy gathered by the condensate particles during the collapse is quite large and so there exists the unique possibility to transfer this energy or the significant amount of it to undercondensate particles radiated from the vicinity of the singularity. It is very important also that while weak collapse the attraction potential will be formed for other non collapsing particles which can be trapped after t_0. It follows, in particular, from asymptotic of $g(\xi)$ for $\xi \to \infty$:

$$g(\xi) \to A/\xi^{1+2i\epsilon} \text{or} \mid \psi \mid^2 \to \mid A \mid^2 r^{-2}. \tag{11}$$

Because during the weak collapse the number of trapped particles is very small the post collapse regime is natural to suppose to be quasi stationary. For search of its distribution we can find simply stationary solution. It is convenient to rewrite eq.(2) for density n and phase Φ. Then simple analysis gives the following solution [7]-[9]:

$$n(r) \simeq 1/2 \log(r/r_0) r^{-2}; \tag{12}$$

$$v(r) = \partial\Phi/\partial r \simeq P/2\pi \log(r/r_0). \tag{13}$$

Here P is the particle flux to the center of this black hole and r_0 the size of the growing drop. The flux , from one side, forms the drop and, from the other, the counter flux of undercondensate particles which carry out the energy of the order of the kinetic energy of falling particles. It is easy to estimate the corresponding energy E loosing by the cavity during the total time of collapse: $E \propto \mid E_{\text{in}} \mid (\beta n_0)^{-1}$ where E_{in} is the initial gas energy. Thus the carried energy occurs to be quite large value. It can be assumed that sufficiently small part of particles (at least, less then half of falling particles) will be emitted. These particles will interact weakly with the condensate particles. Their phases can be considered random and therefore their interaction will be described by four-wave kinetic equation. As known [11], this interaction leads to the formation of power-type Kolmogorov spectrum in the particle energy space realizing the particle flux in this space,i.e., forms the particle condensate. Now we don't know what kinds of the equilibrium is realized in this case.

Now let us estimate the growth rate of the drop. If we assume that approximately half of particles from the cavity with the size $l \propto k_0^{-1}$ absorbs into singularity, i.e.,$N_{dr} \propto n_0 l^3 \propto n_0^{-1/2}$, then the characteristic time of the drop growth has the same order as the inverse growth rate (3): $\Gamma_{\max} N_{dr} \propto l^2 P$. Hence $P \propto n_0^{3/2}$. Substituting this expression in the definition of P one can get for the drop growth $\dot{V} \propto \beta n_0^{3/2}$ where V is the drop volume.

At the end of this part we present the concrete numbers for Cs. For density $10^{11}\mathrm{cm}^{-3}$ the modulation length $l \sim 10^{-3}$cm, the collapse time $\sim$ 0.3 sec that less than sample cooling size and the particle life time in the trap, respectively.It means that the observations of the discussed above effects represent quite real, of course, if the corresponding temperatures will be get.

Conclusion.

In conclusion we would like to discuss briefly other possibilities connected with the repulsion and the spin dependence in atom scattering. For the repulsion case instead of instability (3) we have the stable condensate with the Bogolyubov dispersion law. But, nevertheless, the nonlinear dynamics here is nontrivial. In this case the collapse is also possible that is familiar to the cavitation in fluids [12]. It is very interesting also the question on the formation of vortices in this model that has been studied incompletely. As far as the spin dependence of the exchange interaction the simplest way of its account consists for S-scattering in representation of the potential in the form:

$$V(r) = \left[\frac{G_1}{4}(1 - \vec{\sigma}_1\vec{\sigma}_2) + \frac{G_2}{4}(3 + \vec{\sigma}_1\vec{\sigma}_2)\right]\delta(r) \qquad (14)$$

where $G_{1,2}$ are the constants for singlet and triplet states of two electrons of scattering atoms and $\vec{\sigma}_{1,2}$ are the Pauli matrices for electron spins. This character of the interaction means that the condensate state will be described by two parameters of order $\psi_{1,2}$ the square of modulus of which will be represented the density of particles with spin projections $\pm 1/2$. If $G_{1,2} < 0$, minimum of H will correspond to $n_1 = n_2$ and H will coincide with (1).

One of the authors (E.K.) is greatful to I.B.Khriplovich for useful discussions of the spin dependent nonlinear interaction.

References

[1] C.Monroe, D.Sesko, C.E.Wieman. Phys. Rev.Lett. 64 (1990)408.

[2] A.Kazantsev, G.Surdutovich, V.Yakovlev. Mechanical Action of Light on Atoms. Moscow. Nauka. 1991. (in Russian).

[3] G.Surdutovich. Private communication.

[4] E.P.Gross,Phys.Rev.106 (1957) 161; L.P.Pitaevskii, ZhETF. 40 (1961) 646.

[5] S.N.Vlasov,V.A.Petrischev,V.I.Talanov. Radiofizika, 12 (1970) 1353.

[6] V.E.Zakharov, E.A.Kuznetsov. ZhETF, 91 (1986) 1310.

[7] S.N.Vlasov, L.I.Piskunova, V.I.Talanov. In: Nonlinear and Turb. Proc. in Physics. 2 (1988) 210, Kiev.

[8] V.M.Malkin. Pis'ma ZhETP, 48 (1988) 603.

[9] V.E.Zakharov, N.E.Kosmatov, V.P.Shwets. Pis'ma ZhETP, 49 (1989) 431.

[10] A.A.Kolokolov, N.B.Vakhitov. Radiofizika. 16 (1973) 1020.

[11] V.E.Zakharov. In: Basic Plasma. Handbook. Atomizdat, 1986. Moscow, Vol.2.

[12] E.A.Kuznetsov, S.K.Turitsyn. ZhETF 94 (1988) 119.

THE INTERACTION REPRESENTATION FOR NONLINEAR SCHRODINGER EQUATION PENLEVE PROBLEM

G.M.Fraiman

Department of Plasma Physics and High-Power Electronics
Institute of Applied Physics, Russian Academy of Science
46, Uljanov Street, 603600, Nizhny Novgorod, Russia

Abstract

The interaction between peak and background within self-focusing problem writing by 2d-Nonlinear Schrodinger equation is descussed. The peak-background representation is presented. It is shown that "double logarithm law" for the width of peak is stable to this interaction. The general sceme of pertubation theory is discussed.

This year is marked by the anniversary of the beginning of work on self-focusing (Gurgen Askar'yan, 1962) and the 20th anniversary of publication of the paper by Vladimir Zacharov, 1972 "Lengmuir's wave collapse". So I think it justifies my attempt to give the coverage of basic concepts of collapse theory in the way we see them nowadays.

Being short of time, I will make it on the example of only one problem nonlinear Schrödinger equation in the theory of self-focusing. I shall speak about the tasks which are all reduced to the following equation:

$$-i\frac{\partial u}{\partial t} = \Delta u + |u|^2 u + \hat{\mu}[u]u\,, \qquad \Delta = \frac{\partial^2}{\partial x^2} + \frac{\partial^2}{\partial y^2} \tag{1}$$

R. E. Caflisch and G. C. Papanicolaou (eds.), Singularities in Fluids, Plasmas and Optics, 205–212.
© 1993 *Kluwer Academic Publishers.*

where the perturbation operator is assumed to be small, though structured perturbations of emerging distributions can be rather significant. It can be weak dissipation (the one depending on the field scale included), weak saturating of nonlinearities, weak parametric pumping, etc. It is essential, that such perturbations can, in particular, "switch off" the collapse, i.e. result in the field limitation focal point and, hence, it can be taken into account within the framework of the effective perturbation theory.

It is now understood, that the analysis of tasks of the Eq.1 type is reduced to the choice of a "good" basis in an unperturbed problem

$$-i\frac{\partial u}{\partial t} = \Delta u + |u|^2 u \tag{2}$$

and to "the concept of interaction" for Eq.1 in the framework of which the perturbation theory can be efficiently developed. This, in fact, is the solution of any problem in the theory of the wave collapse.

The manyfold of solutions and some orthogonal perturbations of those corresponding to linearized problem (background) are chosen as this very "good" basis

$$u \simeq u_a(r,t,\vec{\alpha}) + u_\perp(r,t) \tag{3}$$

Here at $\vec{\alpha} = const$ and $u_\perp = 0$, u_a is the exact manifold of initial problem solutions and the division into the peak and the background can always be made so that within the framework of Eq.(2)

$$\frac{d\vec{\alpha}}{dt} = \vec{O}(u_\perp^2)\,, \tag{4}$$

i.e., the parameters change only in the second order over the background.

It is worth noting here that the effectiveness of representation (3-4) is determined by the stability, i.e., by the absence of exponentially growing modes of the background in a linearized problem at fixed parameters of manifold. This is one of the key moments, to our mind, in the construction of the collapse theory. If this stability exists, then in the framework of the perturbed problem 1 we can wait a good description. In the other case we wouldn't "hit the target" and will be forced to return back to the choice of the initial problem of the type (2).

In our case the situation with the manyfold, which is complex enough from the analytical point of view, looks as follows (Fraiman, 1985; Smirnov, Fraiman, 1991)

$$u_\alpha = \frac{1}{a}\varphi(\tau, \frac{r}{a}) \exp i\left(\frac{\dot a}{a}r^2 + \varphi_0\right), \quad \tau = \int_0^+ \frac{dt}{a^2} \tag{5}$$

where $\varphi(\xi, \vec{\alpha, \tau})$ satisfies the equation

$$i\varphi_\tau + (|\varphi|^2 - 1 - c\xi^2)\varphi = 0, \quad C = +\frac{d^2 a}{dt^2}a^3 = -a\frac{d^2}{d\tau^2}\frac{1}{a}, \quad \varphi(r \to \infty) \to 0 \tag{6}$$

Thus, here we are dealing with a four-parametric manifold

$$\vec{\alpha} = (a_0, \dot a_0, \ddot a a^3 = c, \psi_0). \tag{7}$$

The common situation with manifold significantly depends on the sign of C-parameter.

At $C > 0$

Eq.(6) at every time moment τ is the linear Schrödinger equation with the potential

$$U(\xi) = -|\varphi|^2 + c\xi^2 \tag{8}$$

and with the eigenvalue equal unity. Here we can assume that $\partial\varphi/\partial\tau \equiv 0$ and $\varphi(\xi, c)$ exponentially decreases with the growth of distance. In this regime at $c \leq 1$

$$P[\varphi] = \int |\varphi|^2 d\vec{\xi} = P_{cr} - c\xi_{eff}^2 < P_{cr} \tag{9}$$

and

$$a(t) = \left(\dot{a}_0^2 + \frac{c}{a_0^2}\right) t^2 + 2a_0\dot{a}_0 t + a_0^2$$

Consequently at $C > 0$ the peak power is more then critical and the singularity is not possible:

$$a_{\min}^2 = \frac{ca_0^2}{c + a_0^2\dot{a}_0{}^2} \tag{10}$$

At $C < 0$

the "effective" potential

$$U(\xi) = -|\varphi|^2 - |C|\xi^2 \tag{11}$$

is of the open type and we can speak about "leaky" modes only with tunnel luminescence of quanta through the potential barrier. Consequently, analyzing this problem we must understand that the nonstationary character here is of principle importance because the quanta number in the peak region is variated.

Many attempts have been made to solve this problem but I shall restrict myself only to the last one (Smirnov, Fraiman, 1991) which realizes rather a qualitative idea of Alexander Smirnov and is used to analyze resonators and waveguides with luminescence in electrodynamical systems.

The essence of this idea is as follows (the solution is presented as sum of the two parts):

$$\varphi = \varphi_0(\xi, \xi_c) + \delta(\xi, \tau), \tag{12}$$

where φ_0 is the localized mode in the auxiliary potential

$$\bar{U} = -|\varphi_0^2| - \begin{cases} \xi^2, & \xi < \xi_c \\ 0, & \xi > \xi_c \end{cases} \tag{13}$$

Here at $|c| \ll 1$

$$P_0 \equiv P[\varphi_0] = P_{cr} + |c|\xi^2 > P_{cr} \tag{14}$$

and for $\delta(\xi, \tau)$ we get the equation with the source

$$i\frac{\partial \delta}{\partial \tau} + \Delta\delta + \varphi_0^2(2\delta + \delta^*) = q(\xi)\varphi_0 - i\frac{\partial \varphi_0}{\partial \xi_c}\frac{d\xi_c}{d\tau} \tag{15}$$

It means that at the fixed value of c the luminescence depending on the right-hand part of Eq.(15) will be infinite and, thus, perturbations (those in the localization region included) will increase. The condition to avoid it follows:

$$|\delta(\xi, \tau)| < \infty, \quad if \quad \xi \leq 1 \tag{16}$$

It can be shown that it is necessary and sufficient if

$$\frac{dP_0}{d\tau} \equiv -2\pi A_0^2 \exp - \frac{\pi}{2\sqrt{|c|}} \tag{17}$$

This result exactly corresponds to a flow from the peak localization region calculated by us (Fraiman,1985) in the geometry-optical approximation based on qualitative considerations.

It follows from Eqs.(14),(17) "the double logarithm law", i.e., as t tends to singularity moment t_0

$$a(t) \sim \sqrt{\frac{t_0 - t}{\ln|\ln t_0 - t|}} \tag{18}$$

Having determined the peak dynamics of the type (14),(17) let us now turn to the consideration of its interaction with the background. It is worth remembering, that tunnel effects are exponentially weaker (see (17) and thus even minor interaction with the background would break the law (8) even in the framework of Eq.(2). However, in reality it does not happen! Parametric interaction of the peak and the background on the limited stage is reduced to the background spreading and to the partial transfer of quanta from the background to the peak, i.e. to some finite growth of the peak power. Thus, we asymptotically come back to the double logarithm law.

Let us consider this problem in more detail. If we substitute Eq.(3) into Eq.(2) and try to meet condition (4), then in the (ξ, τ)-representation we come to the system of equations for parameters and background. The equations for the peak power and for the background

$$\begin{cases} \dfrac{dP_0}{d\tau} = \xi^2_{eff} \dfrac{d|c|}{d\tau} = -2\pi A_0^2 \exp - \frac{\pi}{2\sqrt{|c|}} + Im \int \varphi_0^{*2} V_\perp^2 d\vec{\xi} \\ i\dfrac{\partial V_\perp}{d\tau} + (\Delta + |\varphi_0^2| + |c|\xi^2 - 1)V_\perp + \varphi_0^2(V_\perp + V_\perp^*) + J_{NL} \end{cases} \tag{19}$$

are considered to be the key ones. Here we restrict ourselves only to the linear approximation for background, because we assume that

$$|\varphi| \gg |V_\perp| \tag{20}$$

i.e., we consider the times which are close to the moment of the singularity formation.

The second equation (19) shows that in the linear approximation

$$V_{\perp} = V_{+}e^{i\tau} + V_{-}e^{+i\tau}\,, \tag{21}$$

where V_+ and V_- are slow functions of τ. It means that the term $V_{\perp}^*$ in Eq.(19) is responsible for the possible parametric instability. In our case this instability is absent due to neutral stability of the automodel solutions manifold. However, it is evident from Eq.(20) that the interaction with the peak is quite possible because in the terms of V_+, V_- the second term in the right-hand side of Eq.(18) is

$$\overline{Jm\int \varphi_0^{*2} V_{\perp}^2 d\vec{\xi}}^{\,T} = 2Jm\int \varphi_0^{*2} V_+ V_- d\vec{\xi} \tag{22}$$

Moreover, if we make use of the equation for V_+ and V_-, following from the equation for the background in Eq.(19) we can express V_- in terms of V_+ (taking into account that V_- is the localized part of the background) and then

$$\left\{ \begin{array}{c} Jm\int \varphi_0^{*2} V_{\perp}^2 d\vec{\xi} \simeq \frac{d}{d\tau} 2\int |\hat{\mathcal{L}}(|\varphi_0|^2 \cdot V_+)|^2 d\vec{\xi} \\ \hat{\mathcal{L}} = |\Delta + \varphi_0^2 - 2|^{-1} \end{array} \right. \tag{23}$$

Thus it is evident that the interaction with the background results in the partial transfer quanta from the background to the peak.

Moreover, let us mark that V_+ is represented in the form

$$\begin{array}{c} V_+ = \sum\limits_{n\geq 0} V_n \Phi_n(\xi, c) \exp - \int \gamma_n(\tau) d\tau \\ \gamma_n \sim \sqrt{|c|} \cdot (n + \frac{1}{2}) \end{array} \,, \tag{24}$$

It corresponds to expansion over radiated modes with complex eigenvalues. Leaving only one term, corresponding to the slowest part of the background instead of Eq.(19) we have

$$\frac{d}{d\tau}\left(\xi_{eff}^2 |c| + B^2 e^{-\int_0^t \frac{\sqrt{|c|}}{2} d\tau}\right) = -2\pi A_0^2 \exp - \frac{\pi}{2\sqrt{|c|}} \tag{25}$$

that again brings us back to the double logarithm law.

Conclusion

1. It is clearly seen that the double logarithm law associated with exponentially weak luminescence turns out to be rather stable. This stability is explained, first of all, by the fact that quanta from the background are transferred into the peak. If we had had a reverse process, we would have got a linear law of singularity $a|\varphi| \sim (t_0 - t)^1$. I know nothing of such an example.

2. At last, the marked detecting of the energy transfer process allows us to easily solve the perturbation problem (1). Thus, conservative perturbations of the saturated nonlinearity type ($\hat{\mu}u \simeq -\mu|u|^u u, \ldots$) lead us only to another expression for the peak power

$$P_0 = P_{cr} - c\xi^2_{iteff} - \frac{Q}{a^2},$$

and, as a consequence, to the transfer to a uniform channel after some oscillations. As for dissipative terms, they only change the law of power losses in the peak. In all the cases, it is enough, as a rule, to take into account the above effects.

References

1. G.A.Askar'yan, Sov.JETF,v.42,p.1566,1962.

2. V.E.Zacharov, Sov.JETF,v.62,p.1745,1972.

3. G.M.Fraiman, Sov.JETF,v.88,p.390 ,1985.

4. A.I.Smirnov, G.M.Fraiman, Physics 2D, v.52,p.2-15,1991.

SELF-SIMILAR REGIMES OF SUBSONIC LANGMUIR WAVE COLLAPSE

V.M. Malkin, and V.N. Khudik
Budker Institute of Nuclear Physics
Novosibirsk 630090 Russia

ABSTRACT. The first examples of regular and singular self-similar solutions of the 3-d nonlinear Schroedinger equation for potential vector field are found explicitly. Regular solutions relate to the process of the "singularities" formation. Singular solutions relate to an early evolution of the "singularities". Together, the solutions describe behavior of Langmuir wave field in the spatiotemporal vicinities of its "singularities" and provide an effective mechanism of energy dissipation from the Langmuir condensate.

In the framework of a weak turbulence theory, the basic nonlinear processes lead to accumulation of long Langmuir waves. As a result so-called Langmuir condensate is formed. The mechanism for the energy dissipation from the Langmuir condensate remained unclear for a long time. Since 1972, the subsonic Langmuir collapse was considered to provide such a mechanism [1]. However, this kind of collapse appeared to be "weak", so that an infinitely small energy gets into arising singularity (where the absorption could take place). An effective mechanism for the energy dissipation was revealed recently in the framework of the scalar collapse model. The singularities of the scalar field were shown not vanish immediately after the absorption occurs of the waves that produced them, but exist a long time and "suck-in" new waves [2–5]. As for the true Langmuir wave field (which is a potential vector and not a scalar), its behavior in the spatiotemporal vicinity of the singularities is unclear still. To advance in the problem, the true equation for Langmuir condensate, namely, the 3-d nonlinear Schroedinger equation for potential vector field is to be studied below. This equation can be written in the form

$$\nabla \left(i\frac{\partial}{\partial t} + \Delta + |\nabla\psi|^2 \right) \nabla\psi = 0. \tag{1}$$

R. E. Caflisch and G. C. Papanicolaou (eds.), Singularities in Fluids, Plasmas and Optics, 213–219.
© 1993 *Kluwer Academic Publishers.*

Substitution

$$\psi(\mathbf{r},t) = f\left(\frac{\mathbf{r}}{a},\tau\right)\exp(\imath\tau), \tag{2}$$

$$\tau \equiv \limsup_{r\to 0} \arg\psi(\mathbf{r},t), \qquad a^{-2} \equiv |\dot{\tau}| \tag{3}$$

turns equation (1) to

$$\nabla\left(\imath\frac{\partial}{\partial\tau} + \Delta + |\nabla f|^2 + \sigma - \imath a\dot{a}\frac{\partial}{\partial\rho}\rho\right)\nabla f = 0. \tag{4}$$

$$\limsup_{\rho\to 0} \arg f(\rho,\tau) = 0, \qquad \sigma \equiv -sign\dot{\tau}. \tag{5}$$

Here the dot above a or t signify the time derivative and the operator ∇ acts on the variable $\boldsymbol{\rho} = \mathbf{r}/a$ already.

Equation (4) has solutions looking as

$$f(\boldsymbol{\rho},\tau) = f^{\sigma}(\boldsymbol{\rho}) = \begin{cases} f^{-}(\boldsymbol{\rho}), & \sigma = -1 \\ f^{+}(\boldsymbol{\rho}), & \sigma = +1 \end{cases}, \tag{6}$$

$$a\dot{a} = \gamma^{\sigma} = const. \tag{7}$$

Time-independent function $f^{\sigma}(\boldsymbol{\rho})$ satisfies the equation

$$\nabla\left(\Delta + |\nabla f^{\sigma}|^2 + \sigma - \imath\gamma^{\sigma}\frac{\partial}{\partial\rho}\rho\right)\nabla f^{\sigma} = 0. \tag{8}$$

The functions $f^{-}(\boldsymbol{\rho})$ and $f^{+}(\boldsymbol{\rho})$ describe the shape of the electrostatic potential $\psi(\mathbf{r},t)$ in the similarity range before and after the singularity arising at the point $\mathbf{r} = 0$. These function are called below as regular and singular self-similar solutions respectively.

While the singularity is formed at the point $\mathbf{r} = 0$, the spatial scale, a, of the field $\nabla\psi(\mathbf{r},t)$ variation decreases. In the range $r \gg a$ the field is "frozen", as the typical time of its variation there is much larger than at $r \leq a$. While a tends to zero, the field is "frozen" at any finite r, corresponding at $a \to 0$ to $\rho \to \infty$. This entails some relationship between asymptoticses of the functions $f^{-}(\boldsymbol{\rho})$ and $f^{+}(\boldsymbol{\rho})$ at $\rho \to \infty$. Basically, the function $f^{\sigma}(\boldsymbol{\rho})$ looks at $\rho \to \infty$ as

$$f^{\sigma}(\boldsymbol{\rho}) \cong F^{\sigma}(\mathbf{n})\rho^{-\imath\sigma/\gamma^{\sigma}}, \qquad \mathbf{n} \equiv \frac{\boldsymbol{\rho}}{\rho}. \tag{9}$$

The "freezing" condition implies that, at the moment t_s of the singularity appearance, the function

$$\psi(\mathbf{r},t_s) = F^{\sigma}(\mathbf{n})r^{-\imath\sigma/\gamma^{\sigma}} \lim_{t\to t_s} a^{\imath\sigma/\gamma^{\sigma}} e^{\imath\tau} \tag{10}$$

do not depend on σ. This is possible at

$$\gamma^{\sigma} = \sigma\gamma \tag{11}$$

only. Then, in view of (3), (5) and (7),

$$a^2 = 2\gamma\,|t - t_s|\,, \qquad \tau = -\frac{1}{2\gamma}\ln|t - t_s| + \tau_0^{\sigma}, \tag{12}$$

where τ_0^{σ} $(\sigma = \pm 1)$ are constants. Hence, the "freezing" condition takes the form

$$F^{+}(\mathbf{n})e^{\imath\tau_0^{+}} \cong F^{-}(\mathbf{n})e^{\imath\tau_0^{-}}. \tag{13}$$

The solutions of equations (8) can be obtained by means of an expansion in associated Legendre polynomials:

$$f^{\sigma}(\boldsymbol{\rho}) = \sum_{l,m} R^{\sigma}_{lm}(\rho)P_l^m(\cos\theta)\exp(\imath m\phi). \tag{14}$$

The radial functions behave at $\rho \to \infty$ as

$$R^{\sigma}_{lm}(\rho) \cong F^{\sigma}_{lm}\rho^{-\imath/\gamma} + C^{\sigma}_{lm}\rho^{-l-1} + D^{\sigma}_{lm}\rho^{-2}\exp\left(\imath\sigma\gamma\rho^2/2\right). \tag{15}$$

Here only the main terms of all kinds are presented, though the first item — corresponding to the "frozen" field — is dominant and small corrections to it may be much larger than a whole other item.

For a regular solution, $f^{-}(\boldsymbol{\rho})$, the last item in the formula (15) corresponds to the converging wave, that can be created only artificially and do not arise at the initial conditions of general kind. Hence,

$$D^{-}_{lm} = 0. \tag{16}$$

The coefficients F^{-}_{lm} and C^{-}_{lm} are to be chosen to secure the regular behavior of the solution at $\rho \to 0$:

$$R^{-}_{lm}(\rho) = A_{lm}\rho^{l} + B_{lm}\rho^{l+2} + O(\rho^{l+4}). \tag{17}$$

At infinitely small perturbation of the parameters F^{-}_{lm} and C^{-}_{lm}, the addition to the regular self-similar solution arises which has the following asymptotics at $\rho \to 0$:

$$\delta R^{-}_{lm}(\rho) \cong \delta A_{lm}\rho^{l} + \delta B_{lm}\rho^{l+2} + E_{lm}\rho^{-l-1} + G_{lm}\rho^{-l+1}. \tag{18}$$

Here again only the main terms of all kinds are presented. The third and forth items give electric field non-analytical at $\rho \to 0$. In the asymptotics of the self-similar solution $f^-(\boldsymbol{\rho})$ both these items are to be absent:

$$E_{lm} = 0, \qquad G_{lm} = 0. \tag{19}$$

The number of conditions (19) coincides with the number of indefinite parameters F_{lm}^- and C_{lm}^- at any finite approximation of the expansion (14). However, the solution differing each other only by a constant phase are in fact identical, and the phase of the function $f^\sigma(\boldsymbol{\rho})$ was fixed above by the first of relationships (5). To satisfy this additional condition, the parameter γ must be used. Hence, the values of γ can constitute only a discreet set.

The actual number of the relevant parameters at any finite approximation of the expansion (14) depends on the symmetry of the solution. Further, axisymmetric solutions odd with respect to the space reflection, $\mathbf{r} \to -\mathbf{r}$, are considered. For such solutions only one value of the "momentum projection" m and only odd values of the "momentum" l are present in the expansion (14).

The most interesting are, probably, the solutions with $m = 0$ and $m = \pm 1$. Those have non-zero pressure of high-frequency field $P(\boldsymbol{\rho}) = |\nabla f^-(\boldsymbol{\rho})|^2$ at $\rho = 0$, in contrast to the solutions with larger values of $|m|$. Noteworthy, that non-zero field $\nabla f^-(\boldsymbol{\rho})$ at the center of the cavity is produced only by dipole terms of the expansion (14). At a relatively small weight of such terms in the expansion, the high-frequency pressure at the center of cavity would be reduced. This looks hardly compatible with the stable collapse to the point $r = 0$ (though the stability problem is not the subject of the current study). That is why the solutions of dipole kind are to be studied first.

Such solutions appear to be well computable by taking into account the first three terms in the expansion (14), namely the terms with $l = 1, 3$ and 5. The basic parameters of thus found solutions of dipole kind are presented in the table 1 (the cases $m = 1$ and $m = -1$ are identical, as they are turned one into another by reflection in a plane containing symmetry axis).

Figure 1 shows isolines of the pressure $P(\boldsymbol{\rho})$, computed for one, two and three terms taken into account in the expansion (14). In the vicinity of the center the pressure is given by the formula

$$P(\boldsymbol{\rho}) = P(0)\left(1 - \frac{\rho_\perp^2}{b_\perp^2} - \frac{\rho_\parallel^2}{b_\parallel^2}\right). \tag{20}$$

The values of the parameters $P(0)$, $b_\perp$ and $b_\parallel$ are also presented in the table 1. As seen from the figure 1 and formula (20), the cavity is oblate at $m = 0$ and prolate at $m = 1$.

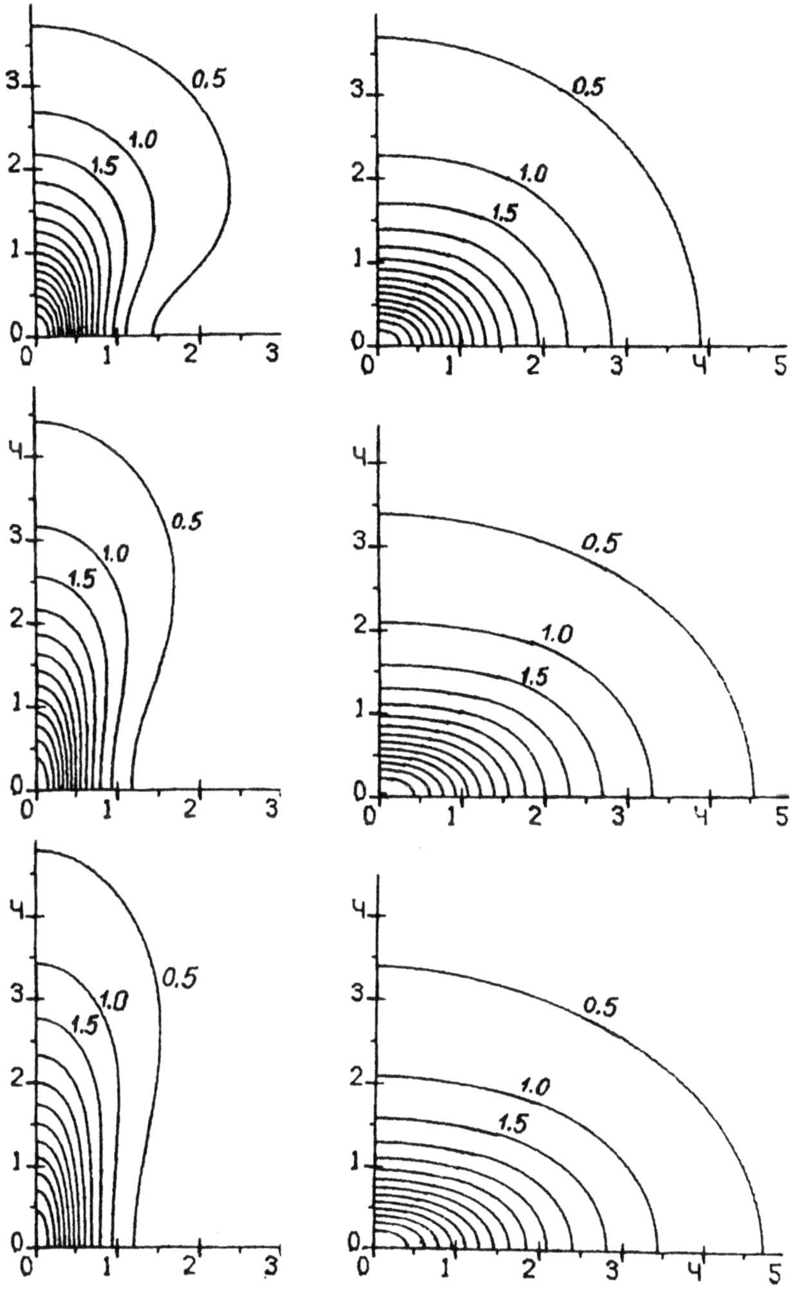

Fig. 1. Isolines of the pressure $P(\rho)$ for regular self-similar solutions of dipole kind in the plane $(\rho_{\parallel}, \rho_{\perp})$. The left figures correspond to $m = 0$ and the right ones — to $|m| = 1$. The upper, middle and lower figures are computed, respectively, for the first, third and fifth harmonics taken into account in the expansion (14).

TABLE I
The basic parameters of regular self-similar solutions of dipole kind.

m	0	1
γ	0.85	0.72
A_{1m}	2.53	1.93
A_{2m}	$-0.34-0.042\imath$	$0.047+0.006\imath$
A_{3m}	$0.059+0.0088\imath$	$0.0027+0.00043\imath$
B_{1m}	$-0.91-0.23\imath$	$-0.63-0.2\imath$
B_{2m}	$0.36+0.073\imath$	$-0.043-0.011\imath$
B_{3m}	$-0.11-0.021\imath$	$-0.0041-0.00094\imath$
F^-_{1m}	$0.79+2.05\imath$	$0.38+1.58\imath$
F^-_{2m}	$-0.18+0.041\imath$	$0.029+0.056\imath$
F^-_{3m}	$0.046+0.068\imath$	$0.0025+0.0027\imath$
$P(0)$	6.3	7.4
$b_\perp$	1.79	0.83
$b_\parallel$	0.58	1.67

After the singularity has formed, the singular self-similar solution $f^+(\rho)$ is relevant. At $\rho \gg 1$ its radial function, $R^+_{lm}(\rho)$, depend on three complex parameters: F^+_{lm}, C^+_{lm} and D^+_{lm} (see (15)). The first of them is linked with the parameter F^-_{lm} by the relationship

$$F^+_{lm} = F^-_{lm} \exp(\imath\Delta\tau_0), \qquad \Delta\tau_0 \equiv \tau_0^- - \tau_0^+, \tag{21}$$

as follows from the "freezing" condition (13). The coefficients C^+_{lm} and D^+_{lm} are to be chosen to secure the proper behavior of the function $f^+(\rho)$ at $\rho \to 0$.

In the region $\rho \ll 1$, where the electric field $\nabla\psi$ is capable of changing in a time $\delta\tau \sim r^2 \ll a^2 \sim t - t_s$, a quasi-stationary state is set in, the quantum flux to the singularity being independent of r. The singular self-similar solution of dipole kind, $f^+(\rho)$, is given there by the formulae

$$f^+(\rho) = \begin{cases} \left(\frac{8}{15}\ln\frac{c}{\rho}\right)^{-1/2}\cos\theta + O\left(\ln\frac{1}{\rho}\right)^{-3/2}, & m=0 \\ \left(\frac{14}{15}\ln\frac{c}{\rho}\right)^{-1/2}\sin\theta e^{\imath\phi} + O\left(\ln\frac{1}{\rho}\right)^{-3/2}, & m=1 \end{cases}. \tag{22}$$

At infinitely small perturbation of the parameters C^+_{lm} and D^+_{lm} , an addition to the solution $f^+(\rho)$ arises which, generally speaking, has bad asymptotics at $\rho \to 0$. For each $l > 1$ two bad asymptoticses in the function δR^+_{lm} arise, proportional to ρ^{-l-1} and ρ^{-l+1} (as in (18)). For $l = 1$ only one bad asymptotics arise, proportional to ρ^{-2} . Hence, the number of indefinite complex parameters, C^+_{lm} and D^+_{lm}, exceeds the number of bad asymptoticses in the function $\delta f^{+}(\rho)$ by the unity. Additional real condition (5) is balanced by additional real parameter $\Delta\tau_0$ in (21). When all the coefficients at bad asymptotic are turned to zeros, one complex, i.e., two real parameters remain free still. For instance, the phase gap $\Delta\tau_0$ and constant c in the formulae (21) and (22) can be treated as such free parameters. Their values depend on what happens inside the "singularity", say, do it absorb or reflect all coming to it quanta. In the latter case, the quantum flux to the singularity is to be absent. However, the physical nature of the singularities is a quite another topic. The results of current study can be formulated by the following way.

The existence of regular and singular self-similar solutions of the three-dimensional Schroedinger equation for potential vector field with attracting cubic nonlinearity is proved. Regular solutions, which describe the singularities formation, are shown to constitute a discreet set. Singular solutions, describing an early evolution of the singularities, contain continuous free parameters. This makes the field outside the singularities compatible with the singularities of different absorption properties. The simplest solutions of dipole kind are presented explicitly. The quasi-stationary asymptoticses (22) of singular self-similar solutions is relevant also for further — not necessarily self-similar — evolution of electric field in spatial vicinities of the singularities. Noteworthy that no direct analytical or numerical evidence on the long-living singularities existence in the Langmuir condensate were given earlier.

References

[1] Zakharov V.E., *Zh. Eksp. Teor. Fiz.* **62**, 1745 (1972) [*Sov. Phys. JETP* **35**, 908 (1972)].

[2] Vlasov S.V., Piskunova L.I. and Talanov V.I., *Zh. Eksp. Teor. Fiz.* **95**, 1945 (1989) [*Sov. Phys. JETP* **68**, n.6 (1989)].

[3] Malkin V.M., *Pis'ma Zh. Eksp. Teor. Fiz.* **48**, 603 (1988) [*Sov. Phys. JETP Lett.* **48**, 653 (1988)].

[4] Zakharov V.E., Kosmatov N.E. and Shvets V.F., *Pis'ma Zh. Eksp. Teor. Fiz.* **49**, 431 (1989) [*Sov. Phys. JETP Lett.* **49**, 492 (1989)].

[5] Malkin V.M. and Shapiro E.G., *Zh. Eksp. Teor. Fiz.* **97**, 183 (1990) [*Sov. Phys. JETP* **70**, 102 (1990)].

ASYMPTOTIC PROFILES OF BLOW-UP SOLUTIONS OF THE NONLINEAR SCHRÖDINGER EQUATION

Hayato NAWA
Department of Mathematics, Faculty of Science
Tokyo Institute of Technology
oh-okayama meguro, Tokyo 152
Japan

ABSTRACT. We consider the blow-up problem for the nonlinear Schrödinger equation with critical power nonlinearity on $\mathbb{R}^N$. We show that every blow-up solution concentrates its L^2 mass at most finite points in some sense. We also estimate the amount of concentrated L^2 mass, and consider the asymptotic profiles of blow-up solutions.

1. Introduction

This article is an collection of author's results of recent papers concerning the nonlinear Schrödinger equation with critical power nonlinearity (see (NSC) below) where the idea of concentrated compactness or of bubble theorems have been emphasized. Although one can find most of the proofs of theorems in this paper in the original articles [24–28], we shall give (or sketch) the proofs of them for reader's convinience. Thoerem B below, however, is a new result in this field of nonlinear shrödinger equations.

We start with a review of the Cauchy problem for the nonlinear Schrödinger equation:

$$\text{(NS)} \qquad 2i\frac{\partial u}{\partial t} + \Delta u + |u|^{p-1}u = 0, \qquad (t,x) \in \mathbb{R}_+ \times \mathbb{R}^N,$$

$$\text{(IV)} \qquad u(0,x) = u_0(x), \qquad x \in \mathbb{R}^N.$$

Here $i = \sqrt{-1}$, $u_0 \in H^1(\mathbb{R}^N)$ and Δ is the Laplace operator on $\mathbb{R}^N$. The unique local existence of solutions of C(p) is well known for $1 < p < 2^* - 1$ ($2^* = \frac{2N}{N-2}$ if $N \geq 3$, $= \infty$ if $N = 1,2$): For any $u_0 \in H^1(\mathbb{R}^N)$, there exist a unique solution $u(t,x)$ of C(p) in $C([0,T_m); H^1(\mathbb{R}^N))$ for some $T_m \in (0,\infty]$ (maximal existence time), and $u(t)$ satisfies the following two conservation laws of L^2 and the energy:

$$\|u(t)\| = \|u_0\|, \tag{1.1}$$

$$E_{p+1}(u(t)) \equiv \|\nabla u(t)\|^2 - \frac{2}{p+1}\|u(t)\|_{p+1}^{p+1} = E_{p+1}(u_0), \tag{1.2}$$

for $t \in [0,T_m)$, where $\|\cdot\|$ and $\|\cdot\|_{p+1}$ denotes the L^2 norm and L^{p+1} norm respectively. Furthermore $T_m = \infty$ or $T_m < \infty$ and $\lim_{t\to\infty} \|\nabla u(t)\| = \infty$. For details, see *e.g.* [11,12,14].

As for the existence and non-existence of global solutions of C(p), the following is well known (see [11,12,13,14]).

R. E. Caflisch and G. C. Papanicolaou (eds.), Singularities in Fluids, Plasmas and Optics, 221–253.
© 1993 *Kluwer Academic Publishers.*

(i) If $1 < p < 1+\frac{4}{N}$, there exists a global solution $u \in C_b(\mathbb{R}; H^1(\mathbb{R}^N))$, for any $u_0 \in H^1(\mathbb{R}^N)$, where $C_b(\mathbb{R}; H^1(\mathbb{R}^N)) = C(\mathbb{R}; H^1(\mathbb{R}^N)) \cap L^\infty(\mathbb{R}; H^1(\mathbb{R}^N))$.
(ii) If $1+\frac{4}{N} \leqq p < 2^* - 1$, there is a subset $\mathcal{B} \in H^1(\mathbb{R}^N)$ such that for any $u_0 \in \mathcal{B}$ the solution of C(p) blows up, *i.e.* the L^2 norm of its gradient explodes in finite time T_m.

As we have seen above, the number $p = 1+\frac{4}{N}$ is the critical number for the existence of blow-up solutions of C(p). In what follows, we refer to (NS) with $p = 1+\frac{4}{N}$ as (NSC), *i.e.*,

$$\text{(NSC)} \qquad 2i\frac{\partial u}{\partial t} + \Delta u + |u|^{\frac{4}{N}}u = 0,$$

and we shall use the notations

$$E(v) = E_\sigma(v), \qquad \sigma = 2 + \frac{4}{N}.$$

The nonlinear Schrödinger equation of the form (NSC) is of physical interest, because (NSC) with $N = 2$ arises in a theory of the stationary self-forcusig of a laser beam propagating along the t-axis in a nonlinear medium (see e.g. [1,2,40]). We may say that the blow-up of solution corresponds to the forcusing of the laser beam.

Recently, many mathematicians have studied the formation of singularities in blow-up solutions of (NSC) near blow-up time (*e.g.*, [7, 20, 21, 22, 23, 24, 25, 26, 27, 28, 29, 35, 37, 38]). Here, it is worth while to note that (NSC) has a remarkable property that it is invariant under the pseudo-conformal transformations (see *e.g.*,[7,29] and (1.7) below). This symmetry seems to be closely related to the structure of solutions of (NSC) (see *e.g.*, Weinstein [37,38], Nawa and M. Tsutsumi [29] and Cazenave and Weisler [7]): in the super critical case ($p > 1+\frac{4}{N}$), Merle [19] suggested that every blow-up solution of (NS) has a strong limit in L^2 at blow-up time; in the critical case ($p = 1+\frac{4}{N}$), Nawa [24,25] and Weinstein [38] showed that every blow-up solution of (NSC) loses its L^2 continuity at blow-up time because of the concentration of its L^2 mass (see also Merle [20], Merle and Y. Tsutsumi [21], and Y. Tsutsumi [35]). Moreover we know how amount the blow-up solution of (NSC) concentrate their L^2 mass. Precisely, we can prove

Theorem A. *Let Q be a nontrivial solution of the elliptic equation*

$$\text{(A.1)} \qquad \Delta Q - Q + |Q|^{\frac{4}{N}}Q = 0$$

such that

$$\text{(A.2)} \qquad \frac{2}{\sigma}\|Q\|^{\frac{4}{N}} = \inf_{\substack{v\in H^1(\mathbb{R}^N)\\ v\not\equiv 0}} \frac{\|v\|^{\frac{4}{N}}\|\nabla v\|^2}{\|v\|_\sigma^\sigma}$$

$$= \inf_{\substack{v\in H^1(\mathbb{R}^N)\\ v\not\equiv 0}} \left\{ \frac{2}{\sigma}\|v\|^{\frac{4}{N}} \ ; \ \|\nabla u(t)\|^2 - \frac{2}{\sigma}\|u(t)\|_\sigma^\sigma \leqq 0 \right\}.$$

Let $u(t)$ be the singular solution of (NSC) such that

$$\text{(A.3)} \qquad \lim_{t\to T_m} \|\nabla u(t)\| = \lim_{t\to T_m} \|u(t)\|_\sigma = \infty$$

for some $T_m \in (0,\infty]$. We put

$$\text{(A.4)} \qquad \lambda(t) = \frac{1}{\|u(t)\|_\sigma^{\sigma/2}},$$

$$\text{(A.5)} \qquad u_\lambda(t,x) = \lambda(t)^{\frac{N}{2}}u(t,\lambda(t)x),$$

$$\text{(A.6)} \qquad A = \sup_{R>0}\left(\liminf_{t\uparrow T_m}\left(\sup_{y\in\mathbb{R}^N}\int_{|x-y|\leqq R}|u_\lambda(t,x)|^2dx\right)\right).$$

(1) Then we have

$$A \geqq \|Q\|^2 \tag{A.7}$$

and we have that, for any $\varepsilon \in (0,1)$, there are constans $K > 0$ and $T_0 > 0$, and there is a right continuous function $\gamma(\cdot) \in L^\infty_{loc}([T_0, T_m); \mathbb{R}^N)$ such that

$$\int_{|x-y| \leqq R\lambda(t)} |u(t, x + \lambda(t)\gamma(t))|^2 dx \geqq (1-\varepsilon)A, \quad t \in [T_0, T_m), \tag{A.8}$$

for any $R \geqq K$. Moreover if $A > \frac{1}{2}\|u(0)\|^2$, then we have $\gamma(\cdot) \in C([T_0, T_m); \mathbb{R}^N)$.
(2) Suppose in addition that $\|u(0)\|^2 < 2\|Q\|^2$. Then by (1), we have $\gamma(\cdot) \in C([T_0, T_m); \mathbb{R}^N)$.

The forcusing of a laser beam could be understood mathematically as "mass concentration" phenomena of blow-up solutions of (NSC).

Remark 1.2. The quantity A measures the "size" of the "largest" singularity, since the blow-up solution, in genaral, has several L^2-cocentration points (see Merle [20] and Nawa [25]).

Remark 1.3. (1) The equation (A.1) is a time-independent version of (NSC) and arises in various domain of physics. See [3,6,32,36] and Proposition 2.5 of this paper for the existence of positive solutions of (A.2) and for the associated minimization problems. The standard argument shows that $Q \in \mathcal{S}$ (the space of C^∞ functions of rapid decreasing). We can also prove that $E(Q) = 0$.
(2) By the first equality in (A.2) and the conservation law (1.2), we see that if $\|u_0\| < \|Q\|$, the corresponding solution exists globally in time. For this, see Weinstein [36, 37]. In this sense, the estimate (A.7) is optimal.
(3) We shall sketch the proof of Theorem A in Sect.2 in this paper.

However the profiles of blow-up solutions have not been investigated so well. Concerning this problem, the following results are known.

(I) Let $u(t)$ be a solution of (NSC) such that $\|u(t)\| = \|Q\|$ and $\|\nabla u(t)\| \to \infty$ as $t \to T_m$ for some $T_m \in (0, \infty]$. Then we have, for $\lambda(t) = \|\nabla u(t)\|^{-1}$,

$$\left\| \lambda(t)^{\frac{N}{2}} u(t, \lambda(t)\cdot - \gamma(t)) e^{i\theta(t)} - Q(\cdot) \right\| \to 0 \quad \text{as} \quad t \to T_m \tag{1.3}$$

for some $\gamma(t) \in \mathbb{R}^N$ and $\theta(t) \in \mathbb{R}$ (Weinstein [37]).

(II) Let $u(t)$ be a solution of (NSC) such that $xu(t) \in L^2(\mathbb{R}^N)$ and $\|\nabla u(t)\| \to \infty$ as $t \to T_m$ for some $T_m \in (0, \infty)$. If $u(t)$ satisfies $\|(x-a)u(t)\| \to 0$ $(t \to T_m)$, then $u(t)$ must be of the form

$$(T_m - t)^{-N/2} \exp\left(\frac{-i|x-a|^2}{2(T_m - t)} \right) V\left(\frac{t}{T_m(T_m - t)}, \frac{x-a}{T_m - t} \right), \tag{1.4}$$

where $V(t)$ is also a solution of (NSC) in $C(\mathbb{R}_+; H^1(\mathbb{R}^N) \cup L^2(\mathbb{R}^N; |x|^2 dx))$ such that $E(V(t)) = 0$ (Nawa and M.Tsutsumi [29]).

(III) For given L points $\{a^1, a^2, \cdots, a^L\} \subset \mathbb{R}^N$, there exist a blow-up solution $u(t)$ of (NSC) such that

$$\left\| u(t) - \sum_{j=1}^{L} Q^j(t) \right\|_\sigma \to 0 \quad \text{as} \quad t \to T_m, \tag{1.5}$$

where

$$Q^j(t,x) = (T_m - t)^{-N/2} \exp\left(\frac{-i|x-a^j|^2}{2(T_m - t)} \right) Q\left(\frac{x-a^j}{T_m - t} \right) e^{it/T_m(T_m - t)} \tag{1.6}$$

for $T_m \in (0,\infty)$ (Merle [20]).

Remark 1.4. $Q(x)e^{it/2}$, which is a standing wave solution of (NSC), is transformed into Q^j by the space-time transformation appearing in the left hand side of (1.4) with $a = a^j$. We call this transformation pseudo-conformal transformation. Since we have $E(Q(\cdot)e^{it/2}) = 0$, Q^j is a blow-up solution of (NSC) such that $\|(x-a^j)u(t)\| \to 0$ $(t \to T_m)$ by virtue of (II).

Theses results require additional conditions on initial data (or solutions): $\|u_0\| = \|Q\|$ for (I); $|x|u_0 \in L^2(\mathbb{R}^N)$ for (II) and (III). Our purpose here is to investigate the asymptotic profile of generic H^1-blow-up soluton of (NSC).

Our main results are:

Theorem B. *If the solution $u(t)$ of (NSC) belongs to $u \in C_b([0,\infty); H^1(\mathbb{R}^N))$, then its energy must be nonnegative, i.e.,*

$$E(u(0)) \geqq 0.$$

In other words, If $E(u(0)) < 0$, then there exists $T_m \in (0,\infty]$ such that the corresponding solution $u(t)$ of (NSC) satisfies

$$\lim_{t\to T_m} \|\nabla u(t)\| = \lim_{t\to T_m} \|u(t)\|_\sigma = \infty,$$

i.e., $u(t)$ blows up (in finite time) or grows up (at infinity).

Remark 1.5. We note that Theorem B does not assert that every negative energy initial datum leads to the blow-up solution of (NSC). There remains a possibility that $T_m = \infty$. Hence, if we establish that all negative energy solution blows up in finite time, Theorem B follows . Unfortunately the blow-up of negative energy solutions has been proved under some conditions: $|x|u_0 \in L^2(\mathbb{R}^N)$ (Glassey [13]); $N \geqq 2$ and u_0 is radially symmetric (Ogawa-Y.Tsutsumi[30]); $N = 1$ and $p = 1 + \frac{4}{N}$ (Ogawa-Y.Tsutsumi[31]). The results of Ogawa-Tsutsumi [31] ensure that Theorem B holds true for the case of $N = 1$. We may say that Theorem B is an improvement of previous results. This theorem will play a crucial role in the proof of Theorem C below.

Theorem C. *Let $u(t)$ be the singular solution of (NSC) such that*

(C.1) $$\lim_{t\to T_m} \|\nabla u(t)\| = \lim_{t\to T_m} \|u(t)\|_\sigma = \infty$$

for some $T_m \in (0,\infty]$. Let $\{t_n\}$ be any sequence such that

(C.2) $$\sup_{t\in[0,t_n]} \|u(t)\|_\sigma = \|u(t_n)\|_\sigma.$$

For this $\{t_n\}$, we put

(C.3) $$\lambda_n = \frac{1}{\|u(t_n)\|_\sigma^{\sigma/2}}$$

and, we consider the rescaled functions

(C.4) $$u_n(t,x) = \lambda_n^{\frac{N}{2}} \overline{u(t_n - \lambda_n^2 t, \lambda_n x)}$$

for $t \in [0, t_n/\lambda_n^2)$. Then there exists a subsequence of $\{u_n\}$ (still denoted by $\{u_n\}$), which satisfies the following properties: there exist

(i) *a finite number of nontrivial solutions $u^1, u^2, \cdots, u^L$ of (NSC) in $C_b(\mathbb{R}_+; H^1(\mathbb{R}^N))$ with $E(u^j) = 0$, and*

(ii) *sequences $\{\gamma_n^1\}$, $\{\gamma_n^2\}$, $\cdots$, $\{\gamma_n^L\}$ in $\mathbb{R}^N$,* *such that, for any $T > 0$,*

(C.5) $$\lim_{n\to\infty}\sup_{t\in[0,T]}\left\| u_n(t,\cdot)-\sum_{j=1}^{L} u^j(t,\cdot-\gamma_n^j)\right\|_\sigma = 0,$$

(C.6) $$\lim_{n\to\infty}\sup_{t\in[0,T]}\left\| \nabla u_n(t,\cdot)-\sum_{j=1}^{L} \nabla u^j(t,\cdot-\gamma_n^j)\right\| = 0,$$

(C.7) $$\lim_{n\to\infty}\sup_{t\in[0,T]}\left\| u_n(t,\cdot)-\sum_{j=1}^{L} u^j(t,\cdot-\gamma_n^j) - \phi_n(t,\cdot)\right\| = 0,$$

where

(C.8) $$\phi_n(t,\cdot) = \exp(\frac{it}{2}\Delta) * \left(u_n(0,\cdot)-\sum_{j=1}^{L} u^j(0,\cdot-\gamma_n^j)\right).$$

Furthermore we have

(C.9) $$\|u_0\|^2 \geqq \sum_{j=1}^{L}\|u^j(t)\|^2 \geqq L\|Q\|^2.$$

where Q is a nontrivial solution of (1.4) and (1.5).

Remark 1.6. (1) If the solution satisfies $\lim_{t\to T_m}\|\nabla u(t)\| = \infty$, then we have, by the energy conservation law $\lim_{t\to T_m}\|u(t)\|_\sigma = \infty$. So, (C.1) is always assured.
(2) We can choose a sequence as in (C.2), since we have (C.1).
(3) The scaled function u_n in (C.4) also solves (NSC), and satisfies $\|u_n(t)\| = \|u(t)\|$ and $E(u_n(t)) = \lambda_n^2 E(u(t))$. This is a special feature of (NSC).

Here the important thing is the finiteness of u^j's. We shall use Theorem B to prove $E(u^j) = 0$ (for any j), which leads us to conclued the finiteness of u^j's.

Next two corollaries give sufficiently conditions that we have $L = 1$ in Theorem C. In the proofs of both corollaries we, however, do not need Theorem B to prove $L = 1$.

Corollary D. *Let $u(t)$, $\{t_n\}$, $u_n(t)$ and $Q(x)$ be as in Thoerem C. Assume in addition $u(0)$ is radially symmetric. Then there exists a subsequence of $\{u_n\}$ (still denoted by $\{u_n\}$), which satisfies the following properties: there exists a nontrivial solution u^1 of (NSC) in $C_b(\mathbb{R}_+; H^1(\mathbb{R}^N))$ with $E(u^1) = 0$, and such that, for any $T > 0$,*

(D.1) $$\lim_{n\to\infty}\sup_{t\in[0,T]}\| u_n(t,\cdot)-u^1(t,\cdot)\|_\sigma = 0,$$

(D.2) $$\lim_{n\to\infty}\sup_{t\in[0,T]}\| \nabla u_n(t,\cdot)-\nabla u^1(t,\cdot)\| = 0,$$

(D.3) $$\lim_{n\to\infty}\sup_{t\in[0,T]}\| u_n(t,\cdot)-u^1(t,\cdot) - \phi_n(t,\cdot)\| = 0,$$

where

(D.4) $$\phi_n(t,\cdot) = \exp(\frac{it}{2}\Delta) * \left(u_n(0,\cdot)-u^1(0,\cdot)\right).$$

Corollary E. *Let $u(t)$, $\{t_n\}$, $u_n(t)$ and $Q(x)$ be as in Thoerem C. Assume in addition $\|u(0)\| = \|Q\|$. Then there exists a subsequence of $\{u_n\}$ (still denoted by $\{u_n\}$), which satisfies the following properties: there exist*

(i) *a nontrivial solution u^1 of (NSC) in $C_b(\mathbb{R}_+; H^1(\mathbb{R}^N))$ with $E(u^1) = 0$, and*
(ii) *a sequences $\{\gamma_n^1\}$, in $\mathbb{R}^N$,*

such that, for any $T > 0$,

$$\lim_{n\to\infty} \sup_{t\in[0,T]} \| u_n(t,\cdot) - u^1(t, \cdot - \gamma_n^1) \|_2 = 0, \tag{E.1}$$

Remark 1.7. (1) Theorem C seems to be closely related to a phenomenon which has been observed in various nonlinear problems by the name of bubble theorem or concentrated compactness theorem (for example, see [4,15,17,18,22,33,34]). In fact, the proof of the above result is inspired by Brézis and Coron [4]. One may find that the underlying idea being the method of concentrated compactness due to Lions [17,18]. However, we do not use the general method of it. Our basic tool is the compactness device as in Lieb [16] (see also Brézis and Lieb [6] and Frölich, Lieb and Loss [9]). We extend Lieb's compactness lemma to space-time one, with which the Ascoli-Arzela theorem plays a crucial role in our analysis (see Proposition 3.1 in section 3 of this paper). The use of the general method of concentrated compactness in the study of blow-up problem for the nonlinear Schrödinger equatuion can be traced back to Weinstein [37]. We can safely say that our analysis investigates how the "dichotomy" (in the terminology of concentrated compactness) occurs in u_n defined by (C.4).
(2) In [28] we have proved Theorem C, Corollaries E and F under the assumption that Theorem B holds true. In this paper we shall sketch the proof of Theorem B in Sect.6. The proof of Therorem B is also closely related to the method of concentrated compactness, and the proof of it, however, is more complicated than the one of Theorem C.

Each u^j can be considered to correspond to the "strong" singularity in blow-up solution, since one has, by (C.7),

$$\lim_{n\to\infty} \sup_{t\in[t_n-\lambda_n^2 T, t_n]} \left\| \overline{u(t,\cdot)} - \sum_{j=1}^{L} u_n^j(t, \cdot - \gamma_n^j) - \widetilde{\phi}_n(t,\cdot) \right\| = 0, \tag{1.7}$$

where

$$u_n^j(t,x) = \frac{1}{\lambda_n^{N/2}} u^j\left(\frac{t_n - t}{\lambda_n^2}, \frac{x - \gamma_n^j \lambda_n}{\lambda_n}\right), \tag{1.8}$$

$$\widetilde{\phi}_n(t,x) = \frac{1}{\lambda_n^{N/2}} \phi_n\left(\frac{t_n - t}{\lambda_n^2}, \frac{x - \gamma_n^j \lambda_n}{\lambda_n}\right). \tag{1.9}$$

We note that there is a possibility that $\widetilde{\phi}_n$ produce "weak" singularities, around which the rate of blow-up is lower than $\|u(t)\|_\sigma$. If $N \geqq 2$, "weak" singularities may form a $N-1$ dimensional manifold as in the case of semilinear heat equations (Giga and Kohn [10]). If $\widetilde{\phi}_n$ produce no singularity, we can safely say that the blow-up set consists of finite number of points as in the case of one dimensional semilinear heat equations (Cheng and Matano [8]). However, there still remains a possibility of ergodic behavior of singularities, *i.e.*, even in the case of $L = 1$, $\lambda_n \gamma_n^1$ may perform a ergodic behavior (in full sequence).

Remark 1.8. As in the proof of [Theroem C;25], we can show that if $u(0) \in H^1(\mathbb{R}^N) \cap L^2(|x|^2 dx)$, then

$$\sup_{n\in\mathbb{N}} |\gamma_n^j \lambda_n| < \infty \quad (j = 1, 2, \cdots, L) \tag{1.10}$$

in Theorem C.

This paper is organized as follows.

In Section 2, we give a lemma concerning the evolution operator for the free Schrödinger equation, and give one proposition which concerns (A.1), (A.2) and Remark 1.3. We also sketch the proof of Threorem A.

In Section 3, we prepare a proposition which is crucial for both proofs of Theorem B and Theorem C.

In Section 4, we prove Theorem C under the assumption that Theorem B holds valid.

In Section 5, we prove Corollaries D and E.

In Section 6, we prove Theorem B.

Throughout this paper we will use the following notations:

Notations. $\partial_t = \dfrac{\partial}{\partial t}$, $\partial_k = \dfrac{\partial}{\partial x_k}$, $\nabla = (\partial_1, \cdots, \partial_n)$; $\langle\cdot,\cdot\rangle$ denotes the scalar product in L^2 and various pairing of dual spaces of functions; μ denotes the Lebesgue measure on $\mathbb{R}^N$; The set $\{x \in \mathbb{R}^N; f(x) > \varepsilon\}$ is simply represented as $[f > \varepsilon]$, which is also used to denote the characteristic function of this set; $B(y;R) = \{x \in \mathbb{R}^N; |x-y| \leqq R\}$; $C(I;F)$ denotes the space of strongly continuous functions from a interval $I \subset \mathbb{R}$ to a Fréchet space F; $L^\theta(I;B)$ denotes the space of measurable functions v from a interval $I \subset \mathbb{R}$ to a Banach space B such that $\|v(\cdot)\| \in L^\theta(I)$; $C_b(I;F) = L^\infty(I;F) \cap C(I;F)$; $U = U(t) = \exp\left(\dfrac{it}{2}\Delta\right)$.

2. Preliminaries

We recall [Proposition D;25] as Proposition 2.1 below. Although we have already proved this in [25] to study the "mass concentration" phenomena in blow-up solutions of (NSC) to obtain a similar formula (A.7), we shall give the proof of Proposition 2.1, since the proof of Theorem C in Sect.4 proceed after the model of it. Using this, we also consider the problem (A.1) and (A.2) in Theorem A (see Proposition 2.5 below). We shall also sketch the proof of Theorem A after [24,25,27].

We start with

Proposition 2.1. *Let $\{f_n(x)\}$ be a bounded sequence of functions in $H^1(\mathbb{R}^N)$ such that, for some positive constants C_σ,*

$$\limsup_{n\to\infty} \|f_n\|_\sigma^\sigma \geqq C_\sigma > 0, \tag{2.1}$$

$$\limsup_{n\to\infty} E_\sigma(f_n) = \limsup_{n\to\infty}\left(\|\nabla f_n\|^2 - \frac{2}{\sigma}\|f_n\|_\sigma^\sigma\right) = E_0. \tag{2.2}$$

Then there exists

(i) *a family of functions in $H^1(\mathbb{R}^N)$: $\mathfrak{A} = \{f^1, f^2, \cdots\}$, and*

(ii) *a family of sequences in $\mathbb{R}^N$: $\mathfrak{B} = \{\{y_n^1\}, \{y_n^2\}, \cdots\}$*

such that we have

$$\lim_{n\to\infty}\left|\sum_{k=2}^{j} y_n^k\right| = \infty \qquad (j \geqq 2), \tag{2.3}$$

and, for some subsequence (still denoted by the same letter), we have

$$f_n^1 \equiv f_n(\cdot + y_n^1) \to f^1 \not\equiv 0, \tag{2.4}$$
$$f_n^j \equiv (f_n^{j-1} - f^{j-1})(\cdot + y_n^j) \to f^j \not\equiv 0 \quad (j \geqq 2), \tag{2.5}$$

weakly in $H^1(\mathbb{R}^N)$ *and strongly in* $L^q(\Omega)$ *for any* $\Omega \Subset \mathbb{R}^N$ *and* $q \in [2, 2^*)$, *and*

$$\lim_{n\to\infty} \int_{\mathbb{R}^N} \left| |f_n^j|^q - |f_n^j - f^j|^q - |f^j|^q \right| dx = 0, \quad q \in [2, 2^*), \tag{2.6}$$
$$\lim_{n\to\infty} \{E(f_n^j) - E(f_n^j - f^j) - E(f^j)\} = 0, \tag{2.7}$$
$$E_0 - \lim_{n\to\infty} E(f_n^j - f^j) = \sum_{k=1}^{j} E(f^k). \tag{2.8}$$

Furthermore, we have: If $L \equiv \#\mathfrak{A} < \infty$,

$$\lim_{n\to\infty} \|f_n^L - f^L\|_{L^\sigma(\mathbb{R}^N)} = 0, \tag{2.9}$$
$$\lim_{n\to\infty} \left\{ \sup_{y\in\mathbb{R}^N} \int_{|x-y|<R} |f_n^L(x) - f^L(x)|^2 dx \right\} = 0; \tag{2.10}$$

if $\#\mathfrak{A} = \infty$,

$$\lim_{j\to\infty} \lim_{n\to\infty} \|f_n^j - f^j\|_{L^\sigma(\mathbb{R}^N)} = 0, \tag{2.11}$$
$$\lim_{j\to\infty} \lim_{n\to\infty} \left\{ \sup_{y\in\mathbb{R}^N} \int_{|x-y|<R} |f_n^j(x) - f^j(x)|^2 dx \right\} = 0, \tag{2.12}$$

for any $R > 0$.

Proposition 2.1 asserts that f_n behaves like a superposition of several functions of the form $f^j(t, x - \sum_{k=1}^{j} y_n^k)$, $(j = 1, 2, \cdots,)$ as $n \to \infty$.

The proof of Proposition 2.1 is inspired by the method of concentrated compactness due to Lions [17,18]. However, we do not use the general method of it. Our basic tool is the same compactness device as in Lieb [16]. We repeatedly use Lieb's lemma to decompose $\{f_n\}$ iteratively into several parts with the help of Brézis-Lieb's lemma [5].

We collect here these results needed for the proof of Proposition 2.1.

Lemma 2.2 (Frölich, Lieb and Loss [9]). *Let* $1 < \alpha < \beta < \gamma$ *and let* $g(x)$ *be a measurable function on* $\mathbb{R}^N$ *such that, for some positive constants* C_α, C_β, *and* C_γ,

$$\|g\|_\alpha \leqq C_\alpha, \quad \|g\|_\beta \geqq C_\beta > 0, \quad \|g\|_\gamma \leqq C_\gamma. \tag{2.13}$$

Then we have

$$\mu([\,|g| > \eta\,]) > C_1 \tag{2.14}$$

for some η, $C_1 > 0$ *depending on* α, β, γ, C_α, C_β, C_γ, *but not on* g.

Lemma 2.3 (Lieb [16]). *(1) In addition to the assumption ofLemma 2.2, we assume*

$$\|\nabla g\|_\alpha \leqq C_2, \tag{2.15}$$

for some positive constants C_2. Then there exists a shift $T_y g(x) = g(x+y)$ such that, for some constant $\delta = \delta(C_1, C_2, \eta)$,

$$\mu\left(B(0;1) \cap \left[|T_y g| > \frac{\eta}{2}\right]\right) > \delta. \tag{2.16}$$

(2) Let $1 < \alpha < \infty$ and let $\{f_n\}$ be a uniformly bounded sequence of functions in $W^{1,\alpha}(\mathbb{R}^N)$ such that $\mu([|f_n| > \eta]) \geqq C$ for some positive constans η and C. Then there exista a sequence $\{y_n\}$ in $\mathbb{R}^N$ such that, for some subsequence (stii denoted by the same letter),

$$f_n(\cdot + y_n) \rightharpoonup f \not\equiv 0, \tag{2.17}$$

weakly in $W^{1,\alpha}(\mathbb{R}^N)$.

Lemma 2.4 (Brézis and Lieb [5]). *Let $\{f_n(x)\}$ be an bounded family in $L^\alpha(\mathbb{R}^N)$ where $\Omega \subset \mathbb{R}^N$. Suppose that $f_n \to f$ a.e. in $\mathbb{R}^N$. Then we have*

$$\lim_{n\to\infty} \int_{\mathbb{R}^N} \left||f_n|^\alpha - |f_n - f|^\alpha - |f|^\alpha\right| dx = 0. \tag{2.18}$$

Remark 2.1. One may find proofs of these results in the next section, since we extend these results to functions of space-time variables in $C_b(I; W^{1,\alpha}(\mathbb{R}^N))$ for $I \subset \mathbb{R}$ in Sect.3. So, Lemmas 2.2 - 2.4 are considered to be a stationary version of lemmas in Sect.3.

Proof of Proposition 2.1. In what follows, We shall often extract subsequences without explicitly mentioning this fact. By the assumption (2.1), Lemma 2.2 and Lemma 2.3 (2), we can shift each f_n so that

$$f_n^1 \equiv f_n(\cdot + y_n^1) \rightharpoonup f^1 \not\equiv 0, \tag{2.19}$$

weakly in $H^1(\mathbb{R}^N)$, strongly in $L^q(\Omega)$ for any $\Omega \Subset \mathbb{R}^N$ and $q \in [2, 2^*)$, and *a.e.* in $\mathbb{R}^N$. Hence we have, by Lemma 2.4,

$$\lim_{n\to\infty} \int_{\mathbb{R}^N} \left||f_n^1|^q - |f_n^1 - f^1|^q - |f^1|^q\right| dx = 0, \quad q \in [2, 2^*), \tag{2.20}$$

Moreover we have

$$\lim_{n\to\infty} \|\nabla f_n^1\|^2 - \|\nabla f_n^1 - \nabla f^1\|^2 - \|\nabla f^1\|^2 = 0, \tag{2.21}$$

since $\nabla f_n^1 \rightharpoonup \nabla f^1$ weakly in L^2. Combining (2.20) with $q = \sigma$ and (2.21), we have

$$\lim_{n\to\infty} \{E(f_n^1) - E(f_n^1 - f^1) - E(f^1)\} = 0. \tag{2.22}$$

Suppose that $\limsup_{n\to\infty} \|f_n^1 - f^1\|_\sigma \neq 0$. Then there exists a sequence $\{y_n^2\}$ in $\mathbb{R}^N$ such that

$$f_n^2 \equiv (f_n^1 - f^1)(\cdot + y_n^2) \rightharpoonup f^2 \not\equiv 0, \tag{2.23}$$

weakly in $H^1(\mathbb{R}^N)$, strongly in $L^q(\Omega)$ for any $\Omega \Subset \mathbb{R}^N$ and $q \in [2, 2^*)$, and *a.e.* in $\mathbb{R}^N$ by Lemma 2.2 and Lemma 2.3 again. Moreover we have, by Lemma 2.4,

$$\lim_{n\to\infty} \int_{\mathbb{R}^N} \left||f_n^2|^q - |f_n^2 - f^2|^q - |f^2|^q\right| dx = 0, \quad q \in [2, 2^*), \tag{2.24}$$

$$\lim_{n\to\infty} \{E(f_n^2) - E(f_n^2 - f^2) - E(f^2)\} = 0, \tag{2.25}$$

$$E_0 - \lim_{n\to\infty} E(f_n^2 - f^2) = \sum_{k=1}^{2} E(f^k), \tag{2.26}$$

since we have

$$\lim_{n\to\infty} \|f_n^1 - f^1\|_q = \lim_{n\to\infty} \|f_n^2\|_q \quad q \in [2, 2^*), \tag{2.27}$$

$$\lim_{n\to\infty} E(f_n^1 - f^1) = \lim_{n\to\infty} E(f_n^2) \tag{2.28}$$

by the translation invariance of $\|\cdot\|_q$ and $E(\cdot)$. The local L^2 convergence of (2.24) and the nontriviality of f^2 yield that $\lim_{n\to\infty} |y_n^2| = \infty$: if not, we have that, for any $K \Subset \mathbb{R}^N$,

$$\lim_{n\to\infty} \int_K |f_n(x + y_n^1) - f^1(x) - f^2(x - y_n^2)|^2 dx = 0, \tag{2.29}$$

from which we conclude $f^2 \equiv 0$ by the local L^2 convergence of (2.19).

Repeating this procedure until the quatity $\limsup_{n\to\infty} \|f_n^j - f^j\|_\sigma$ becomes 0, we obtain desired families $\mathfrak{A}$ and $\mathfrak{B}$. It remains to prove (2.11). Suppose the contrary that, for some positive constant ε_0 and a subsequence $\{j(k)\}$ of $\{j\}$, one has

$$\liminf_{n\to\infty} \|f_n^{j(k)} - f^{j(k)}\|_\sigma^\sigma > \varepsilon_0. \tag{2.30}$$

Hence there is a constant C_0 essentially depends on ε_0 such that

$$\|f^{j(k)+1}\|_\sigma^\sigma > C_0, \tag{2.31}$$

since the size of $\|f^j\|_\sigma$ essentially depends on the lower bound of $\|f_n^{j-1} - f^{j-1}\|_\sigma^\sigma$ by Lemma 2.2, Lemma 2.3 (2) and the construction of f^j. We choose $k \in \mathbb{N}$ large enough (specified latter). Using the formula (2.6) with $j = 1, 2, \cdots, j(k)$, we have by (2.31) that, for enough subsequence with respect to n,

$$\begin{aligned} \limsup_{n\to\infty} \|f_n\|_\sigma^\sigma &> \limsup_{n\to\infty} \left(\|f_n\|_\sigma^\sigma - \|f_n^{j(k)} - f^{j(k)}\|_\sigma^\sigma\right) \\ &= \sum_{j=1}^{j(k)} \|f^j\|_\sigma^\sigma > \sum_{l=1}^{k} \|f^{j(l)}\|_\sigma^\sigma \\ &> kC_0. \end{aligned} \tag{2.31}$$

Thus we reach a contradiction, if we take k as $kC_0 \geqq \limsup_{n\to\infty} \|f_n\|_\sigma^\sigma$.

We next cosider (A.1) and (A.2) in the previous section.

Propositin 2.5. *Let*

$$m = \inf_{\substack{v\in H^1(\mathbb{R}^N)\\ v\not\equiv 0}} \left\{ \|v\| \; ; \; E(v) \equiv \|\nabla v\|^2 - \frac{2}{\sigma}\|v\|_\sigma^\sigma \leqq 0 \right\}, \tag{2.32}$$

$$\frac{1}{C_N} = \inf_{\substack{v\in H^1(\mathbb{R}^N)\\ v\not\equiv 0}} \frac{\|v\|^{\frac{4}{N}} \|\nabla v\|^2}{\|v\|_\sigma^\sigma} \equiv \inf_{\substack{v\in H^1(\mathbb{R}^N)\\ v\not\equiv 0}} J(v). \tag{2.33}$$

There is a function $Q \in H^1(\mathbb{R}^N) - \{0\}$ such that

$$\|Q\| = m, \tag{2.34}$$

$$\Delta Q - Q + |Q|^{\frac{4}{N}} Q = 0, \tag{2.35}$$

$$\frac{2}{\sigma}\|Q\|^{\frac{4}{N}} = \frac{1}{C_N}, \tag{2.36}$$

$$E(Q) = 0. \tag{2.37}$$

Remark.2.1. (1) The constant C_N in (2.33) is the best constant for the Gagliardo-Nirenberg inequality, so that

$$\|v\|_\sigma^\sigma \leqq C_N \|v\|^{\frac{4}{N}} \|\nabla v\|^2 \tag{G-N}$$

holds true for any $v \in H^1(\mathbb{R}^N)$.

(2) The estimate of the best constant C_N in terms of the solution of (2.35) can be traced back to Weinstein [34]. He consider the minimizing sequence for (2.33), and employ the radial rearrangement and radial compatness lemma to obtain the suitable minimizer. We shall give an another proof of Proposition 2.2. Our proof concerns the L^2 minimizing sequence of (2.32) instead of (2.33), and apply Proposition 2.1 to it. Thus our analysis do not rely on the radial rearrangement and radial compactness lemma. We obtain (2.37) without using the fact that $Qe^{i2/t}$ ia a standing wave solution of the nonlinear Schrödinger equation $2i\frac{\partial}{\partial t}u + \Delta u + |u|^{4/N}u = 0$.

Proof of Propositin 2.5. First we note that $m > 0$, more precisely

$$\frac{2}{\sigma} m^{\frac{4}{N}} \geqq \frac{1}{C_N} \tag{2.38}$$

by the Gagliardo-Nirenberg inequality (G-N).

Let $\{v_n\} \subset H^1(\mathbb{R}^N)$ be a minimizing sequence for (2.32), *i.e.*

$$\lim_{n\to\infty} \|v_n\| = m, \tag{2.39}$$

$$E(v_n) \leqq 0 \quad \text{for any } n \in \mathbb{N}. \tag{2.40}$$

It is worth while to note that the boundedness of $\{v_n\}$ in $H^1(\mathbb{R}^N)$ is not known. So we rescale v_n as follows:

$$Q_n(x) = \nu_n^{N/2} v(\nu_n x), \quad \nu_n = \frac{1}{\|v_n\|_\sigma^{\sigma/2}}, \tag{2.41}$$

so that we have

$$\begin{aligned} &\|Q_n\| = \|v_n\| \to m \quad \text{as } n \to \infty, \\ &\|Q_n\|_\sigma = 1, \\ &E(Q_n) = \nu_n^? E(v_n). \end{aligned} \tag{2.42}$$

Hence we get an H^1-bounded minimizing sequence $\{Q_n\}$ for (2.32). We apply Proposition 2.1 to this $\{Q_n\}$ to obtain a subsequence of $\{Q_n\}$ (we still denote it by $\{Q_n\}$) which satisfies

$$Q_n^1 \equiv Q_n(\cdot + y_n^1) \rightharpoonup Q^1 \not\equiv 0 \quad \text{weakly in } H^1(\mathbb{R}^N), \tag{2.43}$$

$$\lim_{n\to\infty} \{E(Q_n^1) - E(Q_n^1 - Q^1) - E(Q^1)\} = 0, \tag{2.44}$$

$$\lim_{n\to\infty} (\|Q_n^1\|^2 - \|Q_n^1 - Q^1\|^2 - \|Q^1\|^2) = 0, \tag{2.45}$$

for some $\{y_n^1\} \subset \mathbb{R}^N$. We Note that Q_n^1 is also a H^1-bounded minimizing sequence of (D.1). Now we suppose that $E(Q^1) > 0$, so that (2,44) and the fact $E_\sigma(Q_n^1) \leqq 0$ yeild that $E_\sigma(Q_n^1 - Q^1) \leqq 0$ for sufficientry large n. Thus we have $\|Q_n^1 - Q^1\| \geqq m$ for large n by the definition of m. Since $\lim_{n\to\infty} \|Q_n^1\| = m$, we get, from (2.45), $\|Q^1\| \leqq 0$, which is a contradiction. Thus we obtain

$$E(Q^1) \leqq 0. \tag{2.46}$$

It follows from (2.46) and the definition of m that $\|Q^1\| \geqq m$, so that we have

$$\|Q^1\| = m, \tag{2.47}$$

since $Q_n^1 \rightharpoonup Q^1$ weakly in $L^2(\mathbb{R}^N)$. Thus we get $\lim_{n\to\infty} \|Q_n^1 - Q^1\| = 0$. (So we have $L = 1$ in the terminology of Proposition 2.1)

Considering the continuous curve $Q_s : (0,\infty) \ni s \mapsto Q^1(\frac{\cdot}{s}) \in H^1(\mathbb{R}^N)$, we have

$$0 \leqq \lim_{s\uparrow 1} E(Q_s) = E(Q^1) \leqq 0, \tag{2.48}$$

since $E(Q_s) > 0$ if $s \in (0,1)$. Hence we have $\lim_{n\to\infty} \|Q_n^1 - Q^1\|_{H^1(\mathbb{R}^N)} = 0$. Thus we obtain

$$E(Q) = 0. \tag{2.49}$$

Let $\{w_n\} \subset H^1(\mathbb{R}^N)$ be a minimizing sequence for (2.33). We rescale w_n as follows:

$$W_n(x) = w_n\left(\frac{x}{\widetilde{\nu}_n}\right), \quad \widetilde{\nu}_n = \sqrt{\frac{\sigma\|\nabla w_n\|^2}{2\|w_n\|_\sigma^\sigma}}. \tag{2.50}$$

Then one has

$$J(W_n) = J(w_n), \tag{2.51}$$

$$E(W_n) = \widetilde{\nu}_n^{N-2}\left(\|\nabla w_n\|^2 - \widetilde{\nu}_n^2 \frac{2}{\sigma}\|w_n\|_\sigma^\sigma\right) = 0, \tag{2.52}$$

so that

$$\frac{1}{C_N} = \lim_{n\to\infty} \frac{2}{\sigma}\|W_n\|^{\frac{4}{N}}, \quad E_\sigma(W_n) = 0. \tag{2.53}$$

Thus by the definition of m, we have $\frac{2}{\sigma}m^{\frac{4}{N}} \leqq \frac{1}{C_N}$. Hence we obtain, by (2.38),

$$\frac{2}{\sigma}m^{\frac{4}{N}} = \frac{1}{C_N}. \tag{2.54}$$

Thus Q^1 is a critical point of $J(\cdot)$. Since $|\nabla|Q^1|| \leqq |\nabla Q^1|$, we may assume $Q^1 \geqq 0$. So we have

$$\frac{d}{dt}J(Q^1 + t\varphi)\bigg|_{t=0} = 0 \tag{2.55}$$

for any $\varphi \in C_0^\infty(\mathbb{R}^N)$. Hence Q^1 satisfies

$$\Delta Q^1 - \left(\frac{2\|\nabla Q^1\|^2}{N\|Q^1\|^2}\right) Q^1 + |Q^1|^{\frac{4}{N}} Q^1 = 0 \tag{2.56}$$

in the sense of distribution.

Taking

$$Q(x) = \widehat{\nu}^{N/2} Q^1(\widehat{\nu}x), \quad \widehat{\nu} = \sqrt{\frac{N\|Q^1\|^2}{2\|\nabla Q^1\|^2}}, \tag{2.57}$$

one can easily verifies that this Q satisfies (2.35) and $\|Q\| = \|Q^1\| = m$.

We briefly sketch the proof of Theorem A. In order to estimate the quantity A defined in (A.6), it is enough to prove

Proposition 2.6. *Let $u(t)$ be the blow-up solution of (NSC) such that*

$$\lim_{t\to T_m}\|\nabla u(t)\| = \lim_{t\to T_m}\|u(t)\|_\sigma = \infty \tag{2.58}$$

for some $T_m \in (0,\infty]$. Let $\{t_n\}$ be any sequence such that $t_n \to T_m$ as $n\to\infty$. Set

$$\widetilde{\lambda}_n \equiv \frac{1}{\|u(t_n)\|_\sigma^{\sigma/2}}, \tag{2.59}$$

$$u_n(x) \equiv \widetilde{\lambda}_n^{N/2} u(t_n, \widetilde{\lambda}_n x). \tag{2.60}$$

Then there exists a subsequence of $\{t_n\}$ (we still denote it by $\{t_n\}$) which satisfies the following properties: one can find a sequence $\{y_n\}$ in $\mathbb{R}^N$ such that, for any ε, there is a positive constant $K > 0$;

$$\liminf_{n\to\infty}\int_{B_n} |u(t_n,x)|^2 dx \geqq (1-\varepsilon)\|Q\|^2, \tag{2.61}$$

where $B_n = \{x \in \mathbb{R}^N; |x - \widetilde{\lambda}_n y_n| \leqq K\widetilde{\lambda}_n\}$.

Proof. We apply Proposition 2.1 to $\{\, u_n\ (= f_n)\,\}$. We have (2.8) with $E_0 = 0$. From this and (2.11), we can see $L < \infty$ or $E(f^k) \leqq 0$ for some $k \in \mathbb{N}$. $L < \infty$ also implies $E(f^k) \leqq 0$ for some $k \in \mathbb{N}$, since we have

$$\sum_{j=1}^{L} E(f^j) \leqq 0, \tag{2.62}$$

by (2.8) and (2.9). Thus Proposition 2.5 yeilds that $\|f^k\| \geqq \|Q\|$, which leads us to (2.61) with the help of Lemma 2.4 with $\alpha = 2$.

As to the "path" $\gamma(t)$ in Theorem A, we have the follwing proposition which immediately give the assertion of Theorem A.

Proposition 2.7. *Let $\rho(t,x)$ be a function defined on $\mathbb{R}\times\mathbb{R}^N$ such that*

$$\rho(\cdot) \in C([0,T); L^1(\mathbb{R}^N)), \quad \text{for some } T > 0, \tag{2.63}$$

$$\rho(t,x) \geqq 0, \tag{2.64}$$

$$\int_{\mathbb{R}^N} \rho(t,x)dx = \rho_0 \quad \text{constant}. \tag{2.65}$$

We define the quantity D_ρ as

$$D_\rho = \sup_{R>0}\left(\liminf_{t\uparrow T_m}\left(\sup_{y\in\mathbb{R}^N}\int_{|x-y|\leqq R}\rho(t,x)dx\right)\right). \tag{2.66}$$

Suppose that $D_\rho > \rho_0/2$. Then we have that, for any $\varepsilon \in (0, 1-\rho_0/2D_\rho)$, there is a constan $K > 0$ and $T_0 > 0$, and there exists a continuous function $\gamma(\cdot) \in C_b([T_0,T);\mathbb{R}^N)$ such that

$$\int_{|x|\leqq R} |\rho(t, x+\gamma(t))|^2 dx \geqq (1-\varepsilon)D_\rho, \quad t\in[T_0,T), \tag{2.67}$$

for any $R \geqq K$.

Proof. See [24,27]. Although we proved this proposition under the assumption that $D_\rho = \rho_0$ in [24,27], we can easily improve the proof to obtain this proposition.

We conclude this section with the estimates of the evolution operator the free Schrödinger equation. We say that a pair (ν, ρ) of indices is ***admissible*** if

$$\frac{1}{2} - \frac{1}{n} < \frac{1}{\rho} \leqq \frac{1}{2}$$

and

$$\frac{2}{\nu} = \frac{n}{2} - \frac{n}{q} \equiv \delta(\rho).$$

Lemma 2.8. *(1) For every $\phi \in L^2$ and for every admissible pair (ν, ρ), the function $t \mapsto U(t)\phi$ belongs to $C(\mathbb{R}; L^2) \cap L^\nu(\mathbb{R}; L^\rho)$ and satisfies*

$$\|U(\cdot)\phi\|_{L^\nu(\mathbb{R};\, L^\rho)} \leqq C\|\phi\|_2,$$

where C is independent of $\phi \in L^2$.
(2) Let I be an interval $I \subset \mathbb{R}$ and let $t_0 \in \bar{I}$. Let (κ, θ) be an admissible pair and let $v \in L^{\kappa'}(I; L^{\theta'})$, where $\dfrac{1}{\kappa'} + \dfrac{1}{\kappa} = \dfrac{1}{\theta'} + \dfrac{1}{\theta} = 1$. Then, for every admissible pair (ν, ρ), the function $t \mapsto \int_{t_0}^{t} U(t-\tau)v(\tau)d\tau$ belongs to $C(\bar{I}; L^2) \cap L^\nu(I; L^\rho)$ and satisfies

$$\left\| \int_{t_0}^{\cdot} U(\cdot - \tau)v(\tau)d\tau \right\|_{L^\nu(I;\, L^\rho)} \leqq C\|v\|_{L^{\kappa'}(I;\, L^{\theta'})},$$

where C is independent of $v \in L^{\kappa'}(I; L^{\theta'})$.

For Lemma 2.1, see [11, 14, 39].

3. A convergence result

The aim of this section is to prove the following propositon, which is crucial for both proofs of Theorem B and Theorem C.

Proposition 3.1. *Let $\{v_n\}$ be an equibounded family in $C([0,T]; H^1(\mathbb{R}^N))$ such that*

$$2i\frac{\partial v_n}{\partial t} + \Delta v_n + |v_n|^{\frac{4}{N}} v_n = g_n, \tag{3.1}$$

$$\sup_{t\in[0,T]} \|v_n(t)\|_\sigma \neq 0. \tag{3.2}$$

Here $\{g_n\}$ is an equibounded family in $C([0,T]; L^{\sigma'}(\mathbb{R}^N))$ such that, for any $R > 0$,

$$\lim_{n\to\infty} \sup_{t\in[0,T]} \|g_n(t,\cdot)\|_{\sigma'} = 0, \tag{3.3}$$

where $\frac{1}{\sigma} + \frac{1}{\sigma'} = 1$. Then there exists

(i) a nontrivial solution v of (NSC) in $C([0,T]; H^1(\mathbb{R}^N))$ and
(ii) a sequence $\{\gamma_n\} \subset \mathbb{R}^N$

such that for $\Omega \Subset \mathbb{R}^N$ and for some subsequence (still denoted by the same letter),

$$\widetilde{v}_n \equiv v_n(\,\cdot\,, \cdot + \gamma_n) \overset{*}{\rightharpoonup} v \quad \text{weakly* in } L^\infty([0,T]; H^1(\mathbb{R}^N)), \tag{3.4}$$

$$\widetilde{v}_n \to v \quad \text{strongly in } C([0,T]; L^\alpha(\Omega)) \text{ for } \alpha \in [2, 2^*) \tag{3.5}$$

as $n \to \infty$.

Furthermore we have

$$|\widetilde{v}_n|^{\alpha-2}\widetilde{v}_n - |\widetilde{v}_n - v|^{\alpha-2}(\widetilde{v}_n - v) - |v|^{\alpha-2}v \to 0 \tag{3.6}$$

strongly in $C([0,T]; L^{\alpha'}(\mathbb{R}^N))$,

$$\lim_{n\to\infty} \sup_{t\in[0,T]} \int_{\mathbb{R}^N} \big||\widetilde{v}_n|^\alpha - |\widetilde{v}_n - v|^\alpha - |v|^\alpha\big| dx = 0, \quad \alpha \in [2, 2^*), \tag{3.7}$$

$$\lim_{n\to\infty} \int_0^T \{E(\widetilde{v}_n) - E(\widetilde{v}_n - v) - E(v)\}\, dt = 0, \tag{3.8}$$

and for any $t \in [0,T]$

$$\lim_{n\to\infty} \{E(\widetilde{v}_n(t)) - E((\widetilde{v}_n - v)(t)) - E(v(t))\} = 0. \tag{3.9}$$

To prove Proposition 3.1, we prepare the following lemmas.

Lemma 3.2. *Let $1 < \alpha < \beta < \gamma$ and $I \subset \mathbb{R}$. Let $g(t,x)$ be a measurable function on $I \times \mathbb{R}^N$ such that, for some positive constants C_α, C_β, and C_γ,*

$$\operatorname*{ess.sup}_{t\in I} \|g(t)\|_\alpha^\alpha \leqq C_\alpha, \quad \operatorname*{ess.sup}_{t\in I} \|g(t)\|_\beta^\beta \geqq C_\beta > 0, \quad \operatorname*{ess.sup}_{t\in I} \|g(t)\|_\gamma^\gamma \leqq C_\gamma. \tag{3.10}$$

Then we have

$$\operatorname*{ess.sup}_{t\in I} \mu([\,|g(t,\cdot)| > \eta]) > C_1 \tag{3.11}$$

for some η, $C_1 > 0$ depending on α, β, γ, C_α, C_β, C_γ, but not on g.

Proof. Simple calculation with (3.10) implies that, for sufficiently small $\eta > 0$,

$$\begin{aligned}
&\int_{\mathbb{R}^N} |g(t,x)|^\beta dx \\
&= \int_{[|g(t,\cdot)|<\eta]} |g(t,x)|^\beta dx + \int_{[\eta<|g(t,\cdot)|<\frac{1}{\eta}]} |g(t,x)|^\beta dx + \int_{[|g(t,\cdot)|>\eta]} |g(t,x)|^\beta dx \\
&\leqq \frac{C_\beta}{4C_\alpha} \int_{[|g(t,\cdot)|<\eta]} |g(t,x)|^\alpha dx + \int_{[\eta<|g(t,\cdot)|<\frac{1}{\eta}]} |g(t,x)|^\beta dx + \frac{C_\beta}{4C_\gamma} \int_{[|g(t,\cdot)|>\eta]} |g(t,x)|^\gamma dx \\
&\leqq \frac{C_\beta}{4C_\alpha} \operatorname*{ess.sup}_{t\in I} \|g(t)\|_\alpha^\alpha + \int_{[\eta<|g(t,\cdot)|<\frac{1}{\eta}]} |g(t,x)|^\beta dx + \frac{C_\beta}{4C_\gamma} \operatorname*{ess.sup}_{t\in I} \|g(t)\|_\gamma^\gamma \\
&\leqq \frac{C_\beta}{2} + \mu([|g(t,\cdot)| > \eta])(\frac{1}{\eta})^\beta
\end{aligned}$$

Thus we have (3.11) with $C_1 = \dfrac{C_\beta}{2}\eta^\beta$.

Lemma 3.3. *Let $1 < \alpha < \infty$ and $I \Subset \mathbb{R}$. Let $g(t,x) \in L^\infty(I; W^{1,\alpha}(\mathbb{R}^N))$ such that*

$$\operatorname{ess.sup}_{t\in I} \|\nabla g(t)\|_\alpha \leqq C_2, \tag{3.12}$$

$$\operatorname{ess.sup}_{t\in I} \mu([|g(t)| > \eta]) \geqq C_3 \tag{3.13}$$

for some positive constants C_2 and C_3. Then there exists a shift $(T_y g)(t,x) = g(t, x+y)$ such that, for some constant $\delta = \delta(C_1, C_2, \eta)$,

$$\operatorname{ess.sup}_{t\in I} \mu\left(B(0;1) \cap \left[|T_y g(t)| > \frac{\eta}{2}\right]\right) > \delta. \tag{3.14}$$

Proof. For simplicity, we suppose that $g \in C_b(I; W^{1,\alpha}(\mathbb{R}^N))$. In general case, we employ Lusin's theorem.

We borrow the idea of Brézis in Lieb [16]. let f be a non trivial function such that $f(\cdot) \in C_b(I; W^{1,\alpha}_{loc}(\mathbb{R}^N))$, $\sup_{t\in I} \|\nabla f(t,\cdot)\|_\alpha \leqq 1$. Let

$$K = 1 + \frac{1}{\sup_{t\in\mathbb{R}} \|f(t)\|_\alpha^\alpha},$$

$$C_y = \text{cube in } \mathbb{R}^N \text{with center } y \text{ and the side length } \frac{2}{\sqrt{N}}.$$

First we claim that there exists a point $(s,y) \in I \times \mathbb{R}^N$ such that

$$\int_{C_y} |\nabla f(t,x)|^\alpha dx < K \sup_{s\in I} \int_{C_{y(t)}} |f(s,x)|^\alpha dx, \tag{3.15}$$

Indeed;suppose the contrary that, for any $(s,y) \in I \times \mathbb{R}^N$,

$$\int_{C_y} |\nabla f(s,x)|^\alpha dx \geqq K \sup_{s\in I} \int_{C_y} |f(s,x)|^\alpha dx. \tag{3.16}$$

We choose $\{y_k\}$ in $\mathbb{R}^N$ such that $C_{y_k} \cap C_{y_j} = \emptyset$ for $j \neq k$ and $\cap_{k=1}^\infty C_{y_k} = \mathbb{R}^N$. For this $\{y_k\}$, we have

$$\int_{C_{y_k}} |\nabla f(s,x)|^\alpha dx \geqq K \int_{C_{y_k}} |f(s,x)|^\alpha dx \tag{3.17}$$

for any $s \in \mathbb{R}$. Summing (3.17) from k equals 1 to ∞, we have

$$1 \geqq \int_{\mathbb{R}^N} |\nabla f(s,x)|^\alpha dx \geqq K \int_{\mathbb{R}^N} |f(s,x)|^\alpha dx \tag{3.18}$$

for any $s \in \mathbb{R}$. Now we take the supremum of the left hand side of (3.18) with respect to $s \in I$, so that we obtain, by the definition of K,

$$1 \geqq 1 + \sup_{t\in\mathbb{R}} \|f(t)\|_\alpha^\alpha > 1, \tag{3.19}$$

which is a contradiction.

By (3.15) we have

$$\int_{C_y} |\nabla f(s.x)|^\alpha + |f(s,x)|^\alpha dx < (K+1) \int_{C_y} |f(s,x)|^\alpha dx. \tag{3.20}$$

On the other hand, by the Sobolev inequlity, we have

$$\int_{C_y} |\nabla f(s,x)|^\alpha + |f(s,x)|^\alpha dx \geqq S\left(\int_{C_y} |f(s,x)|^{\alpha^*} dx\right)^{\frac{\alpha}{\alpha^*}}, \tag{3.21}$$

where $\frac{1}{\alpha^*}+\frac{1}{N}=\frac{1}{\alpha}$ if $\alpha < N$ and, if $\alpha \geqq N$, α^* is arbitrary with $\alpha < \alpha^* < \infty$. S is depends only on α, α^*. Combining (3.20) and (3.21), we have that, for some open, small interval I_s containing s,

$$S\sup_{t\in I_s}\left(\int_{C_y} |f(t,x)|^{\alpha^*} dx\right)^{\frac{\alpha}{\alpha^*}} < (K+1)\int_{C_y} |f(s,x)|^\alpha dx. \tag{3.22}$$

Here we have used the fact that $f(t)$ is continuous in t with values in $W^{1,\alpha}_{loc}(\mathbb{R}^N)$. From (3.22), we have by the Hölder inequlity that, for $\tau \in I_s$ (if necessary, taking smaller interval),

$$S\sup_{t\in I_s}\left(\int_{C_y} |f(t,x)|^{\alpha^*} dx\right)^{\frac{\alpha}{\alpha^*}} < (K+1)\left(C_y\cap \operatorname{supp} f(\tau,\cdot)\right)^{1-\frac{\alpha}{\alpha^*}}\left(\int_{C_y} |f(\tau,x)|^{\alpha^*} dx\right)^{\frac{\alpha}{\alpha^*}}. \tag{3.23}$$

Hence it follows that

$$S < (K+1)\operatorname{ess.sup}_{t\in I}\mu\left(C_y\cap \operatorname{supp} f(t,\cdot)\right)^{1-\frac{\alpha}{\alpha^*}}. \tag{3.24}$$

Now we put $f(t,x)=\max\left(|g(t,x)|-\frac{\eta}{2},0\right)$. For simplicity we assume that $\|\nabla g(t)\|_\alpha \leqq 1$ so that $\sup_{t\in\mathbb{R}}\|\nabla f(t,\cdot)\|_\alpha \leqq 1$. From (3.13), we have

$$\sup_{t\in\mathbb{R}}\|f(t)\|_\alpha^\alpha \geqq \left(\frac{\eta}{2}\right)^\alpha \sup_{t\in\mathbb{R}}\mu\left(\left[\,|g(t,\cdot)| > \frac{\eta}{2}\right]\right) \geqq \left(\frac{\eta}{2}\right)^\alpha C_3, \tag{3.25}$$

and thus $K \leqq 1+\frac{2^\alpha}{\eta^\alpha C_3}$. From (3.24) we deduce (3.14) for some point $y\in\mathbb{R}^N$ and some constant δ depending only on N, α, η, C_2 and C_1.

Remark 3.1. (1) Since we have Lemma 3.2, the condition (3.13) is always assured if g satisfies

$$\operatorname{ess.sup}_{t\in I}\|g(t)\|_\alpha^\alpha \leqq C_\alpha, \quad \operatorname{ess.sup}_{t\in I}\|g(t)\|_\beta^\beta \geqq C_\beta > 0, \quad \operatorname{ess.sup}_{t\in I}\|g(t)\|_{\alpha^*}^{\alpha^*} \leqq C_\gamma,$$

where $\beta=\alpha(1+\frac{\alpha}{N})$, (2) This lemma and Corollary 3.4 below are closely related to the compactness device as in Lieb [16].

Corollary 3.4. *Let $1<\alpha<\infty$ and $I \Subset \mathbb{R}$. Let $\{v_n(t,x)\}$ be a uniformly bounded sequence of functions in $C_b(I;W^{1,\alpha}(\mathbb{R}^N))$ such that $\operatorname{ess.sup}_{t\in I}\mu([|v_n(t)|>\eta]) \geqq C_3$ for some positive constans η and C_3. Furthermore we suppose that $\{v_n(t,x)\}$ is an equi-continuous family in $C_b(I;L^\alpha(\mathbb{R}^N))$. Then there exist a sequence $\{y_n\}$ in $\mathbb{R}^N$ and a nontrivial function $v\in L^\infty(I;W^{1,\alpha}(\mathbb{R}^N))$ such that for $\Omega \Subset \mathbb{R}^N$ and for some subsequence (still denoted by the same letter),*

$$\widetilde{v}_n \equiv v_n(\cdot,\cdot+y_n) \overset{*}{\rightharpoonup} v \quad \text{weakly* in } L^\infty(I;W^{1,\alpha}(\mathbb{R}^N)), \tag{3.26}$$

$$\widetilde{v}_n \to v \quad \text{strongly in } C(I;L^\alpha(\Omega)) \tag{3.27}$$

as $n\to\infty$.

Proof. We note that, for any $\{x_n\}\subset\mathbb{R}^N$, $\{v_n(t,x+x_n)\}$ is also an equi-continuous family in $C_b(I;L^\alpha(\mathbb{R}^N))$. Thus this corollary is a direct consequence of Lemma 3.3 and the Ascoli-Arzelà theorem.

To treat the nonlinear term in (3.1), we need

Lemma 3.5. *Let $1 < \alpha < \infty$. Let $\{f_n(t,x)\}$ be an bounded family in $L^\alpha(I \times \Omega)$ where $I \times \Omega \subset \mathbb{R} \times \mathbb{R}^N$. Suppose that $f_n \to f$ a.e. in $I \times \Omega$. Then we have*

$$|f_n|^{\alpha-2} f_n - |f_n - v|^{\alpha-2}(f_n - f) - |f|^{\alpha-2} f \to 0 \quad \text{in } L^{\alpha'}(I \times \Omega), \tag{3.28}$$

where $\frac{1}{\alpha} + \frac{1}{\alpha'} = 1$, and we have

$$\lim_{n\to\infty} \iint_{I\times\Omega} \big||f_n|^\alpha - |f_n - f|^\alpha - |f|^\alpha\big|\, dtdx = 0. \tag{3.29}$$

Furthermore if I is a compact interval, and if $\{f_n(t,x)\}$ is an equi-continuous family in $C(I; L^\alpha(\Omega))$, then we have

$$|f_n|^{\alpha-2} f_n - |f_n - v|^{\alpha-2}(f_n - f) - |f|^{\alpha-2} f \to 0 \quad \text{in } L^\infty(I; L^\alpha(\Omega)), \tag{3.30}$$

and

$$\lim_{n\to\infty} \sup_{t\in I} \int_\Omega \big||f_n|^\alpha - |f_n - f|^\alpha - |f|^\alpha\big|\, dx = 0. \tag{3.31}$$

Proof. If we establish the assertions of (3.28) and (3.29), one can easily verifies (3.30) and (3.31).

(3.29) is purely a consequence of the Brézis-Lieb lemma [5]. The convergence (3.28) also follows from the same argument performed in [5]. One can easily observed that for any $\varepsilon > 0$, there exists a positive constant C_ε such that

$$\big||f_n|^{\alpha-2} f_n - |f_n - v|^{\alpha-2}(f_n - f) - |f|^{\alpha-2} f\big| \leqq \varepsilon |f_n - f|^{\alpha-1} + C_\varepsilon |f|^{\alpha-1}. \tag{3.32}$$

We note that for $I \Subset \mathbb{R}$,

$$\varepsilon |f_n - f|^{\alpha-1} + C_\varepsilon |f|^{\alpha-1} \in L^{\alpha'}(I \times \mathbb{R}^N). \tag{3.33}$$

Now set

$$W_{\varepsilon,n} = \Big[\big||f_n|^{\alpha-2} f_n - |f_n - v|^{\alpha-2}(f_n - f) - |f|^{\alpha-2} f\big| - \varepsilon |f_n - f|^{\alpha-1}\Big]_+ \tag{3.34}$$

where $[a]_+ = \max(a, 0)$, so that $W_{\varepsilon,n}(t,x) \to 0$ *a.e.* as $n \to \infty$ and by (3.32), $W_{\varepsilon,n} \leqq C_\varepsilon |v|^{\alpha-1}$. Thus the dominated convergence theorem implies

$$\lim_{n\to\infty} \iint_{I\times\mathbb{R}^N} W_{\varepsilon,n}(t,x)^{\alpha'} dt\, dx = 0. \tag{3.35}$$

However we have

$$\big||f_n|^{\alpha-2} f_n - |f_n - v|^{\alpha-2}(f_n - f) - |f|^{\alpha-2} f\big| \leqq W_{\varepsilon,n} + \varepsilon |f_n - f|^{\alpha-1} \tag{3.36}$$

and, thus

$$\limsup_{n\to\infty} \iint_{I\times\mathbb{R}^N} \big||f_n|^{\alpha-2} f_n - |f_n - v|^{\alpha-2}(f_n - f) - |f|^{\alpha-2} f\big|\, dtdx \\ \leqq 2^{\alpha'-1} \varepsilon \limsup_{n\to\infty} \iint_{I\times\mathbb{R}^N} |f_n - f|^\alpha\, dtdx. \tag{3.37}$$

Hence we obtain (3.32).

Now we are in a position to prove Proposition 3.1.

Proof of Proposition 3.1. Since $\{v_n\}$ and $\{g_n\}$ are equi-bounded families in $L^\infty(I;H^1(\mathbb{R}^N))$ and $L^\infty(I;L^{\sigma'}(\mathbb{R}^N))$, respectively, it follows from (3.1) that $\{\frac{\partial v_n}{\partial t}\}$ is an equibounded family in $L^\infty(\mathbb{R};H^{-1})$ where H^{-1} is the dual space of $H^1(\mathbb{R}^N)$. From the identity

$$\|v_n(t)-v_n(s)\|^2 = 2\int_s^t \Re\left\langle \frac{\partial}{\partial\tau}v_n(\tau),\, v_n(\tau)-v_n(s)\right\rangle d\tau, \tag{3.38}$$

we have

$$\|v_n(t)-v_n(s)\|^2 \leqq 4|t-s|\|v_n\|_{L^\infty(\mathbb{R};H^1)}\left\|\frac{\partial}{\partial\tau}v_n\right\|_{L^\infty(\mathbb{R};H^{-1})}, \tag{3.39}$$

where $t,s\in I$ and $H^1 = H^1(\mathbb{R}^N)$. By the Gagliardo–Nirenberg inequality, we also obtain, for $\alpha\in[2,\,2^*)$,

$$\|v_n(t)-v_n(s)\|_\alpha^\alpha \leqq C|t-s|^{\alpha(1-\theta)/2}\|v_n\|_{L^\infty(\mathbb{R};H^1)}^{\alpha(1+\theta)/2}\left\|\frac{\partial}{\partial\tau}v_n\right\|_{L^\infty(\mathbb{R};H^{-1})}^{\alpha(1-\theta)/2}, \tag{3.40}$$

where $\theta = \dfrac{\alpha(\theta)}{2}$. Hence, for any $\{x_n\}\subset\mathbb{R}^N$, $\{v_n(t,x+x_n)\}$ is an equi-continuous family in $C_b(\mathbb{R};L^\alpha(\mathbb{R}^N))$ for $\alpha\in[2,\,2^*)$. On the other hand, we have, by (3.2) and Lemma 3.2, that, for some pocitive constant C independent of n,

$$\operatorname*{ess.\,sup}_{t\in I}\mu([|v_n(t)|>\eta]) \geqq C. \tag{3.41}$$

Thus, by virtue of Corollary 3.4, there exist a nontrivial function v in $C([0,T];H^1(\mathbb{R}^N))$ and a sequence $\{\gamma_n\}\subset\mathbb{R}^N$ such that, for $\Omega\Subset\mathbb{R}^N$ and for some subsequence (still denoted by the same letter),

$$\widetilde{v}_n \equiv v_n(\,\cdot\,,\,\cdot\,+\gamma_n) \overset{*}{\rightharpoonup} v \quad \text{weakly* in } L^\infty([0,T];H^1(\mathbb{R}^N)), \tag{3.42}$$

$$\widetilde{v}_n \to v \quad \text{strongly in } C([0,T];L^\alpha(\Omega)) \text{ for } \alpha\in[\,2,\,2^*) \tag{3.43}$$

as $n\to\infty$.

We claim that the limit function v solves (NSC). Since v_n solves (3.1), it holds that, for any $\chi\in C_0^\infty(\mathbb{R})$ and $\varphi\in C_0^\infty(\mathbb{R}^N)$,

$$-\int_{\mathbb{R}}\langle 2i\widetilde{v}_n,\varphi\rangle\dot{\chi}dt + \int_{\mathbb{R}}\langle\Delta\widetilde{v}_n,\varphi\rangle\chi dt + \int_{\mathbb{R}}\langle|\widetilde{v}_n|^{\frac{4}{N}}\widetilde{v}_n,\varphi\rangle\chi dt = \int_{\mathbb{R}}\langle\widetilde{g}_n,\varphi\rangle\chi dl, \tag{3.45}$$

where $\widetilde{g}_n\equiv g_n(t,x+\gamma_n)$ and $\dot{} = \dfrac{d}{dt}$. Hence, by (3.3), (3.42) and (3.43), we have

$$2i\frac{d}{dt}\langle v,\varphi\rangle + \langle\Delta v,\varphi\rangle + \langle|v|^{\frac{4}{N}}v,\varphi\rangle = 0 \tag{3.46}$$

in $\mathcal{D}'(\mathbb{R})$ (the dual of $C_0^\infty(\mathbb{R})$), so that the standard argument shows

$$2i\frac{\partial v}{\partial t} + \Delta v + |v|^{\frac{4}{N}}u^1 = 0, \quad \text{in} \quad H^{-1} \tag{3.47}$$

By (3.39), we have $v\in C_b(I;L^2(\mathbb{R}^N))$. This fact together with $v\in L_\infty(I;H^1(\mathbb{R}^N))$ implies $v\in C_w(I;H^1(\mathbb{R}^N))$. Hence v has the definite initial value $v(0)\in H^1(\mathbb{R}^N)$. Thus the uniqueness theorem of Kato [14] yeilds $v\in C_b(\mathbb{R};H^1(\mathbb{R}^N))$.

Since $\widetilde{v}_n$ converges to v in $C(I;L^2(\Omega))$ and *a fortiori* in $L^2(I\times\Omega)$ for any $\Omega\Subset\mathbb{R}^N$, we can extract a subsequence from $\{\widetilde{v}_n\}$ (still denoted by $\{\widetilde{v}_n\}$) such that $\widetilde{v}_n\to v$ *a.e.* $I\times\Omega$. Thus (3.6) and (3.7) follow from Lemma 3.5 and the equi-continuity of $\{\widetilde{v}_n\}$. (3.8) follows from (3.6) and the weak convergence of $\{\nabla\widetilde{v}_n\}$ in $L^2(I\times\mathbb{R}^N)$. (3.9) follows from (3.6) and the weak convergence of $\{\nabla\widetilde{v}_n(t)\}$ in $L^2(\mathbb{R}^N)$.

4. Proof of Theorem C

We shall prove Theorem C under the assumption that Theorem B holds valid. Proposition 3.1 plays a crucial role in our analysis.

Let $u(t)$ be the singular solution of (NSC) with initial datum $u(0) = u_0 \in H^1(\mathbb{R}^N)$ such that $u(\,\cdot\,) \in C([0,T_m); H^1(\mathbb{R}^N))$ and

$$\lim_{t\to T_m} \|\nabla u(t)\| = \lim_{t\to T_m} \|u(t)\|_\sigma = \infty \tag{4.1}$$

for some $T_m \in (0,\infty]$. Let $\{t_n\}$ be any sequence such that

$$\sup_{t\in[0,t_n)} \|u(t)\|_\sigma = \|u(t_n)\|_\sigma. \tag{4.2}$$

For this $\{t_n\}$, we put

$$\lambda_n = \frac{1}{\|u(t_n)\|_\sigma^{\sigma/2}} \tag{4.3}$$

and, we consider the rescaled functions

$$u_n(t,x) = \lambda_n^{\frac{N}{2}} \overline{u(t_n - \lambda_n^2 t, \lambda_n x)} \tag{4.4}$$

for $t \in [0, t_n/\lambda_n^2)$. One can easily see that

$$2i\frac{\partial u_n}{\partial t} + \Delta u_n + |u_n|^{\frac{4}{N}} u_n = 0, \tag{4.5}$$

$$\|u_n(t)\| = \|u_0\|, \tag{4.6}$$

$$\sup_{t\in[0,T]} \|u_n(t)\|_\sigma = 1 \quad \text{for any} \quad T > 0, \tag{4.7}$$

$$E(u_n(t)) = \lambda_n^2 E(u_0)\,(\to 0 \quad \text{as}\, n\to\infty). \tag{4.8}$$

From (4.5), (4.6) and (4.7), it follows that $\{u_n\}$ is an equi-bounded family in $L^\infty(\mathbb{R}_+; H^1(\mathbb{R}^N))$. Thus, by Proposition 3.1, there exist

(i) a nontrivial solution u^1 of (NSC) in $C([0,\infty); H^1(\mathbb{R}^N))$

(ii) a sequence $\{y_n^1\} \subset \mathbb{R}^N$

such that for $\Omega \Subset \mathbb{R}^N$ and for some subsequence (still denoted by the same letter),

$$u_n^1 \equiv u_n(\,\cdot\,, \cdot + y_n^1) \overset{*}{\rightharpoonup} u^1 \text{ weakly* in } L^\infty([0,\infty); H^1(\mathbb{R}^N)), \tag{4.10}$$

$$u_n^1 \to u^1 \quad \text{strongly in } C([0,T]; L^\alpha(\Omega)) \tag{4.11}$$

for $\alpha \in [2, 2^*)$ as $n \to \infty$. Furthermore we have that, for $\alpha \in [2, 2^*)$,

$$|u_n^1|^{\alpha-2}u_n^1 - |u_n^1 - u^1|^{\alpha-2}(u_n^1 - u^1) - |u^1|^{\alpha-2}u^1 \to 0 \tag{4.12}$$

strongly in $C([0,T]; L^{\alpha'}(\mathbb{R}^N))$,

$$\lim_{n\to\infty} \sup_{t\in[0,T]} \int_{\mathbb{R}^N} \big||u_n^1|^\alpha - |u_n^1 - u^1|^\alpha - |u^1|^\alpha\big| dx = 0, \tag{4.13}$$

$$\lim_{n\to\infty} \int_0^T \{E(u_n^1) - E(u_n^1 - u^1) - E(u^1)\}\, dt = 0, \tag{4.14}$$

and, for any $t \in \mathbb{R}_+$

$$\lim_{n\to\infty} \{E(u_n^1(t)) - E((u_n^1 - u^1)(t)) - E(u^1(t))\} = 0. \tag{4.15}$$

Here we note that $u^1(t)$ is defined on $\mathbb{R}_+$, and that $u^1 \in L^\infty(\mathbb{R}_+; H^1(\mathbb{R}^N))$. Hence we obtain, by Theorem B, that

$$E(u^1(t)) = E(u^1(0)) \geqq 0 \quad t \in [0,\infty). \tag{4.16}$$

Suppose that $\limsup_{n\to\infty} \|u_n^1 - u^1\|_\sigma \neq 0$. $u_n^1 - u^1$ satisfies

$$2i\frac{\partial(u_n^1 - u^1)}{\partial t} + \Delta(u_n^1 - u^1) + |u_n^1 - u^1|^{\frac{4}{N}}(u_n^1 - u^1) = g_n^1, \tag{4.17}$$

where

$$g_n^1(t,x) = -(|u_n^1|^\sigma - |u_n^1 - u^1|^\sigma - |u^1|^\sigma)(t,x). \tag{4.18}$$

We note here that (4.12) implies that for any $\{x_n\} \subset \mathbb{R}^N$ and for any $T > 0$,

$$\lim_{n\to\infty} \sup_{t\in[0,T)} \|g_n^1(t, \cdot + x_n)\|_\sigma = 0. \tag{4.19}$$

Clearly $\{u_n^1 - u^1\}$ is an equi-bounded family in $L^\infty([0,\infty); H^1(\mathbb{R}^N))$. Thus we apply Proposition 3.1 to $\{u_n^1 - u^1\}$ to obtain

(i) a nontrivial solution u^2 of (NSC) in $C_b([0,\infty); H^1(\mathbb{R}^N))$ with $E(u^2(t)) = E(u^2(0)) \geqq 0$, $t \in [0,\infty)$ (by Theorem B) and

(ii) a sequence $\{y_n^2\} \subset \mathbb{R}^N$

such that for $\Omega \Subset \mathbb{R}^N$ and for some subsequence (still denoted by the same letter),

$$u_n^2 \equiv (u_n^1 - u^1)(\cdot, \cdot + y_n^2) \overset{*}{\rightharpoonup} u^2 \tag{4.20}$$

weakly* in $L^\infty([0,\infty); H^1(\mathbb{R}^N))$, and

$$u_n^2 \to u^2 \quad \text{strongly in } C([0,T]; L^\alpha(\Omega)) \tag{4.21}$$

for $\alpha \in [2, 2^*)$ as $n \to \infty$. Furthermore we have that, for $\alpha \in [2, 2^*)$,

$$|u_n^2|^{\alpha-2}u_n^2 - |u_n^2 - u^2|^{\alpha-2}(u_n^2 - u^2) - |u^2|^{\alpha-2}u^2 \to 0 \tag{4.22}$$

strongly in $C([0,T]; L^{\alpha'}(\mathbb{R}^N))$,

$$\lim_{n\to\infty} \sup_{t\in[0,T]} \int_{\mathbb{R}^N} \left||u_n^2|^\alpha - |u_n^2 - u^2|^\alpha - |u^2|^\alpha\right| dx = 0, \tag{4.23}$$

$$\lim_{n\to\infty} \int_0^T \{E(u_n^2) - E(u_n^2 - u^2) - E(u^2)\}\, dt = 0, \tag{4.24}$$

and, for any $t \in \mathbb{R}_+$,

$$\lim_{n\to\infty} \{E(u_n^2(t)) - E((u_n^2 - u^2)(t)) - E(u^2(t))\} = 0. \tag{4.25}$$

We also have by (4.10) and (4.20) that

$$\lim_{n\to\infty} |y_n^2| = \infty, \tag{4.26}$$

and we can see that $(u_n^2 - u^2)$ satisfies

$$2i\frac{\partial(u_n^2 - u^2)}{\partial t} + \Delta(u_n^2 - u^2) + |u_n^2 - u^2|^{\frac{4}{N}}(u_n^2 - u^2) = g_n^2, \tag{4.27}$$

where

$$g_n^2(t,x) = g_n^1(t, x + y_n^1) - \big(|u_n^2|^\sigma - |u_n^2 - u^2|^\sigma - |u^2|^\sigma\big)(t,x). \tag{4.28}$$

(4.12) and (4.22) imply that for any $\{x_n\} \subset \mathbb{R}^N$ and for any $T > 0$,

$$\lim_{n\to\infty} \sup_{t\in[0,T)} \|g_n^2(t, \cdot + x_n)\|_\sigma = 0. \tag{4.29}$$

It is worthwhile to note that it holds from (4.13), (4.15), (4.23) and (4.25) that, for any $t \in \mathbb{R}_+$ and for $\alpha \in [2, 2^*)$,

$$\lim_{n\to\infty} \big(\|u_n(t)\|_\alpha^\alpha - \|(u_n^2 - u^2)(t)\|_\alpha^\alpha\big) = \sum_{k=1}^{2} \|u^k(t)\|_\alpha^\alpha \tag{4.30}$$

$$\lim_{n\to\infty} \{E(u_n(t)) - E((u_n^2 - u^2)(t))\} = \sum_{k=1}^{2} E(u^k(t)), \tag{4.31}$$

since $\|\cdot\|_\alpha$ and $E(\cdot)$ are invariant under the action of space-translations.

The proof of Theorem 1 consists of iterating the construction of Proposition 3.1. In what follows, we freely take enough subsequences. Repeating the procedure above, we inductively obtain: $(j \geqq 2)$

(i) nontrivial solutions u^j of (NSC) in $C_b([0,\infty); H^1(\mathbb{R}^N))$ with $E(u^j(t)) = E(u^j(0)) \geqq 0 \quad t \in [0,\infty)$ (by Theorem B) and

(ii) a sequence $\{y_n^j\} \subset \mathbb{R}^N$ with $\lim_{n\to\infty} \left|\sum_{k=2}^{j} y_n^k\right| = \infty$

such that for $\Omega \Subset \mathbb{R}^N$ and for any $T > 0$,

$$u_n^j \equiv (u_n^{j-1} - u^{j-1})(\cdot, \cdot + y_n^j) \overset{*}{\rightharpoonup} u^j \tag{4.32}$$

weakly* in $L^\infty([0,\infty); H^1(\mathbb{R}^N))$, and

$$u_n^j \to u^j \quad \text{strongly in } C([0,T]; L^\alpha(\Omega)) \tag{4.33}$$

for $\alpha \in [2, 2^*)$ as $n \to \infty$; and furthermore we have

$$|u_n^j|^{\frac{4}{N}} u_n^j - |u_n^j - u^j|^{\frac{4}{N}}(u_n^j - u^j) - |u^j|^{\frac{4}{N}} u^j \to 0 \tag{4.34}$$

strongly in $C([0,T]; L^{\sigma'}(\mathbb{R}^N))$,

$$\lim_{n\to\infty} \sup_{t\in[0,T]} \int_{\mathbb{R}^N} \big||u_n^j|^\alpha - |u_n^j - u^j|^\alpha - |u^j|^\alpha\big| dx = 0, \tag{4.35}$$

$$\lim_{n\to\infty} \int_0^T \{E(u_n^j) - E(u_n^j - u^j) - E(u^j)\}\, dt = 0, \tag{4.36}$$

and, for any $t \in \mathbb{R}_+$,

$$\lim_{n\to\infty} \{E(u_n^j(t)) - E((u_n^j - u^j)(t)) - E(u^j(t))\} = 0, \tag{4.37}$$

$$\lim_{n\to\infty} \big(\|u_n(t)\|_\alpha^\alpha - \|(u_n^j - u^j)(t)\|_\alpha^\alpha\big) = \sum_{k=1}^{j} \|u^k(t)\|_\alpha^\alpha, \tag{4.38}$$

$$\lim_{n\to\infty} \{E(u_n(t)) - E((u_n^j - u^j)(t))\} = \sum_{k=1}^{j} E(u^k(t)); \tag{4.39}$$

besides, $(u_n^j - u^j)$ satisfies

$$2i\frac{\partial(u_n^j - u^j)}{\partial t} + \Delta(u_n^j - u^j) + |u_n^j - u^j|^{\frac{4}{N}}(u_n^j - u^j) = g_n^j, \tag{4.40}$$

where

$$g_n^j(t,x) = g_n^{j-1}(t, x + y_n^{j-1}) - \big(|u_n^j|^\sigma - |u_n^j - u^j|^\sigma - |u^j|^\sigma\big)(t,x). \tag{4.41}$$

(4.34) implies that for any $\{x_n\} \subset \mathbb{R}^N$ and for any $T > 0$,

$$\lim_{n\to\infty} \sup_{t\in[0,T)} \|g_n^j(t, \cdot + x_n)\|_\sigma = 0. \tag{4.42}$$

We claim that the iteration must terminated at some index $L \in \mathbb{N}$. Suppose the contrary that $L = \infty$. In this case, we have

$$\lim_{j\to\infty} \lim_{n\to\infty} \sup_{t\in[0,T]} \|(u_n^j - u^j)(t)\|_\sigma = 0. \tag{4.43}$$

Indeed; suppose the contrary that, for some positive constant ε_0 and a subsequence $\{j(k)\}$ of $\{j\}$, one has

$$\liminf_{k\to\infty} \lim_{n\to\infty} \sup_{t\in[0,T]} \|(u_n^{j(k)} - u^{j(k)})(t))\|_\sigma > \varepsilon_0. \tag{4.44}$$

Hence there is a constant C_0 essentially depends on ε_0 such that

$$\sup_{t\in[0,T]} \|u^{j(k)+1}(t)\|_\sigma^\sigma > C_0, \tag{4.45}$$

since the size of $\sup_{t\in[0,T]} \|u^j(t)\|_\sigma$ essentially depends on the lower bound of $\sup_{t\in[0,T]} \|(u_n^{j-1} - u^{j-1})(t)\|_\sigma$ by Lemma 3.2, Lemma 3.3 and the construction of u^j. We choose $k \in \mathbb{N}$ large enough (specified latter). Using the formula (4.38) with $\alpha = \sigma$ and $j = j(k)$, we have by (4.45) that, for enough subsequence with respect to n,

$$\begin{aligned} 1 &> \limsup_{n\to\infty} \sup_{t\in[0,T]} \Big(\|u^{j(k)+1}(t)\|_\sigma^\sigma - \|(u_n^{j(k)} - u^{j(k)}(t))\|_\sigma\Big) \\ &= \sup_{t\in[0,T]} \sum_{j=1}^{j(k)} \|u^j(t)\|_\sigma^\sigma > \sup_{t\in[0,T]} \sum_{l=1}^{k} \|u^{j(l)}(t)\|_\sigma^\sigma \\ &> kC_0. \end{aligned} \tag{4.46}$$

Thus we reach a contradiction, if we take k as $kC_0 \geqq 1$. Hence we get (4.43). (4.39) together with (4.7) and (4.43) yields that

$$0 \leqq \lim_{j\to\infty} \sum_{k=1}^{j} E\big(u^k(t)\big) \leqq \lim_{j\to\infty} \lim_{n\to\infty} \sup_{t\in[0,T]} \|(u_n^j - u^j)(t)\|_\sigma = 0, \tag{4.47}$$

so that we have

$$E\big(u^k(t)\big) = 0, \quad j \in \mathbb{N}. \tag{4.48}$$

Then, by Proposition 2.5, we have

$$\|u^j(t)\| = \|u^j(0)\| \geqq \|Q\|, \quad j \in \mathbb{N}. \tag{4.49}$$

This together with (4.38) with $\alpha = 2$ implies

$$j\|Q\|^2 \leqq \sum_{k=1}^{j} \|u^k(t)\|^2 \leqq \|u_0\|^2, \tag{4.50}$$

which is a contradiction. Thus we obtain $L < \infty$, so that

$$\lim_{n\to\infty} \sup_{t\in[0,T]} \|(u_n^L - u^L)(t)\|_\sigma = 0 \tag{4.51}$$

which implies that (C.5) with $\gamma_n^j = \sum_{k=1}^{j} y_n^k$, since we have

$$(u_n^L - u^L)(t,x) \quad = u_n\left(t\,,\,x + \sum_{k=1}^{L} y_n^k\right) - \sum_{j=1}^{L} u^j \left(t\,,\,x + \sum_{k=j+1}^{L} y_n^k\right). \tag{4.52}$$

The formula (C.6) follows from (4.39), (4.51) and (4.52). It remains to prove (C.7). Noting that

$$\lim_{n\to\infty} \sup_{t\in[0,T]} \left\| |u_n^L - u^L|^{\frac{4}{N}} (u_n^L - u^L) \right\|_{\sigma'} = 0, \tag{4.53}$$

we have, by (4.34) with $j = L$ and (4.52), that

$$|u_n|^{\frac{4}{N}} u_n(t,\cdot) - \sum_{j=1}^{L} |u^j|^{\frac{4}{N}} u^j(t, \cdot - \gamma_n^j) \to 0 \tag{4.54}$$

strongly in $C([0,T]; L^{\sigma'}(\mathbb{R}^N))$. Since u_n satsfies

$$u_n(t) = U(t)u_n(0) + \frac{i}{2}\int_0^t U(t-\tau)(|u_n|^{\frac{4}{N}})u_n(\tau)d\tau, \tag{4.55}$$

and $\widetilde{u}^j(t.x) \equiv u^j(t, x - \gamma_n^j)$ satisfies

$$\widetilde{u}^j(t) = U(t)\widetilde{u}^j(0) + \frac{i}{2}\int_0^t U(t-\tau)(|\widetilde{u}^j|^{\frac{4}{N}})\widetilde{u}^j(\tau)d\tau, \tag{4.56}$$

we have

$$u_n(t) - \sum_{j=1}^{L} \widetilde{u}^j(t) - \phi_n(t) = \frac{i}{2}\int_0^t U(t-\tau)\left(|u_n|^{\frac{4}{N}} u_n - \sum_{j=1}^{L} |\widetilde{u}^j|^{\frac{4}{N}} \widetilde{u}^j\right) d\tau, \tag{4.57}$$

where

$$\phi_n(t\,,\,\cdot) = U(t)\left(u_n(0) - \sum_{j=1}^{L} \widetilde{u}^j(0)\right). \tag{4.58}$$

Estimating (4.57) with the help of Lemma 2.8 (2) ($\nu = \infty$, $\rho = 2$ and $\kappa = \theta = \sigma$), we have (C.7) by virtue of (4.54).

5. Proofs of Corollaries

Let $u(t)$, $\{t_n\}$, $u_n(t)$ and $Q(x)$ be as in the previous section. We shall prove Corollaries D and E which give sufficiently conditions that we have $L = 1$ in Theorem C. In the proofs of both corollaries we, however, do not need Theorem B to prove $L = 1$.

Proof of Corollary D. Let $H_r^1(\mathbb{R}^N) = \{v \in H^1(\mathbb{R}^N); v(x) = v(|x|), x \in \mathbb{R}^N\}$. Assume $u(0)$ is radially symmetric and $N \geqq 2$. Then the corresponding solution $u(t)$ is also radially symmetric in space variables. Precisely $u(\cdot) \in C([0, T_m); H_r^1(\mathbb{R}^N))$. We note the following fact:if, in Proposition 3.1, we assume in addition that $\{v_n\} \subset C_b([0,T]; H_r^1(\mathbb{R}^N))$, then we have that $\gamma_n \equiv 0$, and that

$$v_n \to v \quad \text{strongly in } C([0,T]; L^\alpha(\mathbb{R}^N)) \text{ for } \alpha \in (2,\, 2^*) \tag{5.1}$$

as $n \to \infty$, since the embedding $L^\alpha(\mathbb{R}^N) \hookrightarrow H_r^1(\mathbb{R}^N)$ for $\alpha \in (2, 2^*)$ is compact. Thus we have in the same way as in the proof of Theorem C that, for radially symmetric family $\{u_n\}$, there exists a nontrivial solution u^1 of (NSC) in $C([0,\infty); H_r^1(\mathbb{R}^N))$ such that for some subsequence (still denoted by the same letter),

$$|u_n|^{\frac{4}{N}} u_n - |u^1|^{\frac{4}{N}} u^1 \to 0$$

strongly in $C([0,T]; L^{\sigma'}(\mathbb{R}^N))$,

$$\lim_{n\to\infty} \sup_{t\in[0,T]} \|(u_n - u^1)(t)\|_\alpha = 0 \quad \alpha \in (2, 2^*), \tag{5.3}$$

$$\lim_{n\to\infty} \int_0^T \{E(u_n) - E(u_n - u^1) - E(u^1)\}\, dt = 0, \tag{5.4}$$

and for any $t \in [0,T]$

$$\lim_{n\to\infty} \{E(u_n(t)) - E((u_n - u^1)(t)) - E(u(t))\} = 0. \tag{5.5}$$

The strong convergence (5.3) implis $L = 1$ in the terminology in Theorem C. From (5.3) and (5.4), we have

$$E(u(t)) = E(u(0)) \leqq 0. \tag{5.6}$$

Theorem B is only used to prove $E(u(t)) = 0$.

Proof of Corollary E. Assume $\|u_0\| = \|Q\|$. In the same way as in the proof of Proposition 2.5, we can prove that (in the proof of Theorem C) we have

$$|u_n^1|^{\frac{4}{N}} u_n^1 - |u^1|^{\frac{4}{N}} u^1 \to 0 \tag{5.7}$$

strongly in $C([0,T]; L^{\sigma'}(\mathbb{R}^N))$, and we have

$$\lim_{n\to\infty} \sup_{t\in[0,T]} \|(u_n^1 - u^1)(t)\|_\alpha = 0 \quad \alpha \in [2, 2^*), \tag{5.8}$$

$$E(u^1) = 0. \tag{5.9}$$

Thus we have $L = 1$ in the terminology in Theorem C. We note that we do not use Theorem B.

6. Proof of Theorem B

It remains to prove Theorem B.
We begin with

Lemma 6.1. *(1) Let $\Phi \in W^{4,\infty}(\mathbb{R}^N)$, and let $\Psi = \nabla\Phi = (\Psi^1, \Psi^2, \cdots, \Psi^N)$. Then H^1 solution $u(t)$ of $C(p)$ $(p \in (1, 2^*))$ satisfies, for any t, $t_0 \in \mathbb{R}$,*

$$\Im\langle \nabla u(t), u(t)\rangle = \Im\langle \nabla u(t_0), u(t_0)\rangle \tag{6.1}$$

and

$$\langle \Psi^k, |u(t)|^2\rangle = \langle \Psi^k, |u(t_0)|^2\rangle + \Im \int_{t_0}^{t} \langle \nabla\Psi^k(s)\cdot\nabla u(s), u(s)\rangle ds. \tag{6.2}$$

(2) Let $\Phi \in C^1(\mathbb{R}; W^{4,\infty}(\mathbb{R}^N))$, and let $\Psi = \nabla\Phi = (\Psi^1, \Psi^2, \cdots, \Psi^N)$. Then H^1 solution $u(t)$ of $C(p)$ $(p \in (1, 2^*))$ satisfies, for any t, $t_0 \in \mathbb{R}$,

$$\langle \Phi(t), |u(t)|^2\rangle = \langle \Phi(t_0), |u(t_0)|^2\rangle + \Im \int_{t_0}^{t} \langle u(s), \Psi(s)\nabla u(s)\rangle ds + 2\Im \int_{t_0}^{t} \langle \partial_s\Phi(s), |u(s)|^2\rangle ds \tag{6.3}$$

and

$$\begin{aligned} &-2\Im\langle u(t), \Psi(t)\cdot\nabla u(t)\rangle + 2\Im\langle u(t_0), \Psi(t_0)\cdot\nabla u(t_0)\rangle = -2\Im\int_{t_0}^{t}\langle u(s), \partial_s\Psi(s)\cdot\nabla u(s)\rangle ds \\ &+ 2\frac{p-1}{p+1}\int_{t_0}^{t} ds\,\langle \nabla\cdot\Psi(s)|u(s)|^{p+1}\rangle - 2\Re\int_{t_0}^{t} ds\,\langle \partial_j u(s), \partial_k\Psi^j(s)\partial_k u(s)\rangle \\ &+ \frac{1}{2}\int_{t_0}^{t} ds\,\langle \Delta(\nabla\cdot\Psi), |u(s)|^2\rangle. \end{aligned} \tag{6.4}$$

Moreover we have

$$\begin{aligned} &\langle \Phi(t), |u(t)|^2\rangle = \langle \Phi(t_0), |u(t_0)|^2\rangle + 2\Im t\langle u(t_0), \Psi(t_0)\cdot\nabla u(t_0)\rangle + t^2 E_{p+1}(u_0) \\ &+ \Im\int_{t_0}^{t} ds \int_{t_0}^{s} d\tau \langle \partial_\tau\Phi(\tau), |u(\tau)|^2\rangle + 2t\Im\int_{t_0}^{t} ds\int_{t_0}^{s} d\tau\langle u(\tau), \partial_\tau\Psi(\tau)\cdot\nabla u(\tau)\rangle \\ &+ \int_{t_0}^{t} ds\int_{t_0}^{s} d\tau \left\langle \frac{2}{p+1}(\nabla\cdot\Psi(\tau)+2) - \nabla\cdot\Psi(\tau), |u(\tau)|^{p+1}\right\rangle \\ &- 2\int_{t_0}^{t} ds\int_{t_0}^{s} d\tau\,\langle (\delta_{ij} - \Re\partial_i\Psi^j)\partial_i u(\tau), \partial_j u(\tau)\rangle + \frac{1}{2}\int_{t_0}^{t} ds\int_{t_0}^{s} d\tau\,\langle \Delta(\nabla\cdot\Psi), |u(\tau)|^2\rangle. \end{aligned} \tag{6.5}$$

Proof. This lemma is an analogue of Ehrenfest's law in quantum mechanics. We give a formal calculation below. It can be easily justified by approximating the initial datum u_0 with a sequence in $H^2(\mathbb{R}^N)$ (see Kato [14]). We use the following notations: $\mathcal{L}u = (2i\partial_t + \Delta)u$; $F = F(u) = |u|^{p-1}u$; $[A, B] = AB - BA$; $\mathfrak{d} = a(t,x)\partial_t + b(t,x)\cdot\nabla + c(t.x)$; $\langle zf, g\rangle = \overline{z}\langle f, g\rangle$, $z \in \mathbb{C}$.

Let $u(t)$ satisfy $\mathcal{L}u = F(u)$. Then we have (Ehrenfest's law)

$$2i\frac{d}{dt}\langle u, \mathfrak{d}u\rangle = \langle u, \mathfrak{d}F\rangle - \langle F, \mathfrak{d}u\rangle + \langle u, [\mathcal{L}, \mathfrak{d}]u\rangle. \tag{6.6}$$

Putting $\eth = \nabla$ (the generator of space translations), we have (6.1). Putting $\eth = \Psi + it\nabla$ (the generalized generator of Galilean transformations), we have (6.2). One can easily obtain (6.3) from (6.6) with $\eth = \Phi$. In order to obtain (6.4), we put $\eth = 2t\partial_t + \Psi \cdot \nabla + \frac{1}{2}\nabla\Phi$ (the generalized generator of space-time dilations). Combining (6.3) and (6.4), we obtain (6.5) with the help of the energy conservation law (1.2).

Remark 6.1. The identity (6.4) is a variant of the one used in Ogawa - Y.Tsutsumi [30,31].

Now we turn to the proof of Theorem B. We argue by contradiction. We assume that the solution $u(t)$ of C$(1+\frac{4}{N})$ exists globaly in time in the space $C_b([0,\infty); H^1(\mathbb{R}^N))$, and $u(t)$ satisfies

$$M \equiv \sup_{t\in\mathbb{R}_+} \|u(t)\|_{H^1(\mathbb{R}^N)} < \infty, \tag{6.7}$$

$$E(u(t)) = E(u_0) \equiv -E_0 < 0. \tag{6.8}$$

We take any sequence $\{t_n\}$ such that $t_n \to \infty$ as $n \to \infty$, and put

$$u_n(t,x) \equiv u(t+t_n, x). \tag{6.9}$$

As in the preceeding sections, we shall not bother to extract subsequences. We note that $\{u_n\}$ is a bounded sequece in $L^\infty(\mathbb{R}_+; H^1(\mathbb{R}^N))$, and satisfies

$$\|u_n(t)\|_\sigma^\sigma > \frac{\sigma}{2} E_0, \tag{6.10}$$

$$E(u_n(t)) = -E_0 \tag{6.11}$$

for any $t \in [0,\infty)$ by (6.7). In the same way as in the proof of Theorem C and Proposition 2.1, we have

Lemma 6.2. *There exists*

(i) *a family of solutions of (NSC) in $C_b(\mathbb{R}_+; H^1(\mathbb{R}^N))$: $\mathfrak{A} = \{v^1, v^2, \cdots\}$, and*
(ii) *a family of sequences in $\mathbb{R}^N$: $\mathfrak{B} = \{\{y_n^1\}, \{y_n^2\}, \cdots\}$*

such that we have

$$\lim_{n\to\infty} \left| \sum_{k=2}^{j} y_n^k \right| = \infty \qquad (j \geqq 2), \tag{6.12}$$

and, for some subsequence (still denoted by the same letter), we have

$$v_n^1 \equiv u_n(t, \cdot + y_n^1) \to f^1 \not\equiv 0, \tag{6.13}$$

$$v_n^j \equiv \left(v_n^{j-1} - v^{j-1}\right)(t, \cdot + y_n^j) \to v^j \not\equiv 0 \quad (j \geqq 2), \tag{6.14}$$

weakly in $L^\infty(\mathbb{R}_+; H^1(\mathbb{R}^N))$ and strongly in $C([0,T]; L^q(\Omega))$ for any $T > 0$, $\Omega \Subset \mathbb{R}^N$ and $\alpha \in [2, 2^*)$, and*

$$\lim_{n\to\infty} \sup_{t\in[0,T]} \int_{\mathbb{R}^N} \left| |v_n^j|^\alpha - |v_n^j - v^j|^\alpha - |v^j|^\alpha \right| dx = 0, \tag{6.15}$$

$$\lim_{n\to\infty} \int_0^T \left\{ E(v_n^j) - E(v_n^j - v^j) - E(v^j) \right\} dt = 0, \tag{6.16}$$

and, for any $t \in \mathbb{R}_+$,

$$\lim_{n\to\infty} \{E(v_n^j(t)) - E((v_n^j - v^j)(t)) - E(v^j(t))\} = 0, \tag{6.17}$$

$$\lim_{n\to\infty} (\|v_n(t)\|_\alpha^\alpha - \|(v_n^j - v^j)(t)\|_\alpha^\alpha) = \sum_{k=1}^{j} \|v^k(t)\|_\alpha^\alpha, \tag{6.18}$$

$$\lim_{n\to\infty} \{E(u_n(t)) - E((v_n^j - v^j)(t))\} = \sum_{k=1}^{j} E(v^k(t)). \tag{6.19}$$

Furthermore, we have: If $L \equiv \#\mathfrak{A} < \infty$,

$$\lim_{n\to\infty} \sup_{t\in[0,T]} \|(v_n^L - v^L)(t)\|_{L^\sigma(\mathbb{R}^N)} = 0, \tag{6.20}$$

$$\lim_{n\to\infty} \sup_{t\in[0,T]} \left\{ \sup_{y\in\mathbb{R}^N} \int_{|x-y|<R} |(v_n^L - v^L)(t,x)|^2 dx \right\} = 0; \tag{6.21}$$

if $\#\mathfrak{A} = \infty$,

$$\lim_{j\to\infty} \lim_{n\to\infty} \sup_{t\in[0,T]} \|(v_n^j - v^j)(t)\|_{L^\sigma(\mathbb{R}^N)} = 0 \tag{6.22}$$

$$\lim_{j\to\infty} \lim_{n\to\infty} \sup_{t\in[0,T]} \left\{ \sup_{y\in\mathbb{R}^N} \int_{|x-y|<R} |(v_n^j - v^j)(t,x)|^2 dx \right\} = 0, \tag{6.23}$$

for any $R > 0$.

From (6.19), there exists a number $k \in \mathbb{N}$ such that $E(v^k) < 0$. We put

$$u^1(t,x) \equiv v^k(t,x),$$

so that we have

$$u^1 \in C_b([0,\infty); H^1(\mathbb{R}^N)) \tag{6.24}$$

$$2i\frac{\partial u^1}{\partial t} + \Delta u^1 + |u^1|^{\frac{4}{N}} u^1 = 0, \tag{6.25}$$

$$-E_1 \equiv E(u^1(t)) < 0. \tag{6.26}$$

Here we define

$$D_1 = \sup_{R>0} \left(\liminf_{t\uparrow\infty} \left(\sup_{y\in\mathbb{R}^N} \int_{|x-y|\leqq R} |u^1(t,x)|^2 dx \right) \right). \tag{6.27}$$

Next stage we consider

$$u_n^1(t\,,\,x) \equiv u^1(t + t_n\,,\,x). \tag{6.28}$$

Repeating the arguement above, we obtain a sequence $\{u^j\}_{j=1}^L$ (L may be infinite) of solutions of (NSC) such that

$$u^j \in C_b([0,\infty); H^1(\mathbb{R}^N)) \tag{6.29}$$

$$2i\frac{\partial u^j}{\partial t} + \Delta u^j + |u^j|^{\frac{4}{N}} u^j = 0, \tag{6.30}$$

$$-E_j \equiv E(u^j(t)) < 0. \tag{6.31}$$

At this stage, we consider the sequence $\{u^j\}_{j=1}^\infty$ (we put $u^j = u^L$ for $j \geqq L$ if $L < \infty$). We have

Lemma 6.3. *(1) We put*

$$D_j = \sup_{R>0}\left(\liminf_{t\uparrow\infty}\left(\sup_{y\in\mathbb{R}^N}\int_{|x-y|\leqq R}|u^j(t,x)|^2dx\right)\right). \tag{6.32}$$

Then we have

$$\|Q\| \leqq \|u^{j+1}(0)\|^2 \leqq D_j, \tag{6.33}$$

$$\|u^j(0)\|^2 - D_j \downarrow 0 \quad (j\uparrow L). \tag{6.34}$$

(2) There exists a positive constant E^* such that we have

$$E(u^j) < -E^* \tag{6.35}$$

for sufficiently large j.

Proof. (1) is obvious from the construction of u^j's. We shall prove (2). By Proposition 2.5, we know that if $\|v\| < \|Q\|$, then $E(v) > 0$. Hence this fact, (6.32) and the similar formula (6.19) at the level of u_n^j yields the existence of a positive constant E^* satisfying (6.35).

Lemma 6.4. *For any $\varepsilon \in (0,1)$, there is a constant $R_j > 0$ such that for sufficiently large j, there exists a Lipshitz continuous function $\gamma_j(\cdot) \in C([T_j,\infty);\mathbb{R}^N)$ such that*

$$\int_{|x|\leqq R}|u^j(t,x+\gamma_j(t))|^2dx \geqq (1-\varepsilon)D_j, \quad t\in[T_j,\infty), \tag{6.36}$$

for any $R \geqq R_j$ and for some $T_j > 0$.

Proof. Proposition 2.7 together with (6.32) yields the existence of continuous "path " $\gamma_j(t)$. Since $\sup_{t\in[0,\infty)}\|\partial_t u^j(t)\|_{H^{-1}} < \infty$, $u^j(\cdot)$ is L^2-valued 1/2 Hölder continuous function as in the proof of Proposition 3.1, we can take γ_j as a Lipshitz continuous function.

Remark 6.2. If $|\gamma_j(t)| \leqq C$ for some constant $C > 0$, then we have

$$D_{j+1} = \|u^{j+1}(0)\|^2 = D_j,$$

so that we have $L < \infty$. In the case of $u(0)$ being radially symmetric, we have $L = 1$, since we can take $y_n^1(t) \equiv 0$ and $\#\mathfrak{A} = 1$ in Lemma 6.2.

The rest of the proof of Theorem B is closely related to Ogawa - Y.Tsutsumi [30]. Following Ogawa - Y.Tsutsumi [30,31], we introduce

$$\phi(r) = \begin{cases} r, & 0 \leqq r, \\ r-(r-1)^3, & 1\leqq r < 1+\frac{1}{\sqrt{3}}, \\ \text{smooth}, \quad (\phi' \leqq 0) & 1+\frac{1}{\sqrt{3}} \leqq r < 2, \\ 0, & 2 \leqq r, \end{cases} \tag{6.37}$$

and put, for $m > 0$,

$$\Psi_m(x) = \frac{x}{r}\phi_m(r) = \frac{x}{r}m\phi(\frac{r}{m}), \tag{6.38}$$

$$\Phi_m(x) = \int_0^r \phi_m(s)ds, \tag{6.39}$$

where $r = |x|$. Here we note that

$$\left|\frac{\partial^k}{\partial x_k}\phi_m(r)\right| \leqq \frac{C_k}{m^{k-1}}, \quad k = 0, 1, 2, \cdots, \tag{6.40}$$

$$\nabla\Phi_m(x) = \frac{x}{r}\phi_m(r) = \Psi_m(x), \tag{6.41}$$

Now we consider the fine property of $\gamma_j(t)$. We may assume that

$$\Im\langle\nabla u^j(0), u^j(0)\rangle = 0, \tag{6.42}$$

since (NSC) is invariant under the Galilei transformations.

Lemma 6.5. *Let $\varepsilon < \min(1/16, 1/16M^2)$. Then we have that, for u^j (j close enough to L), there are constants $R_j > 0$ and $T_j > 0$, and there is a continuous path $\gamma_j(\cdot) \in C^1([T_j, T_j + 2R_j/\varepsilon_j); \mathbb{R}^N)$ such that*

$$|\dot{\gamma_j}(t)| < \varepsilon_j, \quad t \in [T_j, T_j + 2R_j/\varepsilon_j), \tag{6.43}$$

$$\int_{|x|\leqq 3R_j} |u^j(t, x + \gamma_j(t))|^2 dx \geqq (1-\varepsilon)D_j, \quad t \in [T_j, T_j + 2R_j/\varepsilon_j), \tag{6.44}$$

where $\varepsilon_j = \varepsilon + (\|u^j(0)\|^2 - D_j)/\|u^j(0)\|^2$.

Proof. From Lemma 6.4, we know that there are constant $R_j > 0$ and continuous path γ_j which satisfies

$$\int_{|x|\leqq R_j} |u^j(t, x + \gamma_j(t))|^2 dx \geqq (1-\varepsilon)D_j, \quad t \in [T_j, \infty). \tag{6.45}$$

For simplicity, we suppose $T_j = 0$ and $\gamma_j(0) = 0$. We shall show that, for sufficiently large j, one can choose γ_j to satisfy (6.43) and (6.44). If $B(\gamma_j(t); 2R_j) \subset B(0; 8R_j)$, the path γ_j satisfies

$$\left\langle\Psi_{R_j}(\cdot - \gamma_j(t)), |u^j(t)|^2\right\rangle + \gamma_j(t)\|u^j(0)\|^2 = \langle\Psi_{8R_j}, |u^j(t)|^2\rangle + g_{\varepsilon,j}(t), \tag{6.46}$$

where Ψ_{R_j} and Ψ_{8R_j} are as in (6.38), and $g_{\varepsilon,j}(t) (\in \mathbb{R}^N)$ is a C^1-function of t such that

$$\lim_{j\uparrow\infty}\lim_{\varepsilon\downarrow 0} g_{\varepsilon,j}(t) = 0. \tag{6.47}$$

Here we have used the fact that $\langle\Psi_{8R_j}, |u^j(t)|^2\rangle/\|u^j(0)\|^2$ is almost the center of mass, and that $\left\langle\Psi_{R_j}(\cdot - \gamma_j(t)), |u^j(t)|^2\right\rangle/\|u^j(0)\|^2$ is almost the center of the "cluster". Since we have

$$\left\langle\Psi_{R_j}(\cdot - \gamma_j(t)), |u^j(t)|^2\right\rangle/\|u^j(0)\|^2 < 2R_j,$$

the "center of mass" $\gamma_j^c(t) \equiv \langle\Psi_{8R_j}, |u^j(t)|^2\rangle/\|u^j(0)\|^2$ satisfies

$$\int_{|x|\leqq 3R_j} |u^j(t, x + \gamma_j^c(t))|^2 dx \geqq (1-\varepsilon)D_j \tag{6.48}$$

whenever $B(\gamma_j(t); 2R_j) \subset B(0; 8R_j)$. From (6.2) with $\Psi^k = \Psi^k_{8R_j}$, we have that $\gamma_j^c \in C^1([0,\infty); \mathbb{R}^N)$. Moreover we have from (6.1), (6.42) and (6.2) with $\Psi^k = \Psi^k_{8R_j}$ that,

$$|\dot{\gamma_j^c}(t)| < \varepsilon + (\|u^j(0)\|^2 - D_j)/\|u^j(0)\|^2 \tag{6.49}$$

as far as $B(\gamma_j(t);2R_j) \subset B(\gamma_j^c(t);3R_j)$, since $u^j(t)$ is well-localized in $B(\gamma_j(t);R_j)$. Therefore we can see that

$$\int_{|x|\leqq 3R_j} |u^j(t,x+\gamma_j^c(t))|^2 dx \geqq (1-\varepsilon)D_j, \quad t\in[0,2R_j/\varepsilon_j), \tag{6.50}$$

where $\varepsilon_j = \varepsilon + (\|u^j(0)\|^2 - D_j)/\|u^j(0)\|^2$.

By space-time tranclations, we may assume that $T_j = 0$ and $\gamma_j(0) = 0$. Let

$$\varepsilon < \min\left(\frac{1}{16}, \frac{1}{16M^2}, \frac{E^*}{16}, \frac{E^*}{16(\|u(0)\|^2+M^2)}\frac{E^*}{16\|u(0)\|^2}\right).$$

We take j large enough to obtain $2\varepsilon_j < \varepsilon$. By Lemma 6.5, there is a C^1-path satisfying (6.44) for sufficiently large j. We take Ψ and Φ in (6.5) with $p = 1 + \frac{4}{N}$ as follows:

$$\Psi(t,x) = \Psi_{8R_j}(x),$$
$$\Phi(t,x) = \Phi_{8R_j}(x).$$

Then we have

$$\begin{aligned}
\langle \Phi_{8R_j}, |u^j(t)|^2\rangle &= \langle \Phi_{8R_j}, |u^j(0)|^2\rangle + 2\Im t\langle u^j(0), \Psi_{8R_j}\cdot\nabla u^j(0)\rangle + t^2E(u^j(0)) \\
&+ \int_0^t ds \int_0^s d\tau \left\langle \frac{2}{\sigma}(\nabla\cdot\Psi_{8R_j}(\tau)+2) - \nabla\cdot\Psi_{8R_j}(\tau),\, |u^j(t)|^\sigma\right\rangle \\
&- 2\int_0^t ds\int_0^s d\tau \left\langle (\delta_{ik} - \Re\partial_i\Psi^k_{8R_j})\partial_i u^j(t),\, \partial_k u^j(t)\right\rangle \\
&+ \frac{1}{2}\int_0^t ds \int_0^s d\tau \,\langle \Delta(\nabla\cdot\Psi_{8R_j}),\, |u^j(t)|^2\rangle \\
&= \mathrm{I} + \mathrm{II} + \mathrm{III} + \mathrm{IV}.
\end{aligned} \tag{6.51}$$

The third term III can be estimated by (6.4) with Ψ choosing as follows: $\Phi = (\Psi^1, \Psi^2, \cdots, \Psi^N)$; $\Psi^k(t,x) = \chi^k(x^k)$; $\chi^k \in W^{4,\infty}(\mathbb{R})$; $\chi(x^k) \geqq 0$; $\chi(x^k) = -\chi(-x^k)$; $\mathrm{supp}\chi^k \subset [|x^k| > 6R_j]$; $\chi(x^k) = x^k - 7, x^k \in [8R_j/\sqrt{N} < x^k < 16R_j]$; $\chi(x^k) = 10, x^k \in [x^k > 18R_j]$. Consequently we get

$$2\int_0^t ds \int_{8R_j<|x|<16R_j} |\nabla u^j(t,x)|^2 dx < \frac{1}{8}E^*t + 8R_j(\|u^j(0)\| + M)t^{\frac{1}{2}}. \tag{6.52}$$

Now, from (6.51) and (6.52), we can show that

$$\langle \Phi_{8R_j}, |u^j(2R_j/\varepsilon_j)|^2\rangle < 0$$

for sufficiently small ε (and sufficiently large j). Thus we reach a contradiction.

Remark 6.3. As in the proof of Theorem B, we can show the existence of singular solutions of (NS) with $p > 1 + \frac{4}{N}$

Acknowledgement. The author would like to express his deep gratitude to professors D. Fujiwara, A. Inoue, M. Tsutsumi and N. Nakauchi for having interest in this study and for their helpful discussions. The author is grateful to professors H. Matano and Y.Tsutsumi for their helpful conversations. Finaly the author also grateful to professor R. Iino for his constant encouragement.

References

1. Akhmanov, S.A., Sukhorukov, A.P. and Khokhlov, R.V., *Self-focusing and self-trapping of intence light beams in a nonlinear medium*, Soviet physics JETP **23** (1966), 1025–1033.
2. ———, *Self-focusing and diffraction of light in a nonlinear medium*, Soviet physics USPEKHI **93** (1968), 609–633.
3. Berestycki, H. and Lions, P.L., *Nonlinear scalar field equations.* I, II, Archi. Rat. Mech. Anal. **82** (1983), 313–376.
4. Brézis, H. and Coron, J.M., *Convergence of H−system or how to blow bubbles*, Archi. Rat. Mech. Anal. **89** (1985), 21–56.
5. Brézis, H. and Lieb, E.H., *A relation between pointwise convergence of functions and convergence of functionals*, Proc. Am. Math. Soc. **88** (1983), 485–489.
6. ———, *Minimum action solutions of some vector field equations*, Commun. Math. Phys. **96** (1983), 97–113.
7. Cazenave, T. and Weisler F.B., *The structure of solution to the pseudoconformaly invariaint nonlinear Schrödinger equations*, preprint.
8. Chen, X.-Y. and Matano, H., *Convergence, asymptotic periodicity, and finite-point blow-up in one-dimensional semilinear heat equations*, J. Differential Equations **104** (1986), 251–270.
9. Frölich, J., Lieb, E.H. and Loss, M., *Stability of coulomb systems with magnetic fields I, The one- electron atom*, Commun. Math. Phys. **78** (1989), 160–190.
10. Giga, Y. and Kohn, R.V., *Nondegeneracy of blowup for semilinear heat equations*, preprint.
11. Ginibre, J. and Velo, G., *On a class of nonlinear Schrödinger equations.* I, II, J. Funct. Anal. **32** (1979), 1–71.
12. ———, *The global Cauchy problem for the nonlinear Schrödinger equation revisited*, Ann. Inst. Henri Poincaré, Analyse Nonlinéaire. **2** (1985), 309–327.
13. Glassey, R.T., *On the blowing up solution to the Cauchy problem for nonlinear Schrödinger equations*, J. Math. Phys. **18** (1979), 1794–1797.
14. Kato,T., *On nonlinear Schrödinger equations*, Ann. Inst. Henri Poincaré, Physique Theorique **46** (1987), 113–129.
15. LeMesurier,B., Papanicolau,C., Sulem,C. and Sulem, P.L., *The focusing singularity of the nonlinear Schrödinger equation*, Direction in Partial Differential Equations (Crandall,M.G., Rabinowitz, P.H. and Turner,R.E., eds.), Academic Press, New York, 1987, pp. 159–201.
16. Lieb,E.H., *On the lowest eigenvalue of Laplacian for the intersection of two domains*, Invent. Math. **74** (1983), 441–448.
17. Lions, P.L., *The concentration compactness principle in the calculas of variations. The localy compact case.* I, II, Ann. Inst. Henri Poincaré, Analyse Nonlinéaire. **1** (1984), 109–145, 223–283.
18. ———, *The concentration compactness principle in the calculas of variations. The limit case.* I, II., Riv. Math. Iberoamericana **1** (1985), 45–121, 145–201.
19. Merle, F., *Limit of the solution of the nonlinear Schrödinger equation at the blow-up time*, J. Funct. Anal. **84** (1989), 201–214.
20. ———, *Construction of solutions with exactly k blow-up points for the Schrödinger equation with the critical power nonlinearity*, Commun. Math. Phys. **129** (1990), 223–240.
21. Merle, F. and Tsutsumi, Y., *L^2 concentration of blow-up solutions for the nonlinear Schödinger equation with the critical power nonlinearity*, J. Differential Equations **84** (1990), 205–214.
22. Nagasaki, K. and Suzuki, T., *Asymptotic analysis for two-dimensional ellipticeigenvalue problems with exponentially-dominated nonlinearities*, Asymptotic Analysis **3** (1990), 173–188.
23. Nawa,H., *Blow-up solutions for the nonlinear Schrödinger equations*, unpublished.
24. ———, *"Mass concentration" phenomenon for the nonlinear Schrödinger equation with the critical power nonlinearity*, Funk. Ekva. **35** (1992), 1–18.
25. ———, *"Mass concentration" phenomenon for the nonlinear Schrödinger equation with the critical power nonlinearity.* II, Kodai Math. J. **13** (1990), 333–348.
26. ———, *Convergence theorems for the pseodo-conformally invariant nonlinear Schrödinger equation*, Proc. Japan Acad. **66 (A)** (1990), 214–216; RIMS Kokuroku **755** (1991), 73–92.
27. ———, *Formation of singularities in solutions of the nonlinaer Schrödinger equation*, Proc. Japan Acad. **67 (A)** (1991), 29–34.
28. ———, *Formation of singularities in solutions of the one dimensional nonlinaer Schrödinger equation with critical power nonlinearity*, submitted.
29. Nawa,H. and Tsutsumi, M., *On blow-up for the pseudo-conformally invariant nonlinear Schrödinger equation*, Funk. Ekva. **32** (1990), 417–428.
30. Ogawa,T and Tsutsumi, Y., *Blow-up of H^1-solution for the nonlinear Schrödinger equation*, J. Differential Equations (1991).
31. ———, *Blow-up of H^1-solution for the one dimensional nonlinear Schrödinger equation with critical power nonlinearity*, Proceeding of A.M.S. (1991).
32. Strauss, W. A., *Existence of solitary waves in higher dimensions*, Commun. Math. Phys. **55** (1977), 149–162.

33. Struwe, M., *A global compactness result for elliptic boundary value problem involving limitting nonlinearities*, Math. Z. **187** (1983), 567–576.
34. Takakuwa, S., *Behavior of minimizing sequences for the Yamabe problem*, Tokyo Metropolitan University Mathematics Preprint Series **7** (1990).
35. Tsutsumi, Y., *Rate of L^2 concentration of blow-up solutions for the nonlinear Schödinger equation with critical power nonlinearity*, Nonlinear Anal. T.M.A. **15** (1990), 719–724.
36. Weinstein, M.I., *Nonlinear Schrödinger equations and sharp interpolation estimates.*, Commun. Math. Phys. **87** (1984), 511–517.
37. ———, *On the structure and formation singularities in solutions to nonlinear dispersive evolution equations*, Commun. in Partial Diffrential Equations **11** (1986), 545–565.
38. ———, *The nonlinear Schrödinger equation - Singularity formation, Stability and dispersion*, The Connection between Infinite and Finite Dimensional Dynamical Systems, Contemporary Math.99, 1989, pp. 213–232.
39. Yajima, K., *Existence of solutions for Schrödinger evolution equations*, Commun. Math. Phys. **110** (1987), 415–426.
40. Zakharov, V.E., and Synakh, V.S., *The nature of self-focusing singularity*, Sov. Phys. JETP **41** (1976), 441–448.

Modulational stability of periodic solutions of the Kuramoto-Sivashinsky equation

Demetrios T. Papageorgiou
Department of Mathematics
New Jersey Institute of Technology
Newark NJ 07102
George C. Papanicolaou
Courant Institute
251 Mercer Street
New York NY 10012

and

Yiorgos S. Smyrlis
Department of Mathematics
University of Manchester
Manchester
UK

ABSTRACT. We study the long-wave, modulational, stability of steady periodic solutions of the Kuramoto-Sivashinsky equation. The analysis is fully nonlinear at first, and can in principle be carried out to all orders in the small parameter, which is the ratio of the spatial period to a characteristic length of the envelope perturbations. In the linearized regime we recover a high-order version of the results of Frisch, She and Thual, ([1]), which shows that the periodic waves are much more stable than previously expected.

1. Introduction.

The Kuramoto-Sivashinsky (KS) equation

$$u_t + uu_x + u_{xx} + \nu u_{xxxx} = 0 \tag{1.1}$$

arises frequently as a weakly nonlinear amplitude equation for the evolution of fluid interfaces when viscosity and surface tension are present ([2]; [3]; [4]; [5]; [6]) and in other interfacial phenomena ([7]; [8]; [9]; [10]; [11]; [12]). It is the simplest evolution equation in which a band of unstable wave modes interact through a convective nonlinearity and the long time behvior of the solutions can be effectively described by a finite number of modes, the inertial manifold ([13];

R. E. Caflisch and G. C. Papanicolaou (eds.), Singularities in Fluids, Plasmas and Optics, 255–263.
© 1993 *Kluwer Academic Publishers.*

[14; 15; 16; 17; 18; 19; 20]). It is also one of a few examples of nonlinear partial differential equations whose periodic solutions exhibit period doubling and approach to chaos in a canonical way ([21; 22; 23; 24]). We are interested here in the modulational or long wave stability of periodic solutions and in particular of steady periodic solutions. The linearized, long wave stability of steady periodic solutions was carried out by Frisch, She and Thual ([1]). They found a narrow region of linearized stability for this solution when the parameter ν in (1.1) is near one. We carry out a weakly nonlinear stability analysis to suffiently high order so that the stabilizing fourth derivative term enters fully. As a result we find that the steady periodic solutions are modulationally stable for a much larger range of ν. We outline our analysis here and give full details in ([25]). One of our objectives in this work, in addition to our interest in understanding the steady periodic solutions of KS, (1.1), is to do a high order modulational stability analysis in a model way so that it can applied to other problems as well. In ([26]) we used a similar approach for flows with microstructure but the analysis is much more complicated and less transparent than the one for the Kuramoto-Sivashinsky equation presented here.

2. Formulation.

The steady periodic solutions $U(y) = U(y;\nu)$ of the Kuramoto-Sivashinsky equation which have period 2π and zero mean satisfy

$$UU_y + U_{yy} + \nu U_{yyyy} = 0,$$

$$\overline{U} = \frac{1}{2\pi}\int_0^{2\pi} U(y)dy = 0 \quad , \quad U(y) = U(y+2\pi).$$

Zero mean solutions are easily obtainable for ν near one, from odd parity initial conditions as is done for example in ([21; 22]). Furthermore, given constants c and ρ the unsteady KS equation (1.1) permits the following exact solution which displays its invariance properties

$$u(t,x) = \rho U(\rho(x-ct);\rho^2\nu) + c. \tag{2.1}$$

We will construct general solutions of KS by letting the constants c and ρ vary slowly relative to the period of U, which is what modulated periodic solutions are. We introduce a small parameter ϵ which is the ratio of the period of U to a characteristic length scale for the modulations. We then look for solutions of

$$\frac{1}{\epsilon}u_t + uu_x + u_{xx} + \epsilon^2\nu u_{xxxx} = 0, \tag{2.2}$$

of the form

$$u(x,t) = \frac{\rho(x,t)}{\epsilon} U(\frac{\theta(x,t)}{\epsilon}; \rho^2 \nu) + \frac{c(x,t)}{\epsilon} + w(\frac{\theta(x,t)}{\epsilon}, x, t). \tag{2.3}$$

In (2.2) we have rescaled the time and the viscosity parameter ν so that slowly accumulating effects are allowed to develop and in (2.3) w is a periodic function of its first argument with period 2π and mean zero. We think of the ansatz (2.2) as a change of variables in which a certain family of solutions u is represented by three functions: ρ, c and w, with $\theta_x = \rho$. When ρ and c are constants then $w = 0$ and we have the exact slutions (2.1). In (2.3), as ϵ tends to zero the period of the the oscillations increases and we expect ρ and c to satisfy appropriate conservation equations which are the modulation equations.

We now substitute (2.3) into (2.2) and re-arrange terms to get the following nonlinear system for ρ, c and w, which equivalent to (2.2)

$$\rho_t + (c\rho)_x + 3\epsilon\rho_{xx} + \epsilon^3 \nu \left[\frac{(3(\rho\rho_x)_x + \rho\rho_{xx})_x + \rho\rho_{xxx}}{\rho} \right]_x = 0, \tag{2.4}$$

$$\begin{aligned} c_t + cc_x + \overline{U^2}\rho\rho_x + 2\nu\overline{UU_\nu}\rho^3\rho_x + \epsilon\overline{U(\rho w)_x} + 2\epsilon\nu\overline{U_\nu w}\rho^2\rho_x \\ + \epsilon c_{xx} + \epsilon^2\overline{ww_x} + \epsilon^3 \nu c_{xxxx} = 0, \end{aligned} \tag{2.5}$$

$$\begin{aligned} \rho^2 L(w) + F_0(c,\rho;\nu) + \epsilon F_1(c,\rho,w;\nu) + \epsilon^2 F_2(c,\rho,w;\nu) \\ + \epsilon^3 F_3(c,\rho,w;\nu) + \epsilon^4 \nu w_{xxxx} = 0. \end{aligned} \tag{2.6}$$

In (2.6) the functions F_0, F_1, F_2, F_3 are given explicitly by cumbersome expressions that can be found in ([25]). The operator L in (2.6) is the linearized operator for U and is given by

$$L(\phi) = (U\phi)_y + \phi_{yy} + \rho^2 \nu \phi_{yyyy}. \tag{2.7}$$

In the weakly nonlinear analysis we use L with $\rho = 1$, which is

$$\mathcal{L}(\phi) = (U\phi)_y + \phi_{yy} + \nu \phi_{yyyy}. \tag{2.8}$$

Given ρ and c we consider (2.6) as an equation for w, which is a zero mean, periodic function of its first argument. The solution is then used in calculate the averages in (2.4) and (2.5), which are denoted by overbars. To be able to have zero mean, periodic solutions w for (2.6) we must make sure that the linearized operator can be inverted. First, we exclude terms proportional to U_y from the forcing terms in (2.6), since $L(1) = U_y$ in which case non-zero mean solutions would arise. The elimination of terms multiplied by U_y yields equation (2.4). Second, we ensure that forcing terms of (2.6) have zero mean and this gives condition (2.5). Thus (2.4) and (2.5) are the compatibility conditions for w and are exact to all orders in ϵ. In fact, there are no approximations at all in (2.4)-(2.6). We have just rewritten the KS equation (2.2) in a complicated way, which is, however, convenient for modulation analysis.

3. Asymptotic solutions and linear stability theory.

Now we try to solve for w in (2.6) approximately, given ρ and c. We let

$$w = w_0 + \epsilon w_1 + \epsilon^2 w_2 + \dots . \tag{3.1}$$

Substitution of (3.1) into (2.6) yields the following three problems to leading order:

$$\rho^2 L(w_0) + F_0(c, \rho; \nu) = 0 \tag{3.2a}$$

$$\rho^2 L(w_1) + F_1(c, \rho, w_0; \nu) = 0 \tag{3.2b}$$

$$\begin{aligned}\rho^2 L(w_2) + F_2(c, \rho, w_0; \nu) + ((\rho w_1)_x U - \overline{(\rho w_1)_x U})\\ +2\nu\rho^2\rho_x(wU_\nu - \overline{wU_\nu}) + w_{1t} + (cw_{1x})_x + \rho(w_0 w_{1y} + w_1 w_{0y}) - 2\rho_x w_{1y}\\ +2\rho w_{1xy} + 6\nu\rho^2\rho_x w_{1yyy} + 4\nu\rho^3 w_{1xyyy} = 0\end{aligned} \tag{3.2c}$$

These are still too complicated to analyze so as a first step we linearize ρ and c about the exact solutions

$$\rho(t,x) = 1 \quad , \quad c(t,x) = 0.$$

The functions w_0, w_1, w_2 then have the form

$$w_0 = \rho_x \chi_1(y) + c_x \chi_2(y)$$

$$w_1 = \rho_{xx}\chi_1^{(1)}(y) + c_{xx}\chi_2^{(1)}(y),$$

$$w_2 = \rho_{xxx}\chi_1^{(2)}(y) + c_{xxx}\chi_2^{(2)}(y)$$

where χ_1 etc. are periodic functions of y alone and are solutions of equations of the form $\mathcal{L}(\chi) = f(y)$. For more details see ([25]). Let us define the following scalar average values

$$\alpha_1 = \overline{U\chi_1} \ , \ \alpha_2 = \overline{U\chi_2} \ , \ \alpha_3 = \overline{U\chi_1^{(1)}} \ ,$$

$$\alpha_4 = \overline{U\chi_2^{(1)}} \ , \ \alpha_5 = \overline{U\chi_1^{(2)}} \ , \ \alpha_6 = \overline{U\chi_2^{(2)}}$$

Then the linear system to be solved for the perturbation of ρ and c about one and zero, respectively, becomes

$$\rho_t + c_x + 3\epsilon\rho_{xx} + 5\epsilon^3\nu\rho_{xxxx} = 0, \tag{3.3}$$

$$c_t + (\overline{U^2} + 2\nu\overline{UU_\nu})\rho_x + \epsilon\alpha_1\rho_{xx} + \epsilon(1+\alpha_2)c_{xx}$$
$$+\epsilon^2(\alpha_3\rho_{xxx} + \alpha_4 c_{xxx}) + \epsilon^3\alpha_5\rho_{xxxx} + \epsilon^3(\alpha_6+\nu)c_{xxxx} = 0. \tag{3.4}$$

This system contains linear stability characteristics up to order ϵ^3 which include the essential fourth derivative effects. The crucial parameters of the problem are the coefficients $\alpha_1, ..., \alpha_6$ which are determined numerically. Numerical solutions for the periodic functions χ above are obtained by spectral representation and inversion of the linearized operator ([25]). Normal mode solutions of (3.3) and (3.4) have the form

$$(\rho(t,x), c(t,x)) = (\overline{\rho}, \overline{c})e^{ikx+\omega t} \quad ,$$

where $\overline{\rho}, \overline{c}$ are constants, k is the wavenumber of a particular disturbance and ω is the corresponding growth given by the following dispersion relation

$$\omega^2 + \omega\left[-\epsilon(4+\alpha_2)k^2 + \epsilon^3(6\nu+\alpha_6)k^4\right] + k^2\lambda^2$$
$$+\epsilon^2k^4(3+3\alpha_2-\alpha_3) - \epsilon^4k^6(8\nu+3\alpha_6+5\alpha_2\nu) + 5\epsilon^6k^8\nu(\nu+\alpha_6) \;=\; 0 \,, \tag{3.5}$$

with

$$\lambda^2 = \overline{U^2} + 2\nu\overline{UU_\nu} \quad ,$$

which can be positive or negative. The important feature of our expression is the inclusion of higher order terms in ϵ and their effect on the structural stability of the system. It is easy to see that ω has the expansion

$$\omega = \pm ik\lambda + \epsilon\mu k^2 + \dots \quad ,$$

where

$$\mu = \frac{1}{2}(4+\alpha_2) \quad . \tag{3.6}$$

The parameter μ plays the role of a diffusion coefficient. Clearly the system is (linearly) stable if both $\lambda^2 > 0$ and $\mu < 0$. Such a two-terms result has been obtained by Frisch, She and Thual ([1]). The window of stability predicted by such a result is very small because the stabilizing higher order terms are excluded. They are readily obtainable from our analysis.

The $O(i\epsilon^2k^3)$ to ω is dispersive, while the physically significant correction is obtainable from a large k analysis of the dispersion relation (5.13), and is given by

$$\omega = \epsilon^3k^4\omega_3 \quad , \quad \omega_3 = -\nu - \alpha_6 \;\; or \;\; -5\nu \quad . \tag{3.7}$$

The results (3.6), (3.7) characterize the structural stability of periodic solutions. For stability, a range of ν must be located where $\lambda^2 > 0$ together with the next correction to the frequency $\omega_1 = (1/2)\epsilon k^2(4 + \alpha_2)$ being negative. The region of structural stability is modified according to $\omega_3 = -(\nu + \alpha_6) < 0$ which is found to increase stability boundaries significantly. The results are summarized in Table 1 below which shows the values of λ^2 , μ and $\epsilon^3\omega_3$ for $\epsilon = 0.05$.

ν	λ^2	μ	$\epsilon^3\omega_3$	$\bar{U}$
0.89	-14.59	40.30	-82.91	2.11
0.85	-10.88	21.95	86.48	2.79
0.80	-6.84	10.77	263.75	3.51
0.75	-3.33	4.73	1009.12	4.09
0.70	-0.287	1.07	24.50	4.51
0.65	2.30	-1.29	-77.55	4.79
0.60	4.43	-2.86	-36.49	4.93
0.55	6.14	-3.94	-16.70	4.92
0.50	7.41	-4.69	-7.10	4.74
0.45	8.23	-5.22	-2.53	4.41
0.40	8.70	-5.64	-0.55	3.92
0.35	8.77	-6.09	0.102	3.25
0.30	8.57	-6.97	0.130	2.37

It is clear that at values of ν where μ fails to stabilize the system the fourth derivative term can do so instead. Our computations show the presence of a stability window when $0.365 \leq \nu \leq 0.695$. At values where the $\omega_3 > 0$ the linear problem suffers from a short wave instability which makes the nonlinear problem ill-posed. For values of ν in the enlarged stability window $0.365 \leq \nu \leq 0.695$ the nonlinear problem is well-posed and is considered next.

The figure at the end gives the results of table graphically. The ν interval inside the little triangle around $\nu = .68$ is the stability regime predicted in ([1]). It is much smaller than the actual structural stability regime.

4. Weakly nonlinear theory.

When the size of the perturbations in c and ρ are of order ϵ^3, the equation for w is linear up to order ϵ^3 and the modulation equations contain nonlinear terms of order ϵ^3 also. Consequently the stability constants $\alpha_1, ..., \alpha_6$ are those provided by linear theory and the nonlinear system of conservation laws to be solved is

$$\rho_t + c_x + 3\epsilon\rho_{xx} + \epsilon^3(c\rho)_x + 5\epsilon^3\nu\rho_{xxxx} = 0 \quad , \tag{4.1}$$

$$c_t + \lambda^2 \rho_x + \epsilon(1+\alpha_2)c_{xx} + \epsilon^2 \alpha_3 \rho_{xxx}$$
$$+\epsilon^3(cc_x + \lambda^2 \rho\rho_x) + \epsilon^3(\alpha_6 + \nu)c_{xxxx} = 0 \quad , \tag{4.2}$$

together with initial conditions

$$\rho(x,0) = \rho_0(x) \quad , \quad c(x,0) = c_0(x) \quad .$$

The nonlinear system (4.1)-(4.2) was solved subject to periodic boundary conditions for two values of ν chosen so that in the former case $\mu > 0$ (positive diffusion) and $\omega_3 < 0$ (negative 4th derivative) while in the latter $\mu < 0$ (negative diffusion) and $\omega_3 < 0$. The former case describes the weakly nonlinear dynamics expected from the linear solutions of ([1]). Details of the numerical scheme and further results can be found in ([25]).

References

1. Frisch, U., She, Z.S., Thual, O., *Viscoelastic behaviour of cellular solutions to the Kuramoto-Sivashinsky model*, J. Fluid Mech., **168** (1986) pp. 221-240.
2. Benney, D.J., *Long Waves in Liquid films*, J. Math. and Phys. **45** (1966), pp. 150-155.
3. Shlang, T., Sivashinsky, G.I., *Irregular flow of a liquid film down a vertical column*, J. Phys., **43**, (1982) pp. 459-466.
4. Babchin, A.J., Frenkel, A.L., Levich, B.G., Sivashinsky, G.I., *Nonlinear Saturation of Rayleigh-Taylor Instability in thin films*, Phys. Fluids **26** (1983), pp. 3159-3161.
5. Hooper, A.P., Grimshaw, R., *Nonlinear instability at the interface between two fluids*, Phys. Fluids, **28** (1985) pp. 37-45.
6. Papageorgiou, D.T., Maldarelli, C., Rumschitzki, D.S., *Nonlinear interfacial stability of core-annular film flow*, Phys. Fluids A, **2**, No. 3 (1990) pp. 340-352.
7. Kuramoto, Y., *Diffusion-induced chaos in reactions systems*, Suppl. Prog. Theor. Phys., **64** (1978) pp. 346-367.
8. Kuramoto, Y., Tsuzuki, T., *Persistent propagation of concentration waves in dissipative media far from thermal equilibrium*, Prog. Theor. Phys., **55** (1976) pp. 356-369.
9. Kuramoto, Y., Tsuzuki, T., *On the formation of dissipative structures in reaction-diffusion systems*, Prog. Theor. Phys., **54** (1975) pp. 687-699.
10. Cohen, B.I., Krommes, J.A., Tang, W.M., Rosenbluth, M.N., *Non-linear saturation of the dissipative trapped ion mode by mode coupling*, Nucl. Fusion, **16**, (1976) pp. 971-992.
11. Sivashinsky, G.I., *Nonlinear analysis of hydrodynamic instability in laminar flames, Part I. Derivation of basic equations*, Acta Astronautica, **4** (1977) pp. 1176-1206.
12. Sivashinsky, G.I., *On flame propagation under conditions of stoichiometry*, SIAM J. Appl. Math., **39** (1980) pp. 67-82.
13. Babin, A.V., Vishik, M.I., *Regular attractors and semi-groups of evolution equations*, J. Math. Pures Appl., **62** (1983) pp. 441-491.
14. Manneville, P., *Lyapunov Exponents for the Kuramoto-Sivashinsky Equations*, Proc. Conf. on Turbulence, Nice 1984, Springer-Verlag Lecture Notes in Physics, New York (1985).
15. Constantin, P., Foias, C., Nicolaenko, B., Temam, R., *Integral Manifolds and Inertial Manifolds for Dissipative Partial Differential Equations*, Appl. Math. Sciences, No. 70, Springer-Verlag, New York, 1988.
16. Foias, C., Jolly, M.S., Kevrekidis, I.G., Sell, G.R., Titi, E.S., *On the Computation of Inertial Manifolds*, Phys. Lett. A, **131**, number 7,8 (1988).
17. Temam, R., *Infinite-dimensional Dynamical Systems in Mechanics and Physics*, Springer-Verlag, New York, 1988.
18. Brown, H., S., Jolly, M.S., Kevrekidis, I.G., Titi, E.S., *Use of Approximate Inertial Manifolds in Bifurcation Calculations*, Proc. NATO Adv. Res. Workshop on : Continuation and Bifurcations : Numerical Techniques and Applications, 18-22 September, 1989, Leuven, Belgium.
19. Jolly, M.S., Kevrekides, I.G. and Titi, E.S., *Approximate inertial manifolds for the Kuramoto-Sivashinsky equation: analysis and computations*, Physica D, **44** (1990) pp. 38-60.
20. Kevrekidis, I.G., Nicolaenko, B., Scovel, C., *Back in the saddle again : A computer assisted study of Kuramoto-Sivashinsky equation*, SIAM J. Appl. Math., **50**, No. 3 (1990) pp. 760-790.
21. Papageorgiou, D.T., Smyrlis, Y.S., *The route to chaos for the Kuramoto Sivashinsky equation*, Theoret. Comput. Fluid Dynamics, **3** (1991), pp. 15-42.
22. Smyrlis, Y.S., Papageorgiou, D.T., *Predicting chaos for infinite dimensional dynamical systems: The Kuramoto-Sivashinsky equation, a case study*, Proc. Natl. Acad. Sci. USA, **88** (1991), pp. 11129-11132.
23. Feigenbaum, M., *The transition to aperiodic behavior in turbulent systems*, Commun. Math. Phys., **77** (1980), pp. 65-86.
24. Feigenbaum, M., *The onset spectrum of turbulence*, Phys. Lett., **74**, pp. 375-378.

25. Papageorgiou, D.T., Papanicolaou, G.C., Smyrlis, Y.S., *Modulational stability of periodic solutions of the Kuramoto Sivashinsky equation*, preprint (1992).
26. McLaughlin, D., Papanicolaou, G.C., Pironneau, O., *Convection of Microstructure and Related Problems* SIAM J Appl Math **45** (1985) pp. 780-797.

Dynamic rescaling for tracking point singularities: application to Nonlinear Schrödinger equation and related problems

G. Papanicolaou
Courant Institute of Mathematical Sciences
251 Mercer Street
New York N.Y. 10012
U.S.A
C. Sulem
Department of Mathematics
University of Toronto
Toronto M5S1A1
Canada
P.L. Sulem
CNRS Observatoire de Nice
BP 229
06304 Nice-Cedex 4
France
and
X.P. Wang
Program in Applied Mathematics
University of Colorado
Boulder CO 80309-0526
U.S.A.

ABSTRACT. We present a method with which equations that develop a localized singularity can be integrated numerically up to times very close to the singularity formation. We apply the method, which is based on a dynamic rescaling, to self-focusing solutions of the Nonlinear Schrödinger equation, the Zakharov equations for Langmuir waves in plasmas and the Davey Stewartson equation for capillary–gravity waves at a fluid surface.

1. Introduction

The Nonlinear Schrödinger equation is the simplest example of a wave equation whose solutions may blow up in a finite time because of self-focusing. It describes, in general, the evolution on a long time scale of a fully dispersive, weakly nonlinear wave that satisfies a linear equation on a short time scale ([1]). It appears as the subsonic limit of the Zakharov equations for

R. E. Caflisch and G. C. Papanicolaou (eds.), Singularities in Fluids, Plasmas and Optics, 265–279.
© 1993 *Kluwer Academic Publishers.*

Langmuir waves in plasmas and also as the deep water limit of the Davey–Stewartson equations for capillary–gravity waves at a fluid surface. For all these equations the singularities of the solutions are isolated and correspond to a localized focusing or collapse with a self–similar, or quasi self–similar structure, associated with the invariance properties of the equation.

To accurately compute the singular solutions very close to the blow up time an adaptive mesh algorithm is required. The dynamic rescaling method we discuss here is based on a continuous time rescaling. The idea is to introduce a time–dependent scaling of the dependent and independent variables, so that in the new variables, the solution is smooth and can be solved by a usual numerical method. The singular behavior of the solution in the original variables is determined by the scaling factors. This method enables one to study the detailed structure of the singularities of the above equations.

2. Dynamic rescaling for isotropic NLS

The nonlinear Schrödinger equation (NLS)

$$i\psi_t + \Delta\psi + |\psi|^{2\sigma}\psi = 0 \quad t > 0, \tag{1}$$

$$\psi(0,\mathbf{x}) = \psi_0(\mathbf{x}), \quad \mathbf{x} \in R^d$$

arises in various physical contexts (usually with $\sigma = 1$), as an amplitude equation for weakly nonlinear dispersive waves ([2]). For a certain class of initial conditions, in particular those for which the invariant $H = \int_0^\infty (|\nabla\psi|^2 - \frac{1}{2}|\psi|^2)d\mathbf{x}$ is negative, NLS has solutions that become singular in a finite time when the space dimension d and exponent σ satisfy $\sigma d \geq 2$. This singularity corresponds to a focusing collapse of an initially smooth wavepacket. The local form of this singularity was studied numerically in the special case of radially symmetric solutions in ([3]; [4]), using a dynamic rescaling method suggested by the invariances of the equation and based on the following change of variables ($r = |\mathbf{x}|$):

$$\psi(r,t) = \frac{1}{L(t)}u(\xi,\tau) \quad , \quad \xi = \frac{r}{l(t)} \quad , \quad \tau = \int_0^t \frac{ds}{l^2(s)} \tag{2}$$

where

$$L(t) = \left(\frac{1}{l(t)}\right)^{1/\sigma} . \tag{3}$$

Substituting in (1), we get

$$iu_\tau + u_{\xi\xi} + \frac{d-1}{\xi}u_\xi + |u|^2u + ia\left(\frac{u}{\sigma} + \xi u_\xi\right) = 0 \tag{4}$$

where

$$a = -l\frac{dl}{dt} = -\frac{1}{l}\frac{dl}{d\tau} \ . \tag{5}$$

The scaling factor l should tend to zero fast enough as the singularity time is approached in order to make τ infinite in this limit. It is chosen by demanding that $u(\xi,\tau)$ remain smooth. Various constraints have been used with comparable efficiency. They consist in requiring that one of the spatial norms $|u|_{L^\infty}$, $|\nabla u|_{L^2}$, $|\Delta u|_{L^2}$, $|\nabla u|_{L^4}$, etc., remains constant in time. This enables one to express a as a functional of u. For example, by fixing $|\nabla u|_{L^2}$, we determine the scaling factor by

$$l(t) = (\frac{|\nabla u_0|^2_{L^2}}{|\nabla\psi(\cdot,t)|^2_{L^2}})^p \quad p = 2 + \frac{2}{\sigma} - d \tag{6}$$

and then

$$a(\tau) = -\frac{2\sigma}{p|\nabla u_0|^2_{L^2}} \int_0^\infty |u|^{2\sigma} Im(u\Delta u^*)\xi^{d-1}d\xi \tag{7}$$

is to be substituted in (4). In (6) and (7), the subscript zero refers to the initial condition. Equations (4) and (7) constitute a closed system whose solution remains smooth for all τ. The solution of eq.(1) is recovered by computing the scaling factor l from refeq.3), (7).

In the supercritical case $\sigma d > 2$, the solution near the singularity approaches the self-similar solution which (up to simple rescaling associated with invariances of the problem) is

$$\psi(r,t) = (\frac{1}{2K(t_* - t)})^{1/2\sigma} exp(\frac{i}{2K} ln\frac{t_*}{t*-t}) Q(\frac{r}{(2K(t_* - t))^{1/2}}) \tag{8}$$

where Q is a solution of

$$Q_{\xi\xi} + \frac{d-1}{\xi}Q_\xi - Q + iK(Q + \sigma\xi Q_\xi) + |Q|^{2\sigma}Q = 0 \tag{9}$$

for a unique value $K = K(d,\sigma) > 0$.

In the critical case ($\sigma d = 2$), the self-similarity is weakly broke because in (9), K is no longer a constant but tends to zero very slowly. An asymptotic analysis shows that to leading order, K(which is asymptotically proportional to a defined in (5)) varies like $(\log\tau)^{-1} \sim [loglog(1/(t_* - t))]^{-1}$. The scaling factor $l(t)$ has a log log correction in the form $l(t) = [(t_* - t)/\log\log(t* - t)]^{1/2}$ ([5; 6; 7; 8]).

Remark 1: The method of dynamic rescaling is easily extended to situations where the scale invariance is broken, for example to the case of NLS with a saturating nonlinearity NLS ([4]) In this case, the scaling factor $l(t)$ enters

the scaled equation and should be computed at each time step by solving (5) where $a(\tau)$ is given by a functional of u analogous to (7).
Remark 2: Instead of fixing the scaling factor l by prescribing that a (spatial) norm of u remains constant, one can use the condition

$$\frac{\int r^{2p}|u|^{2p}r^{d-1}dr}{\int |u|^{2p}r^{d-1}dr} = constant \tag{10}$$

where the exponent p is usually taken equal to 2 or 3 in order to ensure fast enough convergence of the integrals. This procedure was used in the case of anisotropic solutions (see Section 3).

3. Dynamic rescaling for anisotropic solutions

When trying to extend the dynamic rescaling to anisotropic solutions corresponding to an initial wavepacket which is not radially symmetric, it is necessary to for change in the location of the maximum of the solution and also different rates of collapse for different directions ([9]). To deal with this situation, the origin of the rescaled coordinates should move in time, the coordinate axes should rotate and different rescaling should be chosen along the various coordinate axes ([10]). For simplicity, we concentrate on NLS with cubic nonlinearity. We thus write

$$\psi(\mathbf{x},t) = \frac{1}{L(t)}u(\boldsymbol{\xi},\tau)\,, \quad \boldsymbol{\xi} = D^{-1}(t)(\mathbf{x}-\mathbf{x}_0)\,, \quad \tau = \int_0^t \frac{1}{L^2(s)}ds\,. \tag{11}$$

where the matrix $D(t)$ can be rewritten as the product $D(t) = O^T(t)\Lambda(t)$ where O is a orthogonal matrix and $\Lambda(t)$ a diagonal matrix whose diagonal elements are $\lambda_i (i = 1, ..., d)$. The scaled equation thus takes the form

$$i[u_\tau - L^{-1}L_\tau u + \mathbf{f}\cdot\nabla u] + \mathcal{L}u + |u|^2 u = 0 \tag{12}$$

where

$$\mathbf{f} = -D^{-1}\frac{dD}{d\tau}\boldsymbol{\xi} - D^{-1}\frac{d\mathbf{x}_0}{d\tau} \tag{13}$$

and

$$\mathcal{L} = L^2 \sum_i \lambda_i^{-2}\frac{\partial^2}{\partial \xi_i^2} \tag{14}$$

The quantities $\mathbf{x}_0$, $D(t)$ and $L(t)$ are determined by the following conditions:
(i) The origin $\mathbf{x}_0$ is chosen to be the centroid of $2p$ power of $|\psi|$:

$$x_0^i = \frac{\int x_i |\psi|^{2p} d\mathbf{x}}{\int |\psi|^{2p} d\mathbf{x}} \tag{15}$$

where p is an integer. Usually $p \geq 3$ suffices to ensure accuracy in the numerical computations of the integrals. In the limit $t \to t_*$ (or equivalently $\tau \to \infty$), $\mathbf{x}_0$ is very likely to be the blow up point.

(ii) The matrix D is chosen so that

$$\frac{\int \xi_i \xi_j |u|^{2p} d\boldsymbol{\xi}}{\int |u|^{2p} d\boldsymbol{\xi}} = \delta_{ij} \; . \tag{16}$$

This choice is the extension of condition (10) to anisotropic solutions and minimizes the deviation of the rescaled solution from isotropy. Coming back to the primitive variables, (16) implies that the matrix D is given by

$$D^{-1} S (D^{-1})^T = I \tag{17}$$

where the matrix $S = (s_{ij})$ is defined as

$$s_{ij} = \frac{\int (x^i - x_0^i)(x^j - x_0^j) |\psi|^{2p} d\mathbf{x}}{\int |\psi|^{2p} d\mathbf{x}} \tag{18}$$

(iii) The factor $L(t)$ is chosen as

$$L(t) = \sqrt{\frac{d}{\sum 1/\lambda_i^2}} \tag{19}$$

in order to ensure that $L^2 \lambda_i^{-2}$ is bounded.

To see how $\mathbf{x}_0, D, L$ vary with time, we take derivatives of (15)-(19) with respect to the scaled time τ. Using NLS together with (12) and (19), we obtain a closed system for the evolution for the transformed solution $u(\boldsymbol{\xi}, \tau)$ and the scaling factors λ_i, which is (refeq.14) together with

$$\frac{d\lambda_i}{d\tau} = -a_{ii} \lambda_i \;\; (i = 1, ..., d) \tag{20}$$

where L is given by (19),

$$L^{-1} L_\tau = -\frac{\sum a_{ii}/\lambda_i^2}{\sum 1/\lambda_i^2} \tag{21}$$

$$a_{ij} = \frac{p \int (\delta_{ij} - \xi_i \xi_j) |u|^{2(p-1)} Im(u \mathcal{L} u^*) d\boldsymbol{\xi}}{\int |u|^{2p} d\boldsymbol{\xi}} \tag{22}$$

$$\mathbf{f} = B\boldsymbol{\xi} - 2\boldsymbol{\beta} \; . \tag{23}$$

The matrix $B = (b_{ij})$ is defined by

$$b_{ii} = -a_{ii} \;\; ; \;\; b_{ij} = \frac{2\lambda_j^2}{\lambda_j^2 - \lambda_i^2} a_{ij} \;\; (i \neq j) \tag{24}$$

and the vector $\boldsymbol{\beta}$ by

$$\beta_j = \frac{p \int \xi_j |u|^{2(p-1)} Im(u\mathcal{L}u^*) d\boldsymbol{\xi}}{\int |u|^{2p} d\boldsymbol{\xi}} \tag{25}$$

Note that the computation of the rotation O and of the centroid $\mathbf{x}_0$ is not required. This is a consequence of the invariance of the equation by rotation and translation. These quantities can be obtained by solving

$$\frac{d\mathbf{x}_0}{d\tau} = 2O^T \Lambda \boldsymbol{\beta} \tag{26}$$

$$\frac{dO}{d\tau} = -GO \tag{27}$$

where

$$g_{ii} = 0 \quad ; \quad g_{ij} = \frac{2\lambda_i \lambda_j}{\lambda_i^2 - \lambda_j^2} a_{ij} \quad (\quad \lambda_i \neq \lambda_j) \tag{28}$$

In practice, it is convenient to express the rotation matrix O in terms of its Euler angles (three angles if $d = 3$ and only one if $d = 2$).

Finally, (12) must be solved in the entire space R^d. However, at large distance, this equation reduces to the hyperbolic equation

$$u_\tau - \frac{L_\tau}{L} u + \mathbf{f} \cdot \nabla u = 0. \tag{29}$$

which can be solved exactly. Equation (12) is then solved in an extended enough but finite domain. The solution is matched at each time step to the exact solution of (29). This approximation allows a substantial reduction of the spatial integration domain. It was introduced in ([11]) in the context of isotropic solutions.

The above approach was used to solve the nonlinear Schrödinger equation with initial conditions corresponding to a wavepacket without rotational symmetry ([10]). We found that after an early transient, during which the anisotropy is significantly amplified, the solution becomes isotropic. Initial conditions corresponding to two peaks with different amplitudes were also considered. We also found that the higher peak becomes stronger and eventually, the lower peak is absorbed by the higher one, leading to a simple isotropic peak.

4. Dynamic rescaling for the Zakharov equations

The Zakharov equations([12]) ([13])

$$i\mathbf{E}_t - \alpha \nabla \times (\nabla \times \mathbf{E}) + \nabla(\nabla \cdot \mathbf{E}) = n\mathbf{E} \tag{30}$$

$$\varepsilon n_{tt} - \Delta n = \Delta |\mathbf{E}|^2 \tag{31}$$

and its scalar version[14]

$$i\psi_t + \Delta\psi - n\psi = 0 \tag{32}$$

$$n_{tt} - \Delta n = \Delta|\psi|^2 \tag{33}$$

are model equations for strong Langmuir turbulence in plasmas. In eq.(31), ε is a parameter which tends to zero in the subsonic limit. For $\varepsilon = 0$, the system (32), (33) reduces to vector NLS.

There is no proof of existence of singular solutions for the Zakharov equations, although numerical simulations in two and three dimensions do indicate that collapse can occur when the energy is negative.

4.1. THE SCALAR MODEL

In order to implement the dynamic rescaling to the scalar model, we introduce the change of variables

$$\psi(\mathbf{x}, t) = \frac{1}{L(t)} U(\boldsymbol{\xi}, \tau), \quad n(\mathbf{x}, t) = \frac{1}{M(t)} V(\boldsymbol{\xi}, \tau), \tag{34}$$

together with

$$w(\mathbf{x}, t) = \frac{1}{H(t)} W(\boldsymbol{\xi}, \tau) \tag{35}$$

where $w = n_t$. The rescaled space variable $\boldsymbol{\xi}$ is defined as for NLS but the rescaled time is now

$$\tau = \int_0^t \frac{1}{\Omega^2(s)} ds, \tag{36}$$

where $\Omega(t)$ is a positive scalar function of time. The form of the equations suggests that we to take

$$L(t) = \Omega^{d/2}(t), \quad M(t) = \Omega^2(t), \quad H(t) = \Omega^{2+d/2}(t) \tag{37}$$

together with

$$\Omega^2(t) = \frac{d}{\sum(1/\lambda_i^2(t))} . \tag{38}$$

It is also convenient to modify (15) and (16) by replacing the amplitudes ψ and u by the densities n and V respectively. Contrary to the original definition, this choice keeps W bounded away from zero near the collapse. We have the following evolution equations

$$i[U_\tau - L^{-1}L_\tau U + \mathbf{f} \cdot \nabla U] + \mathcal{L}U - VU = 0 \tag{39}$$

$$V_\tau - M^{-1}M_\tau V + \mathbf{f}\cdot\nabla V - \frac{M\Omega^2}{H}W = 0 \tag{40}$$

$$W_\tau - H^{-1}H_\tau W + \mathbf{f}\cdot\nabla W - \frac{H}{M}\mathcal{L}V - \frac{H}{L^2}\mathcal{L}(|U|^2) = 0 \tag{41}$$

$$\frac{d\lambda_i}{d\tau} = -a_{ii}\lambda_i\Omega^{2-\frac{d}{2}} \tag{42}$$

$$\frac{d\mathbf{x}_0}{d\tau} = 2\Omega^{2-\frac{d}{2}}O^T\Lambda\boldsymbol{\beta} \tag{43}$$

$$\frac{dO}{d\tau} = -GO \tag{44}$$

The matrices $A = (a_{ij})$ are now defined by

$$a_{ij} = \frac{p\int(\delta_{ij} - \xi_i\xi_j)V^{2p-1}W d\boldsymbol{\xi}}{\int |V|^{2p} d\boldsymbol{\xi}} \tag{45}$$

and G is given by (28). Furthermore,

$$\mathbf{f} = \Omega^{2-\frac{d}{2}}(B\boldsymbol{\xi} - 2\boldsymbol{\beta}) \tag{46}$$

where B is given by (24) and $\boldsymbol{\beta}$ by

$$\beta_j = \frac{p\int \xi_j V^{2p-1}W d\boldsymbol{\xi}}{\int |V|^{2p} d\boldsymbol{\xi}} \tag{47}$$

Again, due to the translation and rotation invariances of the scalar model, (39)–(42) form a closed system for U, V, W and λ_i, while the translation $\mathbf{x}_0$ and the rotation matrix O are secondary quantities which can be determined separately from (43) and (44). The coupled equations (39)–(45) are solved numerically by a finite difference method with the approximate boundary conditions discussed in Section 3.

Computations are reported in ([15]). We summarize the results as follows. In three dimensions, self-similar solutions can only exist asymptotically close to the collapse when the regime is strongly supersonic with the pressure term Δn negligible compared to the ion-inertia term in eq.(33). For a large class of initial conditions corresponding to a single wavepacket, the singular solutions become isotropic and approach asymptotically a universal self-similar solution of the form:

$$\psi(r,t) = \frac{1}{(t_\star - t)}P\left(\frac{r}{(t_\star - t)^{2/3}}\right)e^{i(t_0-t)^{-1/3}} \tag{48}$$

$$n(r,t) = \frac{1}{(t_\star - t)^{4/3}}N\left(\frac{r}{(t_\star - t)^{2/3}}\right) \tag{49}$$

where $P(\eta)$ and $N(\eta)$ are isotropic scalar functions given by

$$\Delta P - \frac{1}{3}P - NP = 0 \tag{50}$$

$$\frac{2}{9}(2\eta^2 N_{\eta\eta} + 13\eta N_\eta + 14N) = \Delta P^2 \tag{51}$$

In two dimensions, the self-similar collapse is no longer supersonic. The pressure term Δn in equation (33) is not negligible. Singular solutions approach a self-similar isotropic solutions of the form:

$$\psi(r,t) = \frac{1}{\alpha(t_\star - t)} P\left(\frac{r}{\alpha(t_\star - t)}\right) e^{i(t_\star - t)^{-1} - i(t_\star - t)^{-1} r^2/4}$$

$$n(r,t) = \frac{1}{(\alpha(t_\star - t))^2} N\left(\frac{r}{\alpha(t_\star - t)}\right)$$

where $P(\eta), N(\eta)$ satisfy

$$\Delta P - P - NP = 0 \tag{52}$$

$$\alpha^2(\eta^2 N_{\eta\eta} + 6\eta N_\eta + 6N) - \Delta N = \Delta P^2 \tag{53}$$

An interesting aspect of the problem in two dimensions is that the parameter α in eq.(53) is not universal but depends on the initial condition. It appears that when the L^2–norm of the initial amplitude approaches its lower bound for collapse given by the L^2–norm $|R|_{L^2}$ of the positive solution of the equation $\Delta R - R + R^3 = 0$, then α tends to zero. From the above equations for P, N, we see that $P \to R$. Therefore, the singular solution of the Zakharov problem tends, up to a simple rescaling, to $\psi(r,t) \approx \frac{1}{t_* - t} R(\frac{r}{t_* - t}) \exp(i(t_* - t)^{-1} - i(t_* - t)^{-1} r^2/4)$. This function is an exact self-similar solution of the two dimensional non–linear Schrödinger equation. As a solution of the latter, it is generally unstable. A possible exception is when the mass is exactly critical. It was recently proved that in this case, it is the only singular solution ([16]; [17]).

4.2. VECTOR ZAKHAROV EQUATIONS AND VECTOR NLS

The main change in implementing the dynamic rescaling to the full vector Zakharov equations is that the rotation also acts on the vector $\mathbf{E}$ which is rewritten

$$\mathbf{E}(\mathbf{x}, t) = \frac{1}{L(t)} O^T \mathbf{U}(\boldsymbol{\xi}, \tau). \tag{54}$$

Our computations performed in three space dimensions([18]) show that when the parameter ε is of order unity, the collapse of solutions of the vector Zakharov equations is supersonic and self–similar with the same rate of collapse as for the scalar model. The intensity distribution which displayed a very strong anisotropy in an early stage, eventually relaxes to an ellipsoidal configuration with moderate aspect ratios. These observations support the argument, based on spherical harmonic decomposition given in ([19]), that weakly anisotropic collapse is far more common than strongly anisotropic collapse.

In contrast, with the vector NLS ($\varepsilon = 0$) and the near subsonic Zakharov equations (ε very small), it appears that the dynamics is sensitive to the initial conditions. For moderately anisotropic initial data, the solution of vector NLS becomes self–similar with a $(t_* - t)^{1/2}$ rate of collapse. Under the same conditions, the solution of the vector Zakharov equation first displays a subsonic scaling. However, as the singularity is approached, the subsonic approximation breaks down and the above supersonic regime emerges, leading to a weakly anisotropic collapse.

For more anisotropic initial data, solutions of the vector NLS and of the near subsonic Zakharov equations contract much faster in one direction than in the other two. After a while, the solutions break apart into two peaks. In its present state the dynamic rescaling is not suitable for studying the subsequent evolution.

5. Dynamical rescaling for the Davey–Stewartson equations

The evolution of a two-dimensional capillary–gravity wave packet at a fluid surface that travels predominantly in one direction is governed by the Davey-Stewartson equations (DS). When written in a non-dimensional form these equations are

$$iA_t + \sigma_1 A_{x_1x_1} + A_{x_2x_2} = \sigma_2 |A|^2 A + \phi_{x_1} A \tag{55}$$

$$\alpha\phi_{x_1x_1} + \phi_{x_2x_2} = -\gamma(|A|^2)_{x_1x_1} \tag{56}$$

Here A denotes the wave amplitude and ϕ the large scale velocity potential produced by the wave. The parameters σ_1, σ_2, α and γ depend on the (dimensionless) fluid depth and surface tension. It is always possible to choose $|\sigma_1| = |\sigma_2| = 1$. The coefficient γ is positive, while α is an arbitrary real number. At infinity, $A \to 0$, while the condition on the potential depends on the sign of α. We concentrate here on the case $\alpha > 0$ where $\phi \to 0$ at infinity. The above equations were derived in ([20]; [21]; [22]) in the absence of surface tension. In the case, $\sigma_1 = -1$, $\sigma_2 = 1$ and $\alpha = 1$ and $\gamma = 2$ and the problem is integrable by inverse scattering ([23]). The effect of surface tension was introduced in ([24]; [25]).

The most dramatic effect of surface tension for surface waves occurs in sufficiently deep water. In this regime, $\sigma_1 = 1$, $\sigma_2 = -1$, $\alpha > 0$, and a wave with a large enough amplitude focuses in a finite time. Proof of existence of such a singularity is similar to that for NLS and is based on a variance identity ([25]; [26]). The solution remains smooth in the large if the initial amplitude is sufficiently small, namely if $|A|^2_{L^2} < \dfrac{1}{1+\gamma/\alpha} |R|^2_{L^2}$. A detailed study of the Cauchy problem is found in ([26]). Here, we report on recent results concerning the nature of the focusing singularity obtained numerically with the dynamic rescaling method and by an asymptotic analysis. A more detailed presentation is given in ([27]). To implement the dynamic rescaling, we rescale the independent variables and the wave amplitude as for NLS. In order to avoid cross terms in the rescaled equation for the potential, we concentrate on initial amplitudes which are symmetric with respect to the coordinate axes. In this case, the rotation matrix O can be taken equal to unity. We denote by $\ell_i^2 = 2L^2/\lambda_i^2$ the normalized scaling factors. The potential is rescaled by setting $v(\xi,\tau) = L(t)\phi(x,t)$. Eqs.(55), (56) become

$$i\{u_\tau + \sum_{i=1}^{2} a_i \ell_i^2 u + \mathbf{f}\cdot\nabla u\} + 2\sigma_1 l_1^2 u_{\xi_1\xi_1} + 2 l_2^2 u_{\xi_2\xi_2} = \sigma_2 |u|^2 u + \sqrt{2} l_1 v_{\xi_1} u \quad (57)$$

$$\alpha l_1^2 v_{\xi_1\xi_1} + l_2^2 v_{\xi_2\xi_2} = -\frac{\gamma}{\sqrt{2}} l_1 (|u|^2)_{\xi_1} \quad (58)$$

$$\frac{l_{i\tau}}{l_i} = -\sum_{j=1}^{2} a_j l_j^2 + a_i \quad (59)$$

In (55)–(59), f_i, a_i and β_i are defined as in eqs.(13-25). Numerical simulations have been done with $\sigma_2 = 1$ and various values of the parameters α and γ, and for $\sigma_2 = -1$ (defocusing nonlinear terms) and "large" γ. For $\gamma = 1$ and $\alpha = 1$, and an initial amplitude $A(x_1,x_2) = 4\exp^{-x_1^2 - x_2^2/4}$, fig.1 shows the evolution of the normalized scaling factors ℓ_1 and ℓ_2. As $\tau \to \infty$, they tend to different limit values while a_1 and a_2 tend to 0 very slowly. The profiles u and and v tend to become stationary (figs 2a-b) while the phase at the origin Arg $u(0,\tau)$ varies linearly with τ. Similar results were observed for other values of the parameters. These observations suggest that the blow up is similar to that of NLS in critical dimension.

As with NLS, the detailed nature of the singularity can be determined by an asymptotic analysis. We rewrite eqs.(55), (56) in terms of A and $\psi = \phi_x$. Since we observe numerically that the anisotropy is moderate (the scaling factors obeying the same law in time), it is convenient to rescale both x_1

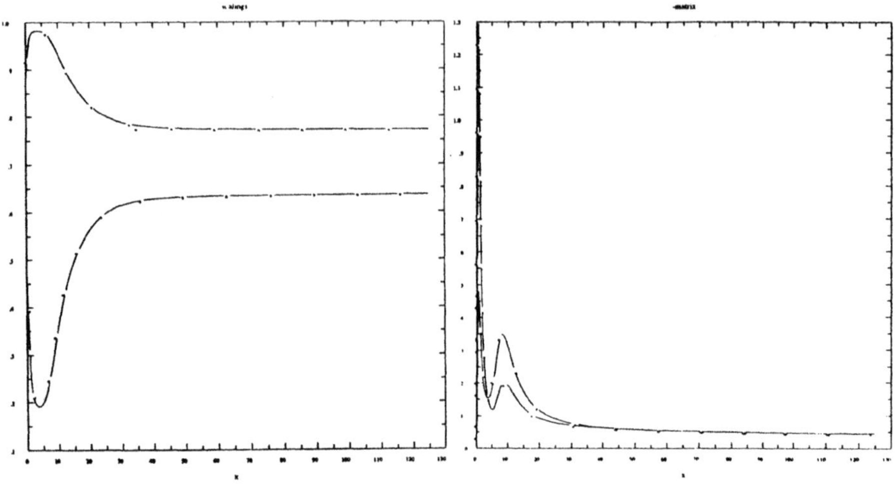

(a) (b)

Fig. 1. Normalized scaling factors l_1 and l_2 (fig. 1a) and functions a_1 and a_2 versus the rescaled time τ.

and x_2 with the same factor $L(t)$. We also write $W(\xi,\tau) = L^2\psi(\mathbf{x},t)$ and introduce $V = e^{-i\tau+ia\xi^2/4}u$. It follows that V and W satisfy

$$iV_\tau - V + \Delta V + \frac{a^2 + a_\tau}{4}|\xi|^2 V + |V|^2 V = VW \tag{60}$$

$$\Delta_\alpha W = -\gamma(|V|^2)_{\xi_1\xi_1} \tag{61}$$

where $\Delta_\alpha = \alpha\frac{\partial^2}{\partial\xi_1^2} + \frac{\partial^2}{\partial\xi_2^2}$. Alternatively to the approach presented in ([6]), singular solutions for critical NLS may be constructed perturbatively starting from the self-similar supercritical solutions obtained by keeping the space dimension equal to 2, but changing the nonlinearity $|u|^2u$ to $|u|^{2\sigma}u$ where the exponent $\sigma > 1$ tends to 1 as the singularity is approached. Here is the main line of the analysis. We introduce the stationary system for (P, Z):

$$\Delta P - P + \frac{a^2}{4}|\xi|^2 P + ia(\frac{1}{\sigma} - 1)P + |P|^{2\sigma}P = PZ \tag{62}$$

$$\Delta_\alpha Z = -\gamma(|P|^{2\sigma})_{\xi_1\xi_1} \tag{63}$$

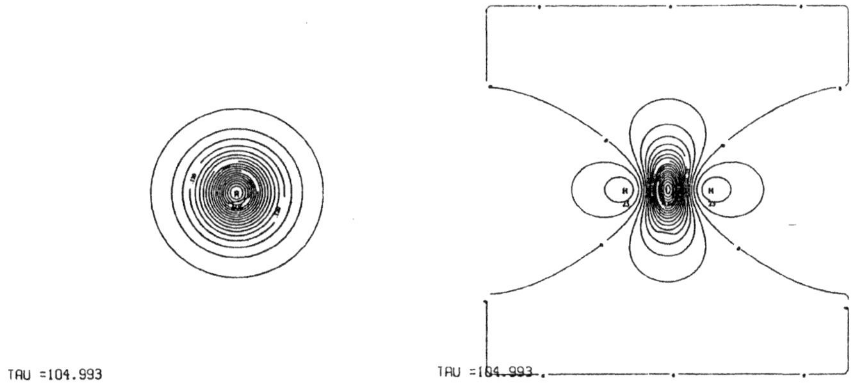

(a) (b)

Fig. 2. Contours of the profiles u (fig.2a) and v (fig.2b) for large τ.

where $\sigma > 1$ and $a > 0$, and write the ansatz:

$$V(\xi, \tau) = P(a(\tau), \xi) + v(\xi, \tau) \tag{64}$$

$$W(\xi, \tau) = Z(\xi) + w(\xi, \tau) \tag{65}$$

with the assumptions $a_\tau << a$ as $\tau \to \infty$ and $\sigma \to 1$ as $a \to 0$. The analysis is more delicate than the one for NLS because we have to deal with a coupled system in fully two dimensions. Also, in the NLS case ($\gamma = 0$), when $a = 0$ and $\sigma = 1$, system (62) (63) reduces to the equation for the ground state for which the mathematical theory is well developed. For the DS system, existence of solutions for (62), (63) is still unproved even in the case $a = 0$ and $\sigma = 1$.

Solvability conditions for the system for v and w (to leading order) give a relation between a and σ of the form

$$a_\tau = C(\frac{1}{\sigma} - 1) \tag{66}$$

where C is a strictly positive constant. In order to solve (66), we need a relation between σ and a, in the limit $a \to 0$, $\sigma \to 1$. It is obtained by a

WKB argument on eq(62) (63) analogous to that given in ([4]) and ([6]). It leads to a relation of the form

$$\frac{1}{\sigma} - 1 \sim \frac{1}{a} e^{-\pi/a}. \tag{67}$$

Eqs (66) and (67) imply

$$a_\tau \sim \frac{1}{a} e^{-\pi/a} . \tag{68}$$

To leading order, this gives $a \sim \dfrac{c}{\ln \tau}$. Coming back to the primitive variables we have a blow-up rate

$$L(t) \sim \frac{(t_* - t)^{1/2}}{\left(\ln \ln \frac{1}{t_* - t}\right)^{1/2}}, \tag{69}$$

which is identical to that of the cubic Nonlinear Schrödinger equation in two space dimensions.

Acknowledgements

Computations were done at the Pittsburg Supercomputiong Center under grant DMS-890012P. We benefitted of support from of a NATO grant No COP 91 0521. The work of G. Papanicolaou was supported also by grants from the National Science Foundation, DMS-9003227, and from AFOSR, F49620-92-J0098. The work of C. Sulem was supported by NSERC operating grant OGPIN16.

References

1. A. Segur, "Solitons as approximate description of physical phenomena", Rocky Mount. J. Math. **8** (1978) 15.
2. A. Newell, Solitons in Mathematics and Physics, CBMS Appl. Math. Series, No **48**, SIAM, 1985.
3. D.W. McLaughlin, G. Papanicolaou, C. Sulem, P.L. Sulem, "The focusing singularity of the cubic Schrödinger equation", Phys. Rev. **A 34** (1986) 1200.
4. B. LeMesurier, G. Papanicolaou, C. Sulem, P.L. Sulem, "Focusing and multifocusing solutions of the Non-Linear Schrödinger equation", Physica **D 31** (1988) 78.
5. Fraiman, G.M., "Asymptotic stability of manifold of self-similar solutions in self-focusing", *Sov. Phys. JETP*, **61** (1986), 228.
6. B. LeMesurier, G. Papanicolaou, C. Sulem, P.L. Sulem, "Local structure of the self-focusing singularity of the Non-Linear Schödinger equation", Physica **D 32** (1988) 210.
7. M. Landman, G. Papanicolaou, C. Sulem, P.L. Sulem, Rate of blow up for solutions of the Non-Linear Schrödinger equation at critical dimension", Phys. Rev. **A 38**(1988) 3837.
8. S. Dyachenko, A.C. Newell, A. Pushkarev, V.E. Zakharov, "Optical turbulence: weak turbulence, condensates and collapsing filaments in the nonlinear Schödinger equation", Physica **D 57** (1992) 96.

9. G. Pelletier, "The asymptotic collapse of Langmuir waves", Physica **D 27** (1987) 187.
10. M. Landman, G. Papanicolaou, C. Sulem, P.L. Sulem and X. Wang, "Stability of isotropic singularities for the Nonlinear Schödinger equation", Physica **47D** (1991) 393.
11.N.E. Kosmatov, I.V. Petrov, V.F. Shvets, V.E. Zakharov, Large Amplitude Simulation of Wave Collapse in Nonlinear Schrödinger equations, Preprint No 1365 (1988), Space Research Institute.
12. V.E. Zakharov, "Collapse of Langmuir waves", Zh. Eksp. Teor. Fiz. **62**, 1745 (1972); Sov. Phys. JETP **35**, 908 (1972).
13. E.A. Kuznetsov, "The collapse of electromagnetic waves in a plasma", Zh. Eksp.Teor. Fiz. **66**, 2037 (1974); Sov. Phys. JETP **39**, 1003 (1975).
14.D.B. Budneva, V.E. Zakharov and V.S. Synakh, "Certain models for wave collapse", Fiz. Plasma **1** (1975) 606; Sov. J. Plasma Phys. **1** (1975) 335.
15. M. Landman, G. Papanicolaou, C. Sulem, P.L. Sulem and X. Wang, "Stability of isotropic self-similar dynamics for scalar wave collapse", Phy. Rev. **A**, in press.
16. F. Merle, "On uniqueness and continuation properties after blow up time of self-similar solutions of Nonlinear Schrödinger equation with critical exponent and critical mass", Comm. Pure Appl. Math. **XLV** (1992) 203.
17. F. Merle, Determination of blow-up solutions with minimal mass for nonlinear Schrödinger equations with critical power. Preprint.
18. G. Papanicolaou, C. Sulem, P.L. Sulem and X. Wang, "Singular solutions of the Zakharov equations for Langmuir turbulence", Phys. Fluids **B** 3, (1991), 969.
19. D.L. Newman, P.A. Robinson and M.V. Goldman, "Field structure of collapsing wave packets in 3D strong Langmuir turbulence", Phys. Rev. Lett. **62** (1989) 2132.
20. Zakharov, V.E., "Stability of periodic waves of finite amplitude on the surface of a deep fluid", Sov. Phys. J. Appl. Mech. Tech. Phys., (1968) 190.
21. Benney, D.J. and Roskes, G.J., "Wave instabilities", Stud. Appl. Math., **48** (1969), 377.
22 Davey, A. and Stewartson, K., "On three-dimensional packets of surface waves", Proc. Roy. Soc., A388 (1974), 191.
23. Ablowitz, M.J. and Haberman, R., "Nonlinear evolution equations-two and three dimensions", Phys. Rev. Letters, **35** (1975) 1185.
24 Djordjevic, V.D. and Redekopp, L.G., On two-dimensional packets of capillary-gravity waves", *J. Fluid Mech.*, **79** (1977), 703.
25. Ablowitz, M.J. and Segur, H., On the evolution of packets of water waves", Fluid Mech., **92** (1979), 691.
26. Ghidalia, J.M. and Saut, J.C., "On the initial value problem for the Davey-Stewartson systems", Nonlinearity, **3** (1990) 475.
27. G. Papanicolaou, C. Sulem, P.L. Sulem and X. Wang, "The focusing singularities of the Davey-Stewartson equations for capilary-gravity waves", Preprint.

DYNAMIC SELF-FOCUSING OF STRUCTURED LASER BEAMS IN PLASMA

Harvey A. Rose and D. F. DuBois
Theoretical Division
Los Alamos National Laboratory
Los Alamos, New Mexico 87545
U.S.A.

ABSTRACT. A statistical theory of laser hot spots produced by optics which incorporates a random phase plate is presented. The ponderomotive self-focusing of isolated intense hot spots can lead to supersonic filament formation.

1. Introduction

Real laser beams have complex variations in intensity and phase in the target plane. Beam smoothing techniques[1], which introduce spatial and/or temporal incoherence, offer a way to control the potentially large intensity fluctuations, "hot spots", which may initiate various instabilities such as Stimulated Raman Scattering (SRS), Stimulated Brillouin Scattering (SBS) and filamentation. While spatial incoherence and temporal incoherence seem to be necessary to suppress filamentation[1,2,3], experiments are commonly done with spatial incoherence alone and in this paper we will use an idealized model of such, the Random Phase Plate (RPP), in order to control the statistical properties of the laser field.

We represent the electric field, E of the laser light by its envelope, ε

$$\mathrm{E} = \mathrm{Real}\ \varepsilon \exp \mathrm{i}\left(k_0 z - \omega_0 t\right), \tag{1}$$

where ω_0 is the laser frequency and k_0 is the wavenumber in plasma. In the paraxial and "adiabatic light" (i.e. ignoring light propagation effects) approximations[1], ε,satisfies

$$\left(i\partial_z + \Delta_\perp\right)\varepsilon = n\varepsilon, \tag{2}$$

where n is the relative fluctuation in the plasma density. The units of length are of the order of a wavelength of light, λ: $2k_0c^2/\omega_p^2$ in the axial (z) direction and c/ω_p in the transverse direction, where c is the speed of light, ω_p is the local plasma frequency and the Laplacian, Δ, is approximated by its transverse part, $\Delta_\perp = \partial^2/\partial x^2 + \partial^2/\partial y^2$.

R. E. Caflisch and G. C. Papanicolaou (eds.), Singularities in Fluids, Plasmas and Optics, 281-299.
© 1993 *All Rights Reserved.*

2. Spatial Incoherence and the Random Phase Plate

The standard model of a RPP[1] can be specified by the following boundary condition for ε,

$$\varepsilon(\mathbf{x}_\perp, z=0, t) = \sum_{\mathbf{k}_\perp} |\varepsilon(\mathbf{k}_\perp)| e^{i(\mathbf{k}_\perp \cdot \mathbf{x}_\perp + \phi_{\mathbf{k}_\perp})}, \qquad (3)$$

where each of the $\mathbf{k}_\perp$, vectors in the x-y plane, is associated with a beamlet of an idealized RPP. We take them to be uniformly distributed in fourier space with an energy spectrum, $|\varepsilon(\mathbf{k}_\perp)|^2$, that cuts off for

$$k_\perp > k_{\max} \equiv \sqrt{\frac{n_c/n_0}{1+F^2}} \quad , \qquad (4)$$

which is a geometric constraint associated with the optics. F=D/2f where D is the RPP width and f is the focal length. n_c/n_0 is the ratio of the critical to the local plasma density. The phases, $\varphi_{\mathbf{k}_\perp}$, are sampled from an ensemble in which: (a) distinct phases are statistically independent; (b) they assume the values 0 or π with equal probability to represent the random, time independent, phase shifts induced by each element of the RPP. This phase ensemble has the property, (c): $\left\langle \exp\left[i\left(\phi_{k'} - \phi_k\right)\right]\right\rangle = \delta_{k'k}$. Any phase ensemble with properties (a) and (c) will yield wave fields, ε, with the same statistical properties as those described below.

2.1 STATISTICAL PROPERTIES OF THE RPP MODEL

2.1.1. *The Two Point Correlation Function.* If the ϕ's are the only random element, then (3) implies that $\langle\varepsilon\rangle = 0 = \langle\varepsilon\varepsilon\rangle$, which in turn implies that $\langle\varepsilon_R^2\rangle = \langle\varepsilon_I^2\rangle$, where $\varepsilon = \varepsilon_R + i\varepsilon_I$. In the absence of plasma density fluctuations (n=0 in equation (2)) the envelope field autocorrelation function, $C_{\varepsilon\varepsilon^*}$, is given by

$$\left\langle \varepsilon(\mathbf{x}_\perp, z, t)\varepsilon^*(\mathbf{x}'_\perp, z', t)\right\rangle = \sum_{\mathbf{k}_\perp} |\varepsilon(\mathbf{k}_\perp)|^2 \exp\, i\left[\mathbf{k}_\perp \cdot (\mathbf{x}_\perp - \mathbf{x}'_\perp) - \mathbf{k}_\perp^2 (z - z')\right]. \qquad (5)$$

For convenience we use the following Gaussian model energy spectrum :

$$|\varepsilon(\mathbf{k}_\perp)|^2 = \left(L_\perp^2 / 2\pi\right)\varepsilon_0^2 \exp\left(-\mathbf{k}_\perp^2 L_\perp^2 / 2\right)\Delta^2, \qquad (6)$$

where Δ represents Δk_x ($=\Delta k_y$). Physically $\tilde{\Delta} = \tilde{k}_0 \tilde{d} / \tilde{f}$ (the superscript "~" signifies the value of a parameter in physical (dimensionfull) units) where d is the width of a RPP element and f is the focal length of the lens1. The choice $L_\perp = 48^{1/8}/k_{max}$ can be shown[4] to ensure that the volume densities of intense hot spots generated by the Gaussian model is the same as that generated by the standard "top hat" spectrum".

If the summation in (5) is replaced by an integral then $\left\langle |\varepsilon(\mathbf{x}_\perp, t)|^2 \right\rangle = \varepsilon_0^2$ and ε_0^2 is the average laser intensity, $\langle I \rangle = I_0$. Equations (5) and (6) imply that

$$C_{\varepsilon\varepsilon^*}(\mathbf{x}_\perp, z, t) \equiv \left\langle \varepsilon(\mathbf{x}_\perp, z, t)\varepsilon^*(0,0,t) \right\rangle = I_0 \frac{L_\perp^2}{L_\perp^2 + 2iz} \exp\left[\frac{-\mathbf{x}_\perp^2}{2\left(L_\perp^2 + 2iz\right)} \right]. \tag{7}$$

2.1.2 *The Gaussian Statistical Limit.* If d is small compared to the size of the RPP then Δ is small compared to k_{max} and there are many terms in the summand of (3) with roughly the same amplitude and independent phases. Therefore in this regime where the RPP has many elements $\varepsilon(\mathbf{x}_\perp)$ is effectively a Gaussian random variable by application of the central limit theorem, with the real and imaginary parts of ε statistically independent at a given spatial location. In addition, because of the independent phases, the multipoint correlation function, $< \psi(\mathbf{x}_\perp^a)\psi(\mathbf{x}_\perp^b)... >$, with ψ representing either ε or ε^* factors into the product of all possible two point correlation functions with corrections that are negligible in the many element limit. Therefore the joint distribution function for ε at different points in space is multivariate Gaussian.

The many element regime has the property that the collection of hot spots obtained from a single realization of the random phase ensemble will be a good representation of the entire ensemble. There are two reasons why this is the case: (1) Since each beamlet of a RPP has an independent phase all local configurations of beamlet phases will be explored in a single realization as the number of beamlets grows; (2) The volume of plasma in which the beamlets of a finite RPP overlap scales as[5] $\left(F^4\lambda^3\right)N^{3/2}$ where N is the number of beamlets. As shown in section 2.1.3 below, a hot spot volume scales as $F^4\lambda^3$ and therefore the number of hot spots in the beamlet overlap region scales as $N^{3/2}$. The number of hot spots is large and therefore even rare intense hot spots will be adequately sampled in a single RPP realization if N is sufficiently large. This is experimentally significant because typically a unique RPP is available in a given experimental environment.

2.1.3. *Hot Spot Statistics In the Gaussian Limit.* If filamentation is initiated in hot spots then an a priori knowledge of the geometry of hot spots would be useful to estimate the properties of "typical filaments". While the statistics of ε

are known (Gaussian), the statistics of "hot spots" - defined as those regions that surround local maxima of $|\varepsilon|^2(\mathbf{x}_\perp,z,t)$ - are different as we will show below.

Since ε $(=\varepsilon_R+\varepsilon_I)$ is Gaussian and $\langle\varepsilon\varepsilon\rangle=0$, the probability distribution of ε at a given point, $P(\varepsilon_R,\varepsilon_I)$, is given by

$$P(\varepsilon_R,\varepsilon_I)=\frac{1}{2\pi I_0}\exp\left[-\left(\varepsilon_R^2+\varepsilon_I^2\right)/I_0\right]. \tag{8}$$

The probability distribution for the intensity,

$$I=\varepsilon_R^2+\varepsilon_I^2, \tag{9}$$

is therefore

$$P(I)=P(I)=e^{-I/I_0}/I_0. \tag{10}$$

We will show that compared to the exponential distribution, hot spot intensities are much more likely to assume values that are several times greater than I_0 because of the biased nature of the samping process: only thoses points at which the intensity is a local maximum are considered.

Certain aspects of the hot spot histogram can be understood theoretically as the theory of the local maxima of Gaussian random fields is well developed[6].

2.1.4. *Statistics of Intense Hot Spots.* If ε is expressed in terms of its real and imaginary parts,

$$\varepsilon=\varepsilon_R+i\varepsilon_I, \tag{11}$$

then (7) and the fact that ε is Gaussian imply that ε_R and ε_I are Gaussian random fields with

$$\langle\varepsilon_R(\mathbf{x}_\perp,z,t)\varepsilon_R(\mathbf{0},0,t)\rangle=\langle\varepsilon_I(\mathbf{x}_\perp,z,t)\varepsilon_I(\mathbf{0},0,t)\rangle=\frac{1}{2}\operatorname{Re}C_{\varepsilon\varepsilon^*}(\mathbf{x}_\perp,z,t) \tag{12}$$

$$\langle\varepsilon_R(\mathbf{x}_\perp,z,t)\varepsilon_I(\mathbf{0},0,t)\rangle=\frac{1}{2}\operatorname{Im}C_{\varepsilon\varepsilon^*}(\mathbf{x}_\perp,z,t). \tag{13}$$

Although equations (7) and (13) imply that ε_R and ε_I are uncorrelated for zero axial separation, i.e., $z=0$, they are not independent fields. We are not aware of any theory of the local maxima of the sum of the squares of dependent Gaussian fields. Estimates for hot spot statistics will now be obtained using the fact that at a point ε_R and ε_I are independent.

Because of this independence, if the point $(\mathbf{x}_\perp,z)$ is a strong local maximum of ε_R^2, i.e. $\varepsilon_R^2>>\langle\varepsilon_R^2\rangle=I_0/2$, then an intense hot spot ("hot spot" refers only to local intensity maxima) will occur near $(\mathbf{x}_\perp,z)$ because the contribution of ε_I^2

to the intensity at that point will typically be of order $I_0/2$ and is therefore a perturbation. Similarly there will be a hot spot near every strong local maximum of ε_I^2. The RPP phase ensemble for the boundary condition of ε, however, is phase (or gauge) invariant: the boundary condition ε, occurs with the same probability as $\varepsilon\exp(i\phi)$ for arbitrary ϕ. Therefore the hot spot phase is equally likely to assume any value. Thus we are led to our basic hot spot ansatz that the relative frequency of intense hot spots with a given intensity is the same as the relative frequency of intense local maxima of ε_R^2.

There are two key theoretical results, concerning local maxima of Gaussian fields, which we shall first quote and then apply to RPP hot spots. The first result, equation (14), concerns an asymptotic expression for the number of local maxima of a Gaussian field per unit volume. The second equation (17), concerns the local structure of high maxima. This will be useful in formulating models to study plasma instabilities in isolated hot spots (see equation (24)).

Let B(**x**) be a statistically homogeneous, real, zero mean, Gaussian field with $\mathbf{x}=(x_1,x_2,\dots,x_d)$, and let M_u^B be the volume density of local maxima of B above the level u. Then Adler[6] shows (theorem 6.3.1) that

$$\langle M_u^B\rangle = \frac{|\Lambda|^{1/2}u^{d-1}}{(2\pi)^{(d+1)/2}\sigma^{2d-1}}\exp\left(-\frac{u^2}{2\sigma^2}\right)\left[1+O\left(\frac{1}{u}\right)\right], \tag{14}$$

where σ^2 is the variance of B and Λ is the matrix

$$\Lambda_{ij} = \langle \partial B(\mathbf{x})/\partial x_i\, \partial B(\mathbf{x})/\partial x_j\rangle = -\partial^2 R(\mathbf{x})/\partial x_i \partial x_j\big|_{\mathbf{x}=\mathbf{0}}, \tag{15}$$

where

$$R(\mathbf{x}) = \langle B(\mathbf{x})B(\mathbf{0})\rangle,\ R(\mathbf{0})=\sigma^2 \tag{16}$$

is the two point correlation function. Note that equations (14) and (15) imply that the density of maxima varies inversely with the correlation volume, $|\Lambda|^{-1/2}$ (the product of the correlation lengths in each of d dimensions which, by virtue of equation (24) below, is a hot spot volume). In general, transverse correlation lengths of a laser after passing through a RPP are linear in the f number, F, and axial correlation lengths are quadratic in F (in the paraxial approximation). Therefore the volume density of maxima will vary as $1/F^4$ for d=3.

Let B have a high local maximum with height $u >> \sigma$ at $\mathbf{x}=\mathbf{0}$. Then[6] near x=0 (Adler[6], Lemma 6.7.3)

$$B(\mathbf{x}) \approx u R(\mathbf{x})/\sigma^2 . \tag{17}$$

These results can be applied to both the real and imaginary parts of ε Let B be either ε_R or ε_I. Then $\sigma^2 = I_0/2$ and in three dimensions (d=3) equations (7), (12) and (15) imply that

$$|\Lambda|^{1/2} = I_0^{3/2}/L_\perp^4, \; d=3. \tag{18}$$

Therefore the volume density of maxima of ε_R (or ε_I) above the level u is given by

$$\left\langle M_u^{\varepsilon_R} \right\rangle = \frac{2^{5/2}\left(u^2/I_0\right)}{(2\pi)^2 L_\perp^4} \exp\left(-\frac{u^2}{I_0}\right)\left[1+O\left(\frac{1}{u}\right)\right]. \tag{19}$$

Because ε_R has zero mean this is also the volume density of minima below -u and therefore the volume density of maxima of ε_R^2 above the level u^2 is simply $2\left\langle M_u^{\varepsilon_R} \right\rangle$

Our hot spot ansatz implies that the volume density of hot spots above the level I , $\langle M_I \rangle$, is given by the sum of the densities of maxima of ε_R^2 and ε_I^2:

$$\langle M_I \rangle \sim 4\left\langle M_{\sqrt{I}}^{\varepsilon_R} \right\rangle \sim \frac{(I/I_0)^{((d-1)/2)}}{L_\perp^{d+1}} \exp\left(-\frac{I}{I_0}\right)\left[1+O\left(\sqrt{\frac{I_0}{I}}\right)\right]. \tag{20}$$

The probability distribution function for hot spots, $P(I)$, is proportional to $-d\langle M_I \rangle/dI$. For $I/I_0 >> 1$

$$P(I) \sim \frac{(I/I_0)^{((d-1)/2)}}{I_0} \exp\left(-\frac{I}{I_0}\right)\left[1+O\left(\sqrt{\frac{I_0}{I}}\right)\right]. \tag{21}$$

We will now use equation (17) to predict the local structure of the laser field, ε, in the neighborhood of an ***intense*** hot spot. Let the hot spot be located at the origin and let its value there be given by $\varepsilon(\mathbf{0}) = |\varepsilon(\mathbf{0})| e^{i\phi}$. Because of the gauge invariance of the RPP ensemble, ε is statistically equivalent to the field obtained at a real hot spot, i.e. $\phi = 0$, multiplied by $e^{i\phi}$. Our strategy then is to find the local structure of a real valued hot spot and then multiply by $e^{i\phi}$ to obtain the actual hot spot structure. For a real hot spot which is a strong local maximum of ε_R equations (12), (16) and (17) imply that

$$\begin{aligned} \varepsilon_R(\mathbf{x}_\perp, z, t) &= \varepsilon_R(\mathbf{0}, 0, t) \operatorname{Re} C_{\varepsilon\varepsilon^*}(\mathbf{x}_\perp, z, t)/I_0(t) \\ &= \varepsilon_R(\mathbf{0}, 0, t) \sum_{\mathbf{k}_\perp} \left|\varepsilon(\mathbf{k}_\perp, t)\right|^2 \cos\left(\mathbf{k}_\perp \cdot \mathbf{x}_\perp - \mathbf{k}_\perp^2 z\right)/I_0(t). \end{aligned} \tag{22}$$

The imaginary part of ε in the neighborhood of a real hot spot can be obtained by using (22) for z=0 as a boundary condition for equation (2) (with n=0) and integrating in z Note that if the operator $(i\partial_z + \Delta_\perp)$ is applied to the right hand side of (5) zero is obtained which means that the complex field

$$\varepsilon(\mathbf{x}_\perp, z, t) = \varepsilon_R(\mathbf{0}, 0, t) C_{\varepsilon\varepsilon^*}(\mathbf{x}_\perp, z, t) / I_0(t) \tag{23}$$

satisfies the paraxial wave equation and its real part coincides with that given by equation (22) so that equation (23) gives both the real and imaginary parts of ε in the neighborhood of a real hot spot. Reinserting the phase factor $e^{i\varphi}$ it follows that in the neighborhood of an intense hot spot (now taken to be at $(\mathbf{x}'_\perp, z')$), and before the plasma density responds,

$$\varepsilon(\mathbf{x}_\perp, z, t) = \varepsilon(\mathbf{x}'_\perp, z', t) C_{\varepsilon\varepsilon^*}(\mathbf{x}_\perp - \mathbf{x}'_\perp, z - z', t) / I_0(t). \tag{24}$$

The intensity variation near a hot spot varies as $\left|C_{\varepsilon\varepsilon^*}\right|^2$ and for the gaussian spectrum equation (7) implies (for a hot spot at the origin),

$$\begin{aligned} I(\mathbf{x}_\perp, z)/I(\mathbf{0}, 0) &= \frac{L_\perp^4}{L_\perp^4 + 4z^2} \exp\left[\frac{-\mathbf{x}_\perp^2 L_\perp^2}{L_\perp^4 + 4z^2}\right] \\ &= \frac{1}{1 + (z/\Lambda_z)^2} \exp\left[\frac{-(\mathbf{x}_\perp/\Lambda_\perp)^2 \ln 2}{1 + (z/\Lambda_z)^2}\right], \end{aligned} \tag{25}$$

where Λ_z is the half width at half maximum intensity in z ,$\Lambda_z = L_\perp^2/2$, and the half width at half maximum intensity in the perpendicular direction, $\Lambda_\perp$, is given by $\Lambda_\perp = L_\perp \sqrt{\ln 2}$. While the peak amplitude and phase are stochastic the detailed variation of the laser envelope field in the neighborhood of an intense hot spot is deterministic. It is remarkable to us that these hot spot features are not stochastic. The mathematics behind this has been known for some time[6].

3. Self-focusing of Intense Hot Spots

The paraxial wave equation, equation (2), with the RPP plate boundary condition, equations (3) and (6), together with an as yet unspecified model dynamics for the plasma density fluctuation, n, will be the framework in which to study dynamic self-focusing of structured laser beams in plasma.

If I_0 is small enough then self-focusing will only be initiated in intense hot spots. Since the volume density of hot spots decreases with the hot spot intensity, the very intense hot spots are well separated from each other and we will assume that each such hot spot evolves independently. The evolution of an isolated hot spot requires an effective "hot spot" boundary condition for ε that is discussed next.

3.1 HOT SPOT BOUNDARY CONDITIONS

Equation (5) implies that

$$(i\partial_z + \Delta_\perp)C_{\varepsilon\varepsilon^*}(\mathbf{x}_\perp, z, t) = 0 \tag{26}$$

and therefore the hot spot electric field given by equation (24), which we take to be located at the origin of coordinates without loss of generality, may be generated by the following boundary condition for ε

$$\varepsilon(\mathbf{x}_\perp, z = z_0, t) = \varepsilon_{hot}(t) C_{\varepsilon\varepsilon^*}(\mathbf{x}_\perp, z_0, t)/I_0(t), \tag{27}$$

in the plane $z = z_0$, z_0 arbitrary, if the plasma is quiescent, i.e., $n = 0$. If the laser light propagates in the positive z direction then for $z_0 < 0$ and $|z_0|$ "large enough" the nonlinear development of self focusing will be independent of z_0. An estimate of what "large enough" means is obtained by noting that the $n\varepsilon$ term may be regarded as a perturbation to the dispersive term in (2) if

$$|n| << k_{\max}^2. \tag{28}$$

We will show below that in general $|n| \le |\varepsilon|^2$ and therefore $|\varepsilon(z)|^2 << k_{\max}^2$ or by (25)

$$I_{hot} \frac{L_\perp^4}{L_\perp^4 + 4z_0^2} << k_{\max}^2. \tag{29}$$

Assume that (29) is violated for $z_0 = 0$ and therefore for values of z_0 for which (29) is well satisfied

$$I_{hot} \frac{L_\perp^4}{4k_{\max}^2} << z_0^2. \tag{30}$$

Equations (4) and (30) imply that, approximately,

$$I_{hot} F^6 (n_0/n_c)^3 << z_0^2, \tag{31}$$

is the condition for adequate separation between the boundary condition, and initial focal, planes.

3.2 PLASMA DENSITY DYNAMICS IN THE PONDEROMOTIVE REGIME

If the electron temperature is high enough, typically of the order of several KeV in the regime of short (~micron) wavelength lasers, it can be shown[7] that ponderomotive effects upon the plasma dynamics dominate those due to thermal effects caused by the collisional absorption of laser energy. In this ponderomotive regime the plasma dynamics is modeled by the following linearized hydrodynamic equation:

$$\frac{\partial^2}{\partial t^2}n + 2\hat{v}_i\frac{\partial}{\partial t}n - \Delta_\perp n = \Delta_\perp\left|\varepsilon^2\right|. \tag{32}$$

$\hat{v}_i$ is a linear operator which in fourier space is $k_\perp v_i$ and represents both electron and ion contributions to Landau (collisionless) damping of ion acoustic waves. One is usually interested in regimes where $ZT_e/T_i >> 1$ and then the numerical value of $v_i << 1$. Z is the ionic charge in units of the electron charge, $\tilde{e}$, and $\tilde{T}_{e(i)}$ is the electron (ion) temperature. In addition we will emphasize the inertial regime of plasma dynamics where the $\partial^2 n/\partial t^2$ term dominates the left hand side of (32) and the precise value of v_i is irrelevant. Time is measured in terms of the unit of time, $\hat{t}$, given by $\hat{t} = (c/c_s)/\tilde{\omega}_p$, where c_s is the ion acoustic speed given by $\tilde{c}_s = \sqrt{(Z\tilde{T}_e + \tilde{T}_i)/\tilde{m}_i}$, where $\tilde{m}_{e(i)}$ is the electron (ion) mass. The unit of electric field, $\hat{\varepsilon}$, which relates $\tilde{\varepsilon}$ to the dimensionless , $\tilde{\varepsilon} = \hat{\varepsilon}\,\varepsilon$,is given by $(2\tilde{\omega}_0/\tilde{e})\sqrt{\tilde{m}_e(Z\tilde{T}_e + \tilde{T}_i)/Z}$.

3.3 SYMMETRIES OF THE PONDEROMOTIVE MODEL

Solutions of the ponderomotive self-focusing model , equations (2) and (32), are transformed into other solutions under the transformation:

$$\varepsilon(\mathbf{x}_\perp, z, t) \rightarrow \beta\varepsilon(\beta\mathbf{x}_\perp, \beta^2 z, \beta t) \tag{33}$$

$$n(\mathbf{x}_\perp, z, t) \rightarrow \beta^2 n(\beta\mathbf{x}_\perp, \beta^2 z, \beta t). \tag{34}$$

The adiabatic ion model, in which (32) is replaced by

$$n = -|\varepsilon|^2, \tag{35}$$

which together with (2) implies that

$$(i\partial z + \Delta_\perp)\varepsilon = -|\varepsilon|^2\varepsilon, \tag{36}$$

is also consistent (33). Under this transformation the total power, P, which by (2) is independent of z,

$$P=\int\left|\varepsilon(\mathbf{x}_\perp,z,t)\right|^2 d\mathbf{x}_\perp, \tag{37}$$

is invariant so that except for changes in scale, P is the only parameter that is needed to specify a solution if initially the plasma is quiescent, i.e. $n(t=0)=\partial n(t=0)/\partial t=0$, Note that the boundary conditions for ε must also transform according to (33). For example the family of boundary conditions whose elements correspond to the hot spot given by (27) for different focal length optics yield evolving solutions that differ only by their length and time scales if they have the same power.

For times short compared to an ion acoustic transit time across a hot spot the following supersonic ponderomotive model,

$$\partial^2 n/\partial t^2=\Delta_\perp|\varepsilon|^2, \tag{38}$$

is a good approximation to (32). Solutions of (2) and (38) are transformed into other solutions under a two parameter family of scale transformations:

$$\varepsilon(\mathbf{x}_\perp,z,t)\rightarrow\beta\varepsilon\left(\gamma\mathbf{x}_\perp,\gamma^2 z,\beta t\right) \tag{39}$$

$$n(\mathbf{x}_\perp,z,t)\rightarrow\gamma^2 n\left(\gamma\mathbf{x}_\perp,\gamma^2 z,\beta t\right). \tag{40}$$

While for $\gamma=\beta$ equations (33) and (34) are recovered, if γ and β are varied independently then a solution with a given power is transformed into a solution with another power. This implies, for example, that in the supersonic regime different focal length optics generate evolving filaments which differ only by their length and time scales even if they have different power.

3.4 SINGULAR SOLUTIONS OF DYNAMIC SELF FOCUSSING[8]

If solutions of the dynamic ponderomotive self-focusing model tend to approach those of the adiabatic ion model, (36), then we expect that for a given hot spot boundary condition the dynamic solution of the former will not blow up unless the power exceeds a certain critical amount[9,10], $P\geq P_c\approx 11.7$, the condition that the adiabatic ion model blowup. We will first provide additional theoretical motivation for this conjecture and then present numerical evidence that dynamic blow up requires $P\geq P_c$.

Assume that a solution to (2) and (32) blows up and let τ_B be the blow up time. As $t\uparrow\tau_B$ $|\varepsilon|^2\rightarrow\infty$ at some z, say z_B. If $z_f(t)$ is the time dependent depth at which $|\varepsilon|^2$ is a maximum then as $t\uparrow\tau_B$ $z_f(t)\rightarrow z_B$. $|\varepsilon|^2$ can blow up only for $r\equiv|\mathbf{x}_\perp|<\delta(t)$ with $\delta(t)\rightarrow 0$ as $t\uparrow\tau_B$ otherwise $\int|\varepsilon|^2 d\mathbf{x}_\perp$ would diverge, violating the invariance of the total hot spot power. Equation (2) implies that as $t \uparrow \tau_B$ $n \to -\infty$ at $z = z_B$, (otherwise n is finite in some neighborhood of z_B when $t = \tau_B$ and $|\varepsilon|^2$ could not change by an infinite amount in that neighborhood as it must for $|\varepsilon|^2$ to blow up as $z \to z_B$) and therefore at $z_f(t)$ (or some z which approaches $z_f(t)$ as $t \uparrow \tau_B$) n is a local minimum and $\partial n/\partial z = 0$.

If n were independent of z then the solution of (2) is naturally expressed as a sum over the eigenfunctions of the operator $-\Delta_\perp + n$ with constant coefficients. This motivates the following representation of ε for z near $z_f(t)$, where $\partial n/\partial z$ is small, and $r \equiv |\mathbf{x}_\perp| < \delta$:

$$\varepsilon(r,z,t) \equiv \sum_i \varepsilon_i = \sqrt{P_0(z,t)}\phi_0(r,z,t)\exp\left[-i\int^z \sigma_0(z',t)dz'\right] + \sum_{i>0}\sqrt{P_i}\phi_i\ldots, \tag{41}$$

where ϕ_0 is the normalized ground state of $-\Delta_\perp + n$ (taken to be real without loss of generality) with eigenvalue σ_0,

$$\begin{aligned} &\sigma_0(z,t)\phi_0(r,z,t) = [-\Delta_\perp + n(r,z,t)]\phi_0(r,z,t) \\ &2\pi\int_0^\infty \phi_0^2(r,z,t)r\,dr = 1 \end{aligned}, \tag{42}$$

and the summation in (41) represents the contribution from excited states. Only the eigenfunctions with zero slope at $r=0$ need be considered because $\partial\varepsilon/\partial r|_{r=0} = 0$ (in order that $\Delta_\perp\varepsilon$ be finite at $r=0$). Since distinct eigenfunctions are mutually orthogonal each one may be thought of as contributing a definite amount to the total hot spot power with P_0 representing the ground state contribution. As blow up is approached we expect that only the contributions from localized states in (41) will be significant for $r < \delta$. The separate contribution of each localized state to $\Delta_\perp|\varepsilon|^2$, $\Delta_\perp|\varepsilon_i|^2$, in (32) is estimated as

$$\Delta_\perp|\varepsilon_i|^2 \sim P_i\,\phi_i^2/\delta_i^2 \tag{43}$$

where P_i is the power in the ith state and δ_i is a measure of the radius of the ith eigenfunction, ϕ_i. Equation (42) implies that $\phi_i^2 \sim 1/\delta_i^2$ and therefore

$$\Delta_\perp|\varepsilon_i|^2 \sim P_i/\delta_i^4. \tag{44}$$

For a given amount of power in localized states the greatest ponderomotive force would obtain if it were all in the ground state because it is the most tightly localized ($\delta_i > \delta_0$ if $i>0$) and therefore the most efficient collapse would occur if all the localized power were in the ground state.

If, in fact, dynamic blow up requires a minimum amount of power in the localized states then the conclusion of the preceding paragraph implies that a lower bound for this power is obtained by demanding that this power is entirely in the ground state. We will now make this ansatz and explore the consequences.

Substitute (41), with only the ϕ_0 term retained, into (32):

$$\frac{\partial^2}{\partial t^2} n + 2\,\hat{v}_i \frac{\partial}{\partial t} n - \Delta_\perp n = P\Delta_\perp \phi_0^2(r,t). \tag{45}$$

The z dependence of ϕ_0 has been dropped because with the replacement of (2) by (42) and (32) by (45) self focusing at different depths of the plasma decouple. Alternatively equations (42) and (45) describe the evolution of a (uniform in z) cylindrical filament. If P is independent of time and it can be shown that solutions to (42) and (45) cannot blow up for arbitrarily small P then, according to the discussion that follows equation (44), this will establish a lower bound for dynamic blowup (solutions to the general ponderomotive model, (2) and (32), which blow up). Looking ahead it appears that $P > P_c$ is required for dynamic blowup, where P_c is the critical power required to obtain singular solutions to the adiabatic ion model.

Some information about a critical value of P for dynamic blow up is obtained from the static solutions to (42) and (45). $n = -P\phi^2$ is a time independent solution to (45) which, when substituted in (42), yields a nonlinear eigenvalue problem whose solution is a (radially localized) soliton solution of (36), the nonlinear Schröedinger equation in two dimensions ($\mathbf{x}_\perp$ has two components):

$$\begin{aligned} &\sigma\phi(r) = -\left[\Delta_\perp + P\phi^2(r)\right]\phi(r) \\ &2\pi\int_0^\infty \phi^2(r)\, r dr = 1\,. \end{aligned} \tag{46}$$

It is well known[10], that soliton solutions of the nonlinear Schröedinger equation in two dimensions require $P = P_c$. Since the $-P\phi^2$ term in (46) plays the role of a self consistent potential that determines the eigenfunction ϕ we can say that P_c is a nonlinear localization threshold.

Now consider the possibility of collapsing solutions to (42) and (45). By this we mean solutions whose width, δ, decreases to zero in a finite time, τ_B. Physically a filament cannot collapse to zero radius but τ_B would estimate the time at which saturation effects become important. Let $-(d/dt)\delta$ be identified as a collapse speed, v_c. As $v_c \downarrow 0$ a solution to (45) is $n = -P\phi^2$. For finite v_c (finite collapse time) we expect that the response of n will be weaker than for $v_c = 0$ because of the inertial terms in (45) and therefore for fixed P the instantaneous potential, n in (42), will be less effective than the $-P\phi^2$ term in (46). Therefore in order that a filament collapse in a finite time through a succession of localized ground states of (45) it is required that $P > P_c$ since P_c is a localization threshold when inertial effects are ignored and as $P \downarrow P_c$ v_c must go to zero (collapse time goes to infinity) because otherwise inertial effects would remain finite and an amount of power equal to P_c would not be adequate to yield a localized solution.

3.5 SELF SIMILAR DYNAMIC SOLUTIONS WHOSE POWER EXCEEDS CRITICAL

The basic ponderomotive symmetries, (33)-(34), and the results of numerical simulations (see section 3.6.2) suggest that an analysis of the self similar solutions to (42) and (45) would be useful. If a self similar solution scales with a time dependent length parameter, δ(t), then the eigenvalue equation, (45), implies that $n \sim 1/\delta^2$ and the normalization of ϕ implies that $\phi \sim 1/\delta$ This motivates the following ansatz:

$$\begin{aligned} n(r,t) &= \rho[r/\delta(t)]/\delta^2(t) \\ \phi(r,t) &= \psi[r/\delta(t)]/\delta(t) \\ \sigma(t) &= \omega/\delta^2(t) \\ \delta(t) &= v_c t\,. \end{aligned} \tag{47}$$

ω, v_c and the functions ρ and ψ depend parametrically upon P which we take to be a constant $> P_c$.

We now show that the assumed linear dependence of δ upon time, i.e., a constant collapse speed, is the only consistent choice. Substitute (47) in the four terms of (45) and obtain the relative ordering:

$$(d\delta/dt)^2 : (d\delta/dt) : 1 : 1\,. \tag{48}$$

If $\delta \sim t^\alpha$ with $\alpha > 1$ then the first two terms in (45) are ignorable as $t \to 0$ and the solution to (45)-(45) reduces to the solution of (46). The ground state solution of (46), however, has the unique power[10] P_c and we are looking for collapsing solutions to (45)-(45) with fixed power greater than P_c. If $\delta \sim t^\alpha$ with $\alpha < 1$ then the last two terms in (45) can be ignored as $t \to 0$ which then has a solution which varies exponentially with time, contradicting (47).

Substitution of the self similar[11,12] ansatz in equations (45)-(46) yields (for $V_i = 0$):

$$\begin{aligned} v_c^2\left(6\rho + 6u\, d\rho/du + u^2\, d^2\rho/du^2\right) - \Delta_u \rho &= P \Delta_u \psi^2 \\ \omega\psi &= \left[-\Delta_u + \rho\right]\psi \end{aligned}, \tag{49}$$

where u is the scaled radial variable, $r/\delta(t)$, and $\Delta_u = (1/u)d/du + d^2/du^2$.

3.5.1 *The Supersonic Regime.* In the supersonic collapse regime ($v_c >> 1$) the $\Delta_u \rho$ term in (49) can be ignored and the self similar solution is asymptotically invariant to changes in P except for the speed of collapse: $v_c \sim \sqrt{P}$. This is consistent with (39), which is a general property of self focussing in the supersonic regime which implies that the time scale for evolution varies inversely with the square root of P. Therefore, in the comparison of collapsing filaments with the same instantaneous radius, δ, but different power, those with larger power have larger intensity, $P(\psi/\delta)^2$, but the same density fluctuation, ρ/δ^2. Since the latter may saturate and stop collapse in a physically more accurate model, collapsing filaments can attain much higher intensities in the supersonic regime than in the adiabatic regime. The same conclusion follows directly from the supersonic transformation in equations (39)-(40).

3.6 DYNAMIC SIMULATIONS OF HOT SPOT SELF FOCUSSING

Even if there are self similar collapsing solutions of the form presented in (47), i.e., if the nonlinear eigenvalue problem, equation (49), had physically acceptable solutions for $P > P_c$, it would not necessarily follow that solutions to the initial-boundary value problem, (2) and (32), would even be locally (i.e. in a neighborhood of z_B) attracted to such a solution. There are a variety of mechanisms that might prevent such an attraction: these solutions are translationally invariant in z while hot spots have no such symmetry; these solutions might be unstable to perturbations that break cylindrical symmetry.

In section 3.4 collapse (or blow up) was characterized by a collapse time and a collapse power. While our simulation results do not conclusively show that filaments which evolve from hot spot boundary conditions are locally attracted to the self similar collapsing cylindrical filament solutions, it appears that there are well defined collapse times and collapse powers over a finite range of z, $z_c \leq z \leq z_B$, (where z_c is the blow up depth for the solution to the adiabatic ion model) and we are led to the following conjectures.

For $t \geq \tau_B$, blow up will occur at $z_B(t)$, a monotonically decreasing function of time (i.e., the collapse plane propagates toward the laser[13]) with

$$\lim_{t \to \infty} z_B(t) = z_c \ . \tag{50}$$

Equivalently for each $z_c \leq z \leq z_B$, there is a blow up time $\tau_B(z)$, a monotonically decreasing function of z which has the property

$$\lim_{z \downarrow z_c} \tau_B(z) = \infty, \tag{51}$$

and a collapse power, $P_B(z)$, a monotonically increasing function of z, with the property

$$\lim_{z \downarrow z_c} P_B(z) = P_c. \tag{52}$$

Therefore the unique blow up of the adiabatic ion model takes an infinite time to evolve. Physically, the time for saturation effects to become important may be very large at the depth z_c

3.6.1 *Numerical Technique.* The radial dependence of ε and n is represented as a sum over Bessel functions of order zero. For example,

$$\varepsilon(r,z,t)=\sum_{j=0}^{j_{max}}\varepsilon_j(z,t)J_0(k_j r). \tag{53}$$

We shall call the ε_j's the Bessel space transform. The k_j are chosen such that $\partial/\partial r\, J_0(k_j r)\big|_{r=R}=0$ where the boundary radius, R, is chosen large enough, about $15\Lambda_\perp$, so that ε and n are small at the radial boundary. j_{max} is chosen large enough to resolve all scales that are relevant up to a particular time of evolution.

Given $n(r,z,t)$ and a hot spot boundary condition for ε at a depth z_0 consistent with (30), equation (2) is developed in z using a split step technique with the $\Delta\varepsilon$ term evaluated in Bessel space and the $n\varepsilon$ term in r space. The right hand side of (32) is now evaluated in Bessel space and treated as a constant for purposes of evolving n and $\partial n/\partial t$ from time t to time $t+dt$.

The simulation results presented below are for a hot spot whose power is ten times critical. Time, t, and the radial coordinate, r, are normalized to $\Lambda_\perp$ while the axial coordinate, z, is normalized to Λ_z. The basic ponderomotive symmetry, equations (33) and (34), imply that the results expressed in these units are independent of the optics (F number). The choice $z_0 \approx 10\Lambda_z$ ensured that condition (28) was well satisfied.

3.6.2 *Eigenfunction Resolution.* The eigenfunction resolution of a given simulation is a useful diagnostic for estimating collapse times and powers. The radius of the ground state eigenfunction, ϕ_0 (equation (42)), which we shall now denote simply by $\delta(z,t)$, is defined by

$$\delta(z,t)=\int_0^R r\phi_0^2(r,z,t)r\,dr. \tag{54}$$

In figure 1 $\delta(z,t)/\Lambda_\perp$ is shown as a function of $c_s t/\Lambda_\perp$[14] for $\left(z-z_f^0\right)/\Lambda_z=-3.0$. If it is assumed that the collapse is self similar then a linear extrapolation of the graph in figure 1 yields a collapse time. The ground state eigenvalue, $\sigma_0(z,t)$, scales as $1/\delta^2$. Its numerical determination is less susceptible to noise than $\delta(z,t)$ and is in fact used to estimate the collapse time τ_B.

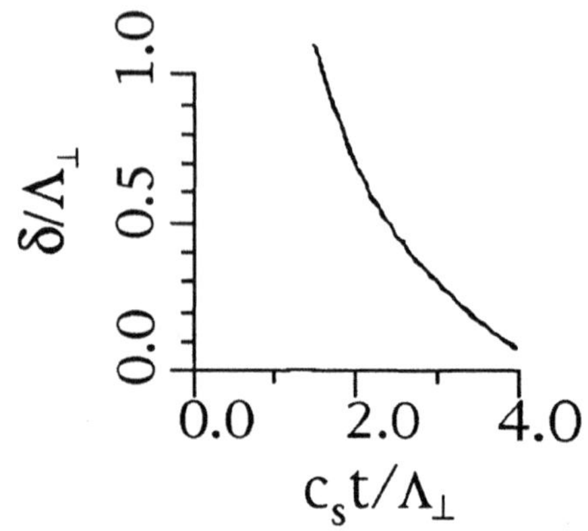

Figure 1. Evolution of normalized filament radius at a depth of $3\Lambda_z$ before the hot spot focal plane.

The projection of ε on ϕ_0 determines the ground state power, P_0. As the collapse time is approached P_0 varies slowly and its value is identified as the collapse power P_B. In figure 2 τ_B and P_B are shown.

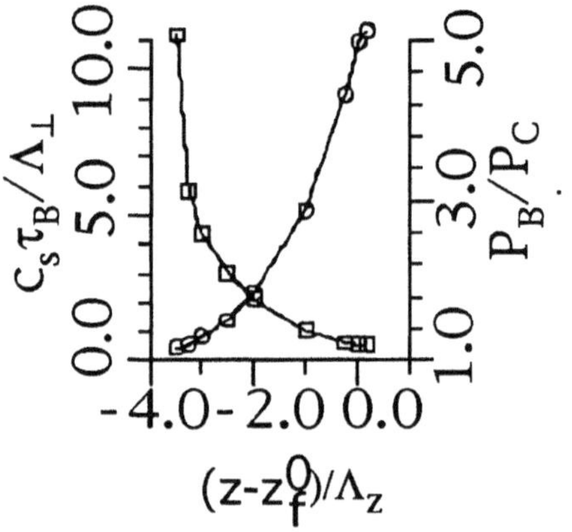

Figure 2. Normalized collapse times and powers vs. normalized depth with respect to the hot spot focal plane.

The evolution was not followed far enough (the intensity increased only by a factor of about 100) to say with confidence that the evolution approached a self-similar solution of the form indicated by equation (49). Nevertheless the qualitative resemblance between the simulation results and the self-similar solutions (length scales going to zero linearly with time, ground state power approaching a constant, divergence of the collapse time as the critical power is approached) strongly suggests to us that the self-similar solutions are at least a good qualitative guide to the actual solutions of the ponderomotive self-focusing model.

In figure 3 $\delta(z,t)$ is shown for selected times over part of the simulation volume. Note the distinguished nature of the earliest collapse where $\partial\delta/\partial z = 0$.

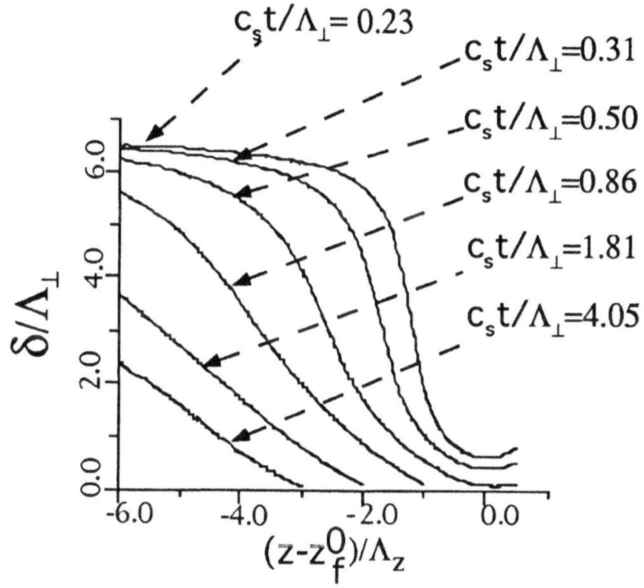

Figure 3. Normalized filament radius at different times.

As discussed before collapse times decrease with depth over a range of z, $z_c \le z \le z_B$, and therefore one must ask what sense it makes to continue in time past the earliest collapse since, strictly speaking, the dynamic self-focusing equations cease to have meaning once collapse has occurred. Because equation (2) may be thought of as propagating a hot spot boundary condition in the positive z direction, and equation (32) is strictly local in z, it is possible to continue the evolution up to $z_B(t)$ even if the dynamic model ceases to have meaning at greater depths.

3.6.3 *Axial Sound Wave Propagation Effects.* Several physical approximations have been employed[7] to obtain the self-focusing model, equations (2) and (32). Here we will discuss only one, the omission of axial sound waves.

This omission is manifested in (32) where the Laplacian is approximated by its perpendicular component, $\Delta_\perp$. Since the collapse location, $z_B(t)$, is a monotonically decreasing function of time, with the first collapse in the neighborhood of the initial hot spot focal plane, it is plausible that sound wave propagation from that region to shallower depths is more important than sound wave propagation in the opposite direction.

Inspection of figure 2 allows an estimate of $|dz_B/dt|$ which seems to diverge at the first collapse. The scaling of the axes shows that $|dz_B/dt|$ varies as $c_s \Lambda_z/\Lambda_p \sim Fc_s$ and is easily supersonic, i.e., exceeds the speed of sound. In this regime of supersonic axial motion of the collapse location axial sound wave propagation may have a small effect.

4. Conclusions

We have predicted the detailed structure of intense laser hot spots produced by random phase plate optics. Dynamic numerical simulations suggest that a sequence of catastrophic self-focusing events may be initiated by such hot spots if they contain more than a certain critical power. This sequence covers a finite range of depths, of order F^2 wavelengths of light, and an indefinite time interval since the time to collapse diverges at the shallowest depth that corresponds to the self-focusing depth based upon the adiabatic ion model (the nonlinear Schröedinger equation). Self-focusing may develop supersonically in two senses: (a)the radial collapse speed will exceed the speed of sound if the hot spot power is large compared to critical; (b)the propagation speed of the collapse location toward the laser initially exceeds the speed of sound.

[1]A. J. Schmitt, Phys. Fluids **31**, 3079 (1988).

[2]S. P. Obenschain, C. J. Pawley, A. N. Mostovych, J. A. Stamper, J. H. Gardner, A. J. Schmitt and S. E. Bodner, Phys. Rev Lett. **62**, 768 (1989).

[3]O. Willi, T. Afshar-rad, S. Coe and A. Giulietti, Phys. Fluids **B2**, 1318 (1990).

[4]H. A. Rose and D. F. DuBois, "Statistical Properties of Laser Hot Spots Produced by a Random Phase Plate", submitted to Phys. of Fluids **B**.

[5]M. H. Emery, J. H. Gardner, R. H. Lehmberg and S. PL Obenschain, Phys. Fluids **B3**, 2640 (1991).

[6]J. Adler, "*The Geometry of Random Fields*", John Wiley & Sons Ltd., (1981).

[7]H. A. Rose and D. F. DuBois, "The Development of Ponderomotive Filaments in Plasma from Intense Hot Spots Produced by a Random Phase Plate", submitted to Phys. of Fluids **B**.

[8]H. A. Rose, D. F. DuBois and D. Russell, Sov. J. Plasma Phys. **16**, 537 (1990).

[9]R. Y. Chiao, E. Garmire and C. H. Townes, Phys. Rev. Lett. **13**, 479 (1964).

[10]M. I. Weinstein, Comm. Math. Phys. **87**, 567 (1983).

[11]L. Bergé, P. Doussear, G. Pelletier and D. Pesme, Phys. Rev. A. **42**, 4952 (1990).

[12]M. Landman, G.C. Papanicolaou, C. Sulem, P.L. Sulem, and X.P. Wang, "Stability of Isotropic Self-Similar Dynamics for Scalar Wave Collapse", submitted to Phys. Rev. A.

[13]G. Litvak, V. A. Mironov and A. M. Sergeev, Physica Scripta T30, 57 (1990).

[14]c_s, the ion acoustic speed, has the value unity in our system of units.

Experimental Manifestation of Strong Langmuir Turbulence*

A.M. Rubenchik†

Plasma Physics Research Institute
Livermore, CA 94550

Abstract

The review of experiments in strong Langmuir turbulence is presented. The existence of Langmuir collapses, singularities of the electric field is demonstrated. It is shown that conception of collapse explains a number of experimental data in laboratory and ionospheric situations.

† Permanent address:
Institute of Automation and Electrometry
Siberian Division, Russia Science Academy
Novosibirsk 630090 Russia

*Some of this work was performed under the auspices of the U.S. Department of Energy by the Lawrence Livermore National Laboratory under contract no. W-7405-Eng-48. Work also supported by IAE.

R. E. Caflisch and G. C. Papanicolaou (eds.), Singularities in Fluids, Plasmas and Optics, 301–308.
© 1993 *Kluwer Academic Publishers.*

Nowadays, it is understood that very often the turbulence is not homogeneous. Often the crucial role for turbulence is having the field singularities bursting randomly. As examples, Langmuir collapse, surface waves wavebreaking, vortices line coalescence can be mentioned. The singularities play important roles in turbulence. Particularly, they define the energy absorption and hence the turbulence level. The singularities are studied, mainly theoretically. The proof of their experimental existence is a subject of great importance.

Recently, the essential results were obtained for plasma Langmuir wave turbulence. The strong Langmuir turbulence (SLT) is the ensemble of the collapsing cavities, compressing density wells filled by plasma waves. It is a self-accelerated process with terminal stage duration ~ 5÷10 ω_{pi} and minimal cavity size ~ 15÷20 r_d.[1]

Here $\omega_{pi}=\frac{4\pi n_o e^2}{M}$ - ion plasma frequency, r_d - is electron Debye radii.

Theoretically, Langmuir collapse was predicted by Zakharov[2] in 1972 and, in detail, was studied both analytically and numerically (see review e.f. [1,3])

The evolution of the isolated cavity was observed by Cheung and Wang. [4] This is a unique experiment, presented in a special machine for very special parameter values. Principally, they deal with the only cavity fixed by the boundary conditions, not turbulence.

As a rule, due to the small size of the cavity on the terminal stage of the evolution and small time of this stage the direct collapse observation is very difficult. But it is possible to record the indirect, but some independent consequences of collapse, which all together present the solid proof of collapse existence.

The first essential manifestations of SLT in nature were recognized by Dubois et al. [See [5] and reference therein] They studied Langmuir turbulence excited near the reflection layer of the ionosphere by powerful radio transmitters. This provides a large plasma turbulence volume with no boundary effects.

Radar Thomson scattering was used to measure the power spectra of turbulent fluctuations. Dubois et al (see [5] and reference therein) have shown that there are unique spectral signatures that can only be explained by SLT concepts.

The most bright result is presented on fig. 1. Experimentally, the Langmuir waves power spectrum for $K=2K_oSin(\theta/2)\equiv K_m$ is measured. Here K_o - is the wavenumber of probe radar radiation and θ - is the scattering angle.

One can see that spectra for fixed K is broad, which is absolutely inconsistent with weak turbulence theory (WLT). The spectra is expanded to the red range. It corresponds to plasmons trapped in the collapsing cavities. Plasma is expelled from them, resulting in the decrease of local plasma frequency. The narrow blue peak represents so called free mode, the waves escaped from the cavities and propagated out of the density wells. The frequency of free mode ω_f exactly corresponds to linear dispersion law $\omega_f = \omega_p(1+\frac{3}{2}K_m^2 r_d^2)$. Let us emphasize that ω_f is larger than pump frequency, which is impossible to explain without SLT. Dubois et al were able to explain the very fine details of the experimental observation.

The next experiment, specially aimed for SLT studies [6], deals with turbulence excited by the electron beam in rarefied plasma. The low density poses to use the probes for diagnostics. Unfortunately, one can not use more than a few probes simultaneously, due to their mutual influence. Hence, it is impossible to measure the electric field spatial structure.

Authors of [6] used the probe's loop antennae to measure electromagnetic radiation with frequencies close to ω_p. When beam density (growth rate of beam-plasma instability) exceed some value n there appear bursts with a frequency which increases with increasing beam density. We note that this kind of radiation bursts have been observed in a number of experiments.

It is natural to identify these radiation bursts with their generation in collapsing cavitons. However, the non-stationarity of the radiation can be connected with the relaxation oscillations of the beam, its bunching, and the passage of electron bunches through the observation region. To elucidate the mechanism of the bursts, the correlation measurements were carried out.

Measurements showed that the radiation sources are localized. For large distances between the probe-antennae, L>0.1 cm, authors did not observe correlations between the bursts. For small distances between the antennae, authors registered a mutual correlation for a significant number of bursts. Figure 2 illustrates the correlation of signals for two radiation bursts. The size of the sources is, clearly less than the distance between the antennae for which the correlation between bursts disappeared. The results of the measurements showed that the size of the source is l<0.1 cm ($100r_{De}$). The length of the bursts also depends weakly on the beam parameters.

The simple estimates [6] demonstrates that electromagnetic radiation from the collapsing cavities is peaked on the terminal stage. It is natural to associate the bursts with collapse phenomena. The duration of the bursts is about $10\omega_{pi}^{-1}$ which fits well with numerical simulation results. Experiments with different plasmas were done with an increase of burst length $T \alpha M^{1/2}$, with an increase in ion mass being observed.

The creation of singularities means energy transfer in large k range and generation of shortwave sound radiation. Fig. 3 demonstrates ion-sound waves frequency spectra. Due to linear dispersion $\omega = c_s k$ Fig. 3 is directly related to spectra in k space. One can see that the appearance of radiation bursts (collapses) correspond to the growth of the shortwave fluctuation.

It is natural to relate the peak of spectra k_{max} with the minimal size of the cavity. Experimentally, $\ell_{min} = \frac{\pi}{k_{max}} \sim 7r_d$, fits well with minimal size ~15-20r_d from simulation.[1] More important, that K_{max} doesn't change with the beam density and ion mass.

Also in [6], the direct measurement of electron distribution evolution is presented. The collapse, energy dissipation on the terminal stage means the acceleration of background electrons, and at moderate turbulence intensity, is the only mechanism of their acceleration. Fig. 4 demonstrates the fast electron tail formation and the gap between fast electrons and beam deacceletated electrons.

The third experiment I would like to mention deals with Laser-plasma interaction.[7] The process of the electromagnetic wave decay into two plasmons were studied ($2\omega_{pe}$ instability).

$$\omega_o = \omega_t(æ_o) = \omega_e\left(\frac{\vec{æ}_o}{2} - \vec{k}\right) + \omega_e\left(\frac{\vec{æ}_o}{2} - \vec{k}\right) \qquad (1)$$

Here $\omega_t = \sqrt{\omega_p^2 + æ^2 c^2}$ -electromagnetic wave dispersion law. The $2\omega_{pe}$ instability is easy to monitor by measurement of electromagnetic radiation with frequency $\sim \frac{3}{2}\omega_o$. This radiation is generated by coupling of the primary laser light and plasmon

$$\omega_t(æ) + \omega_e(k) = \omega_t(-\vec{k} + \vec{æ}) \qquad (2)$$

From (2) one can see that only longwave, $k \sim \frac{\omega_p}{C}$ plasmons can participate in (2).

Such a plasmon can directly produce via (1), if plasma density is close to $\frac{nc}{4}$, $(\omega_p(n_c)=\omega_o$. When density goes down, the wave number k for plasmons in (1) grow and finally reach Landau cutoff value(kr_d)~0.3. In WLT approximation, longwave plasmons can arise after a number of cascading. In this case it must be a big, $\Delta\omega\sim\frac{3}{2}\omega_p k^2 r_a^2$, red shift of radiation.

Experimental data presented on fig. 5. The experiment was done with decayed plasma. Density decrease terminates $\frac{3}{2}\omega_o$ emission. Data of [7] demonstrates that radiation cutoff coincides with Landau cutoff density, but the red shift is much smaller than WLT prediction. The result can be naturally explained in terms of SLT.[8] The modulation instability development expands turbulent spectra to the small k range without essential frequency shift. This longwave part of the spectra can participate in radiation.

Physically, it is dipole radiation from localized cavities. $3/2\omega$ radiation spectra have a specific two-hump structure. We see from (1), instability produced plasmons with blue and red shifts relatively $\omega_o/2$. The frequency difference of $\delta\omega\sim 3\omega_p(\vec{æ}\,\vec{k})r_d^2$ is larger than modulational instabilities increment and is manifested in two-hump-like spectra. To explain such a structure within WLT is possible only if one will take into account reflection from critical surface(s?). But, maximum plasma density was below n_c, in experiment [7].

Finally, let us discuss some future development. As was mentioned above, diagnostics are the crucial point. For dense plasmas, a very powerful diagnostic tool is laser scattering. But, in traditional experiments, due to the high plasma density, singularity development time is less than picosecond, and scale less than 10^{-5}cm. It is extremely difficult to resolve them.

In the rarefied plasma evolution, times and scales growth $\alpha\frac{1}{\sqrt{n}}$ but in laboratory plasmas, density is very small for laser scattering.

Experiments can be very promising with beat-wave excitation, when the beating of two electromagnetic waves resonates with plasma waves. In experiment [9], beating of two N_d lasers excited waves in plasma with density $10^{17}cm^{-3}$, which corresponds to times ~10^{-10} and scales ~10^{-3}cm, which is possible to resolve easy enough in laser scattering measurement.

These schemes have some other advantages. The plasmon excitation is "soft", it is "forced", not parametrical excitation. It is easy to control turbulence level, also the size of the turbulent zone can be smoothly changed. It is possible to have one collapsing cavity within it, two, three.... As a result, beat wave experiments look very attractive for fundamental studies of SLT.

Conclusion

The presented results demonstrate that SLT rising of singularities in plasma is not the game of theorists, but corresponds to some real plasma phenomena. We demonstrated that SLT is useful from a pragmatic point of view, helping to understand and explain a lot of plasma turbulence manifestations.

References

1. Dyachenko, A.I., Pushkarev, A.N., et al, "Computer Simulation of Langmuir Collapse", *Physica D,*, **52**, 78, 1991.

2. Zakharov, V.E., "Collapse of Langmuir Waves", *Sov. Phys. GETP*, **35**, 908, 1972.

3. Goldman, M., "Strong Langmuir Turbulence", *Rev. Mod. Phys.*, **56**, 709, 1984.

4. Cheung, P., Wang, A., "Nonlinear Evolution of Electron-beam-plasma Interaction", *Phys. Fluids*, **28**, 1538, 1985.

5. Dubois, D.F., Harvey, A.R. and Russel, D., "Excitation of Strong Langmuir Turbulence in Plasmas near Critical Density: Application to HF Heating of the Ionosphere", *Journal of Geophysical Research*, **95**, 21221, 1990.

6. Karfidov, D.M., Rubenchik, A.M., "Strong Langmuir Turbulence Excited by an Electron Beam in Plasma", *Sov. Phys. GETP*, **71**, 892, 1991.

7. Seka, W., Bahr, R.E., et al, "Nonlinear Laser-matter Processes in Long-scale-length Plasmas", *Phys. Fluids B*, **4**, 2232, 1992.

8. Rubenchik, A.M., Zakharov, V.E., "Strong Langmuir Turbulence in Laser Plasma", Handbook of Plasma Physics, ed. by M. Rosenbluth and R. Sagdeev, Elsevier, New York, 1991.

9. Amiranoff, F., Laberge, M. et al, "Observation of Modulational Instability in N_d-Laser Beat-wave Experiments, *Phys. Rev. Lett.*, **68**, 3710, 1992.

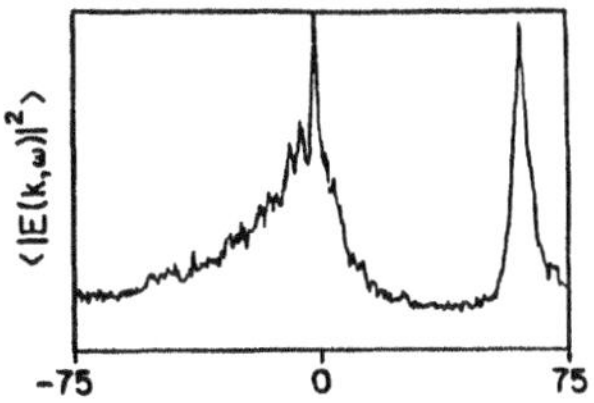

Fig. 1 Experimental plasma waves spectra [5] obtained by Radar scattering. O - corresponds to the heater frequency.

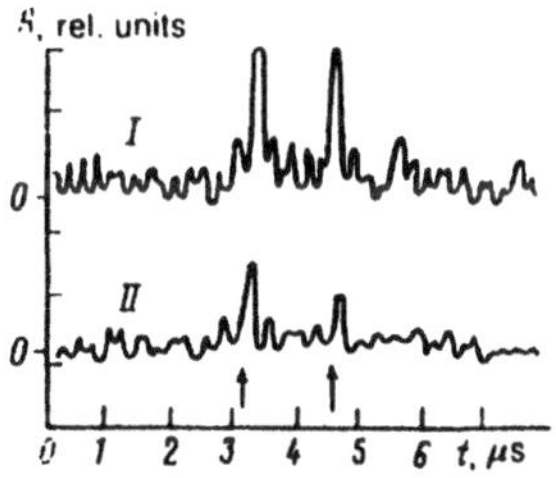

Fig. 2 Oscillograms of plasma field "bursts" obtained from two identical loop antennae with their loop planes separated by a distance of 0.1 cm [6].

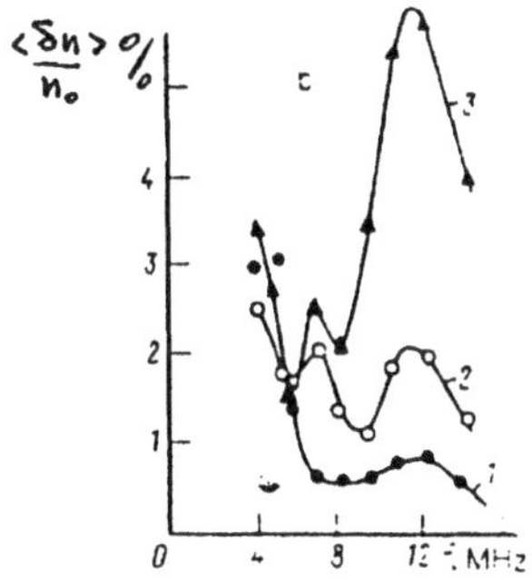

Fig. 3 Ion-sound oscillation spectra for three values of the electron beam parameters.[6] 1. $n_b=n_{bth}$ 2. $n_b=2n_{bth}$ 3. $n_b=12n_{bth}$. n_{bth} related to the threshold of the radiation bursts.

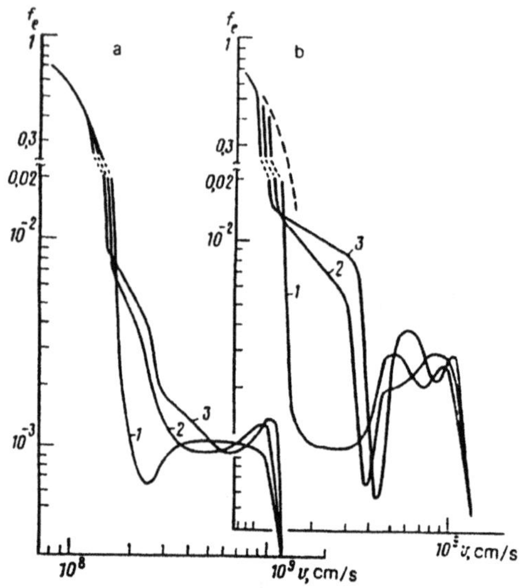

Fig. 4 Electron distribution function for three consequent time moments[6]. a) $n_b=3n_{bth}$ b) $n_b=6n_{bth}$. For case b the background electron acceleration and gap between them and beam's electrons are demonstrated.

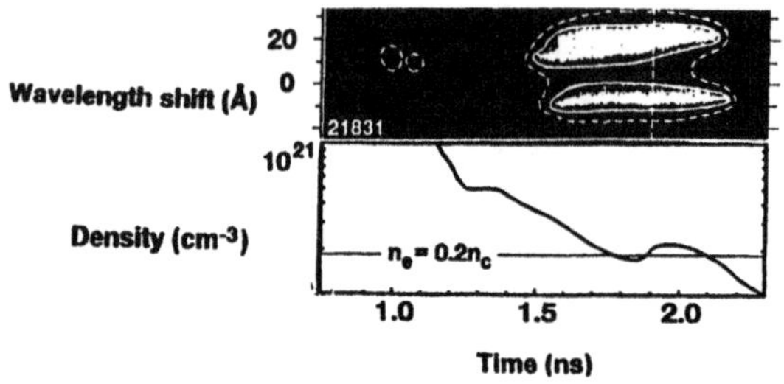

Fig. 5 Time resolved 3/2ω radiation spectra and plasma density evolution[7] $n=0.2n_e$ corresponds to Landau cutoff density.

ON LONG-LIVED SINGULARITIES IN THE NONLINEAR SHROEDINGER EQUATION

V. F. SHVETS, N. E. KOSMATOV
Department of Physics
Auburn University
Auburn, AL 36849-5311, USA

B. J. LeMESURIER
Australian National University
GPO Box 4 Canberra
ACT 2601, Australia

ABSTRACT. This article presents qualitative and computational analysis of the energetic effectiveness of superstrong wave collapse and associated long-lived singularities in the Nonlinear Schroedinger Equation. It is shown that in cases of higher dimensions the amount of energy absorbed by medium can be arbitrarily larger than critical energy. The validity of the point singlularity approach along with self-similar description is demonstrated at the early stage of the process of absorption.

1. Introduction

Singularities of wave fields, resulting from the development of wave collapses, determine, in many cases, the nature of strong wave turbulence (plasma, optical, etc.), which is responsible for numerous macroscopic displays of nonlinear wave phenomena. The theory of wave collapses is highly developed (e.g. see reviews [1,2] and references therein) due to their important role in the processes of transport of energy to small dissipative scales in turbulent media. The main goal of wave collapse theory is to determine the energetic effectiveness of the collapse as a channel of wave dissipation and to estimate the amount of energy transferred to medium.

Wave collapses have been studied intensively in the fundamental Nonlinear Schroedinger Equation (NLS)

$$i\psi_t + \Delta_r\psi + |\psi|^s\psi = 0, \quad \Delta_r = \frac{d^2}{dr^2} + \frac{d-1}{r}\frac{d}{dr}; \quad 0 \le r < \infty, \quad \psi(r,t)\underset{r\to\infty}{\to} 0, \tag{1}$$

that describes, in particular, various types of d-dimensional isotropic singularities in media with positive dispersion and local nonlinearity (for recent references see, in particular, [2-11]). Collapse (singularity formation in a finite time $t{=}t_0$) occurs in Eq. (1) at dimensions $d \ge d_c = 4/s$. Stability of isotropic singularities in the inertial range $t<t_0$ has been demonstrated numerically in [12]. At the pre-collapse stage Eq. (1) conserves the integral of wave energy

R. E. Caflisch and G. C. Papanicolaou (eds.), Singularities in Fluids, Plasmas and Optics, 309–316.
© 1993 *Kluwer Academic Publishers.*

$$N = \int_0^\infty |\psi|^2 r^{d-1} dr\,, \quad \frac{dN}{dt} = 0, \quad t < t_0, \tag{2}$$

some portion of which flows into developing singularity. In supercritical case, $d>d_c$, this quantity vanishes as the collapse point is approached, while in critical case, $d=d_c$, the energy captured approaches the finite critical value, $N_c{\sim}1$, that is determined by Townes stationary soliton[13,14]. For cubic nonlinearity, $s=2$, critical case $d=d_c=2$ corresponds to stationary self-focusing of optical beams, and at higher dimensions, $d\geq3$, Eq. (1) describes and/or models nonstationary self-focusing of light and various plasma collapse problems [13-17]. In general wave collapse theory, parameters s and d are considered to be arbitrary real quantities with restrictions $s\geq2$ and $d\geq1$.

Wave energy trapped by singularity in inertial range is transferred to medium through dissipation. It is said that collapse is strong or weak depending on whether the amount of energy absorbed, N_{abs}, is finite or infinitesimal when damping coefficient decreases. Until recently [7-9], it has been generally accepted that supercritical collapse is weak and that maximum possible amount of energy, which is restricted by critical energy, can be absorbed in the process of strong collapse in critical case. However, recent discovery of the phenomenon of a long-lived singularity, which continuously absorbs energy from the surrounding space [7-11,16-18], changed the picture of absorption dramatically. In the model of damped NLS [2,7,9,10,17], it has been numerically established that, while for parameters $d_c\leq d<d_s=2+2/s$ ($2\leq d<3$ in case of cubic nonlinearity) wave collapse has a short lifetime, breaking up with abrupt fall of field amplitude after the dissipation begins, at higher dimensions $d\geq d_s>d_c$ ($d\geq3$ for cubic NLS) collapse carries on in a quasistationary manner after the stabilization, and the small-size long-lived dissipative zone is formed. Although this effect received various names - 'distributed collapse' [7,10], 'bi-self-similar collapse' [8], 'singular collapse' [11], and 'superstrong wave collapse' (SWC) [2,9,17,18] - the latter name seems to be most pertinent, since it emphasizes a probable high energetic effectiveness of this type of collapse in comparison with ordinary strong wave collapse process.

Important to note that previous computations were performed for relatively narrow time intervals $t-t_0$ of post-collapse dynamics. Correspondingly, quantitative estimates of energetic effectiveness of SWC, which require long computational runs, have not been obtained. In this paper, we consider this critically important problem. In particular, we demonstrate by direct computations that in higher dimensions the amount of energy absorbed is determined basically by initial wave energy and can be arbitrarily larger than critical energy. The applicability of the model of mathematical singularity and self-similar description is also analyzed.

2. Physical and Mathematical Singularities

To go through the singularity and to study the process of absorption, one should take into account wave dissipation. An adequate model is NLS with multiphoton absorption localized in the small vicinity $0\leq r<r_{min}<<1$ around the singularity [7,9,10]:

$$i\psi_t + \Delta_r\psi + |\psi|^s\psi + i\beta|\psi|^m\psi = 0, \quad \psi_r(0,t) = \psi(\infty,t) = 0, \quad \beta <<< 1\,, \; m > s. \tag{3}$$

From wave energy balance

$$\frac{dN}{dt} = -2\beta \int_0^\infty |\psi|^{m+2} r^{d-1} dr \tag{4}$$

an estimate of the amount of energy absorbed is given by

$$N_{\rm abs} \sim 2\beta \int_0^T dt \int_0^\infty |\psi|^{m+2} r^{d-1} dr \,, \tag{5}$$

where T is a characteristic lifetime of a collapse that determines its energetic effectiveness. In cases $d \geq d_s$, when this time is much larger that corresponding time for the range of parameters $d_c \leq d < d_s$, a long-lived physical singularity, supported by energy flux

$$j = 2|\psi|^2 r^{d-1} (\arg\psi)_r, \tag{6}$$

is formed, starting with the time $t \approx t_0$ and the level of the amplitude $\psi_{st} \sim \beta^{1/(s-m)}$. Characteristic size of a thin dissipative layer is then given by $r_{\min} \sim [(d-d_c)\beta]^{s/2(m-s)}$ [2].

In order to determine critically important quantities (4) and (5), one should perform long-time computer simulations of SWC dynamics in Eq. (3). A number of results of such computations will be presented in the next section. Preliminary analytical estimates can be obtained considering an approximation of mathematical or point singularity, $r_{\min} \to 0$. Corresponding limit model of post-collapse behavior is Eq. (1) with nonzero negative energy flux at the center

$$j(0,t) \equiv j_0 < 0, \quad t_0 < t < \infty. \tag{7}$$

Due to this boundary condition, the solutions turn out to be singular as $r \to 0$ [7-9]. Corresponding asymptotical expressions have been found for arbitrary s and d with the additional approximation of quasi-stationary state in spatio-temporal vicinity of the singularity [9]. It has been shown that depending upon parameters SWC can be developed in four different regimes which correspond to the following ranges of parameters: $d=d_s$, $d_s<d<2d_s-2$, $d=2d_s-2$, and $2d_s-2<d$. In particular, for $d_s \leq d < 2d_s-2$ main asymptotical terms do not depend on the flux (7) while at higher dimensions $d \geq 2d_s-2$ the flux prevails. In case of $d>2d_s-2$, quasiclassical description (e.g., see [16] and references therein) is valid.

In the model of point singularity the expression (4) for the rate of absorption is replaced by

$$\frac{dN}{dt} = j_0(t), \quad t_0 < t < \infty. \tag{8}$$

In order to estimate time dependence of the flux at the origin, consider the generalization of well-known self-similar solution in supercritical case (e.g., see [2-4,6,15]), suggested in [8,11],

$$\psi(r,t) = [l(t)]^{-2/s-2i\kappa} g(\xi), \quad \xi = \frac{r}{l(t)}, \quad l(t) = |t_0-t|^{1/2}, \tag{9}$$

that is valid in some restricted regions in both space and time. From this solution, "frozen" field asymptotical distribution

$$\psi(r,t) \to Cr^{-2/s-2i\kappa} \text{ as } l \to 0 \tag{10}$$

leads to the expression for wave energy flux density at collapse point [11]

$$j(r,t_0) = -4\kappa C^2 r^{d-4/s-2}. \tag{11}$$

After collapse point, the flux at the origin is given from (9) by

$$j_0(t) = c(t-t_0)^{d/2-2/s-1}, \quad c = \lim_{\xi \to 0}[|g|^2\xi^{d-1}(\arg g)_\xi] < 0, \quad t_0 < t << t_1, \tag{12}$$

where t_1 is characteristic time of validity of self-similar description. Thus, in the model of point singularity wave energy drops at an early stage of absorption according to a power law

$$N(t) = N_0 - 2|c|\delta^{-1}(t-t_0)^{\delta/2}, \quad \delta \equiv d-d_c, \quad t_0 < t << t_1, \tag{13}$$

where N_0 is the initial energy of wave packet.

We expect these expressions to be also valid for real cases of physical singularities during short periods of time and in some regions outside dissipative layer. However, since lifetime of SWC is typically much larger than characteristic time t_1, final conclusions and quantitative estimates of the enegetic effectiveness of SWC can be obtained only from long-time computer simulations.

3. Computational Results and their Discussion

Such computations have been performed in the model of damped NLS (3) with Gaussian initial conditions, $\psi(r,0)=\psi_0\exp(-r^2/a^2)$. We note that adequate formulation of computer simulation of SWC is a difficult problem. Since the simulations are not appropriate in cases of both relatively thick and very thin dissipative layers, there is a high parameter sensitivity of the whole process of computational solution of the problem. The results presented below correspond to parameters s=2, m=6, $\beta=10^{-6}$, with two cases of higher dimensions: d=4 and d=5. For most of our runs, amplitude and semi-width of initial distribution were equal ψ_0=1 and a=4, respectively. Using one of the modifications of adaptive computational algorithm [3,6], we were able to follow post-collapse dynamics for sufficiently long period of time.

Consider first the case d=4. It is seen from Fig. 1(a) that blow-up-like growth of the amplitude in inertial range converts into slow process of absorption at the time $t_0 \approx 1.66$; the physical singularity formed is, obviously, long-lived. The dynamics of absorption (Figs. 1(b) and 1(c)) is governed by energy flux directed into small-size dissipative zone. The energy transferred to medium during the time t=5.05 is much larger than the critical energy, $N_{abs} \approx 14 >> N_c \sim 1$, and it is about 44% of initial energy, N_0=32. Clearly, we observe the picture of a superstrong collapse. In this particular case, an energy flux, estimated from mathematical singularity model, turns out to be uniform and time-independent. We find, after substitution the values of constants κ and C, which have been

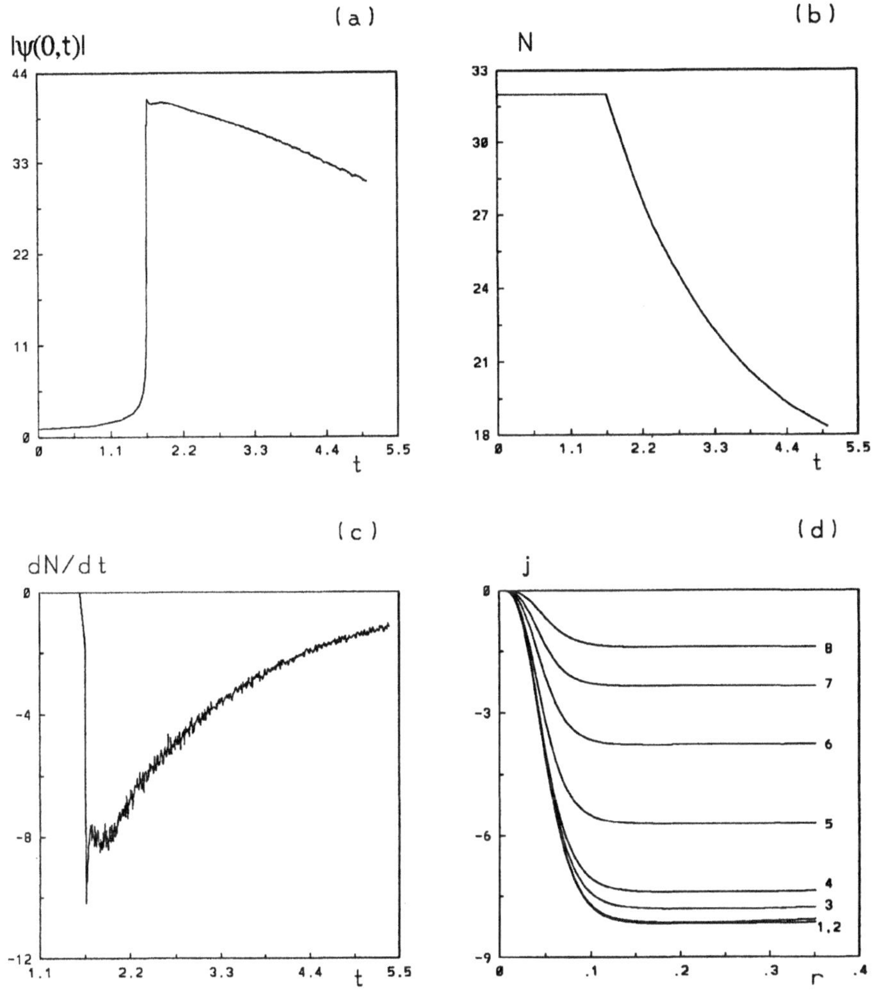

Figure 1. The results of computer simulation for d=4. (a) Time dependence of the maximum field amplitude. (b) Evolution of the absorption of wave energy. (c) Rate of absorption versus time. (d) Spatial structure of energy flux density in the vicinity of the singularity at different moments of time: 1, t=1.71; 2, t=1.98; 3, t=2.05; 4, t=2.11; 5, t=2.50; 6, t=3.25; 7, t=4.11; 8, t=5.05.

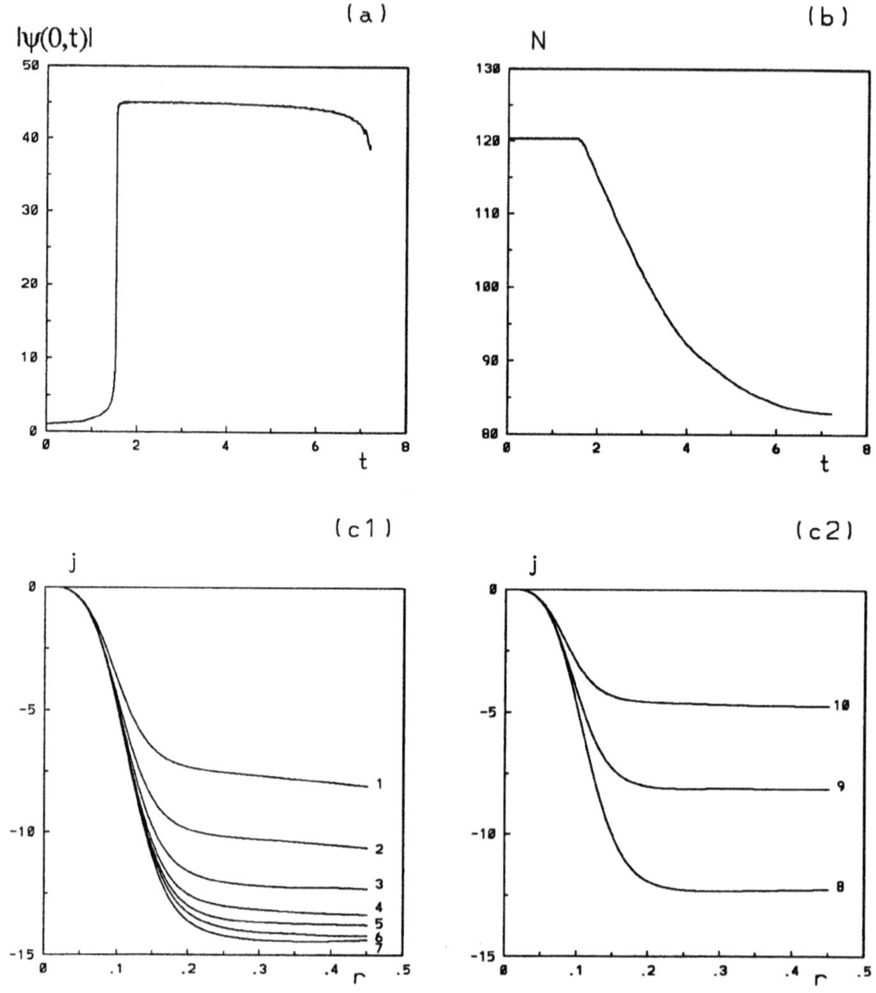

Figure 2. The results of computer simulation for d=5. (a) Time dependence of the maximum field amplitude. (b) Evolution of the absorption of wave energy. (c1) and (c2) Spatial structure of energy flux density in the vicinity of the singularity at different moments of time: 1, t=1.58; 2, t=1.63; 3, t=1.68; 4, t=1.72; 5, t=1.77; 6, t=1.82; 7, t=1.87; 8, t=2.99; 9, t=3.96; 10, t=5.13.

calculated from pre-collapse dynamics, into Eq. (11), that $j_0 \approx -7.54$. In Fig. 1(c), one can see a remarkable agreement between this value and the one found from computer simulation. This agreement holds, however, during narrow time interval (until $t \sim 2$), beyond which the rate of absorption slowly decreases. The same tendency appears in Fig. 1(d), which also confirms the uniform flux distribution outside dissipative layer, $r >> r_{\min} \approx 0.04$.

With the dimension $d=5$, nonlinear effects become stronger[16-18]. An estimated flux (12) is now growing function of time, which has to be stabilized inevitably by the effect of weakening of absorption that is outside the self-similar approach. Due to joint influence of two competitive effects, quasistationary behavior of maximum field amplitude and non-monotonic change of energy flux are expected. This preliminary picture is completely confirmed by computer simulation. Starting at the time $t_0 \approx 1.57$, we observe quasistationary state of the amplitude (Fig. 2(a)) that continues almost the whole time of the run, $t \sim 7$. The amount of energy absorbed, $N_{\text{abs}} \approx 37$, is about 31% of initial wave energy (Fig. 2(b)). The magnitude of the flux in the vicinity of physical singularity reaches its maximum around $t \sim 2$ and then drops continuously (Fig. 2(c)). Again, the whole picture perfectly corresponds to SWC phenomenon with high absorptive ability. In both cases the point singularity model is adequate on the initial stage of the absorption.

4. Conclusions

We have demonstrated that in higher dimensions the energy transferred to medium is determined essentially by the intensity of initial wave packet and that it does not relate to critical energy. Since long-lived singularities absorb considerable part of initial energy (that can exceed substantially the critical value), the process of superstrong wave collapse is very effective and, possibly, the main mechanism of nonlinear dissipation of waves in many strongly turbulent processes. The self-similar approach in the model of point singularity describes adequately an early stage of post-collapse dynamics and gives, in particular, a correct estimate of the initial rate of absorption; however, it fails to provide proper evaluation of total energetic effectiveness.

5. References

[1] Shvets, V. F. and Zakharov, V. E. (1990) 'Computer simulation of wave collapses and wave turbulence', in V. G. Bar'yakhtar et al. (eds.), Nonlinear World, World Scientific, Singapore, vol. 1, pp. 671-692.

[2] Kosmatov, N. E., Shvets, V. F., and Zakharov, V. E. (1991) 'Computer simulation of wave collapses in the nonlinear Schroedinger equation', Physica D 52, 16-35.

[3] McLaughlin, D. W., Papanicolaou, G.C., Sulem, C., and Sulem, P.L. (1986) 'Focusing singularity of the cubic Schroedinger equation', Phys. Rev. A 34, 1200-1210.

[4] Kosmatov, N. E., Petrov, I. V., Shvets, V. F., and Zakharov, V. E. (1988) 'Large - amplitude simulation of wave collapses in nonlinear Schroedinger equation', Space Research Institute, Moscow, Preprint No. 1365.

[5] Zakharov, V .E. and Shvets, V. F. (1988) 'Nature of wave collapse in the critical case', JETP Lett. 47, 275-278.

[6] LeMesurier, B. J., Papanicolaou, G. C., Sulem, C., and Sulem, P. L. (1988) 'Focusing and multi-focusing solutions of the nonlinear Schroedinger equation', Physica D 31, 78-102; 'Local structure of the self-focusing singularity of the nonlinear Schroedinger equation', Physica D 32, 210-226.

[7] Vlasov, S. N., Piskunova, L. V., and Talanov, V. I. (1988) 'Distributed wave collapse', in Nonlinear and Turbulent Processes in Physics, Naukova Dumka, Kiev, vol. 2, pp. 210-213.

[8] Malkin, V. M. (1988) 'Bi-self-similar wave collapse', JETP Lett. 48, 653-655.

[9] Zakharov, V. E., Kosmatov, N. E., and Shvets, V. F. (1989) 'Ultrastrong wave collapse', JETP Lett. 49, 492-495.

[10] Vlasov, S. N., Piskunova, L. V., and Talanov, V. I. (1989) 'Three-dimensional wave collapse in the nonlinear Schroedinger equation model', Sov. Phys. JETP 68, 1125-1128.

[11] Malkin, V. M. and Shapiro, E. G. (1990) 'Singular wave collapse', Sov. Phys. JETP 70, 102-107.

[12] Landman, M. J., Papanicolaou, G. C., Sulem, C., Sulem, P. L., and Wang, X. P. (1991) 'Stability of isotropic singularities for the nonlinear Schroedinger equation', Physica D 47, 393-415.

[13] Chiao, R. Y., Garmire, E., and Townes, C. H. (1964) 'Self-trapping of optical beams', Phys. Rev. Lett. 13, 479-482.

[14] Zakharov, V. E. and Synakh, V. S. (1975) 'The nature of self-focusing singularity', Sov. Phys. JETP 41, 465-468.

[15] Zakharov, V. E. (1972) 'Collapse of Langmuir waves', Sov. Phys. JETP 35, 908-922.

[16] Kuznetsov, E. A. and Turitsyn, S. K. (1990) 'Quasiclassical Langmuir wave collapse in a magnetic field', Sov. J. Plasma Phys. 16, 524-527.

[17] Shvets, V. F. 'On the superstrong wave collapse', in Proceedings of the III Potsdam - V Kiev International Workshop on Nonlinear Processes in Physics, to be published.

[18] Shvets, V. F. (1990) 'On superstrong wave collapse in lower dimensions', Phys. Lett. A 150, 74-78.

ON COLLAPSING SOLUTIONS OF THE NONLINEAR SCHROEDINGER EQUATION IN SUPERCRITICAL CASE

V. F. SHVETS, N. E. KOSMATOV
Department of Physics
Auburn University
Auburn, AL 36849-5311, USA

B. J. LeMESURIER
Australian National University
GPO Box 4 Canberra
ACT 2601, Australia

ABSTRACT. Long distant behavior of the mode of supercritical collapse in the Nonlinear Schroedinger Equation is analyzed both analytically and numerically. The boundary of region of self-similarity of a weak collapse is determined. Exponential decay of the solution at large distances and convergence of wave energy integral are demonstrated.

1. Introduction

The problem of wave collapse, described by isotropic Nonlinear Schroedinger Equation (NLS)

$$i\psi_t + \Delta_r\psi + |\psi|^s\psi = 0,\ \ \Delta_r = \frac{d^2}{dr^2} + \frac{d-1}{r}\frac{d}{dr};\ \ 0 \leq r < \infty,\ \psi_r(0,t) = \psi(\infty,t) = 0 \tag{1}$$

in supercritical case, $d>d_c=4/s$, that includes, in particular, variuos multi-dimensional focusing phenomena in cubic media, has been the subject of many studies (see [1-7] and references therein). It has been established that the process of approaching collapse point $t=t_0$ is self-similar:

$$\psi(r,t) = [l(t)]^{-2/s-2i\kappa}g(\eta),\quad \eta = \frac{r}{l(t)},\quad l(t) = (t_0-t)^{1/2}. \tag{2}$$

This self-similar law leads to "frozen" field asymptotical distribution

$$\psi(r,t) \approx Cr^{-2/s-2i\kappa} \text{ as } r >> l(t), \tag{3}$$

where parameters $\kappa(s,d)$ and $C(s,d)$ are eigenvalues of nonlinear boundary value problem that determines the function $g(\eta)$ from Eqs. (1) and (2).

The self-similar solution (2) corresponds to the process of a weak collapse with infinitesimal wave energy flowing into the singularity as $t\rightarrow t_0$ [2,4,5,7]. Actual realization of solution (2) and

R. E. Caflisch and G. C. Papanicolaou (eds.), Singularities in Fluids, Plasmas and Optics, 317–321.
© 1993 *Kluwer Academic Publishers.*

stability of a weak self-similar collapse have been repeatedly proved in adaptive computations [3-7] performed in the vicinity of the singularity.

However, since self-similar solution leads to divergency of the integral of wave energy

$$N = \int_0^\infty |\psi|^2 r^{d-1} \mathrm{d}r = N_0, \tag{4}$$

it is inconsistent with wave energy conservation and is not valid beyond some finite region $r<r_0$. Although this inconsistency was already noted [2,7], important questions concerning characteristic size of self-similarity region and solution outside this region have not been considered previously.

This paper is devoted to analysis of these questions. Along with analytical estimates, we performed a number of computational studies of Eq. (1) in adaptive variables

$$\xi = \frac{r}{L(t)}, \quad \frac{\mathrm{d}\tau}{\mathrm{d}t} = [L(t)]^{-2}, \quad L(t) \propto (t_0 - t)^{1/2} \text{ as } t \to t_0 \tag{5}$$

(the similar way as in [3-7]), with the resolution of both central and periferal parts of field distribution, using one of modifications of algorithm [3,6]. In section 2, the boundary of self-similarity region is found. The analysis of long distant behavior of the solution is given in section 3.

2. Boundary of Region of Self-Similarity

For arbitrary finite region $r<r_1<r_0$, wave energy integral is given by [2,7]

$$N_1(t) = l^{\delta} \int_0^{r_1/l} |g|^2 \eta^{d-1} \mathrm{d}\eta, \tag{6}$$

where $\delta = d - d_c$. Near the singularity, as $t \to t_0$, this becomes a time-independent value

$$N_1 \approx C^2 \delta^{-1} r_1^{\delta}. \tag{7}$$

In order to estimate the boundary point of region of self-similarity r_0, we further assume that the nature of the behavior of field amplitude changes abruptly in the vicinity of this point, making the intergal (4) converge as $r \to \infty$. This supposition leads to a time-independent boundary point

$$r_0 \sim (C^{-2} N_0 \delta)^{1/\delta}, \tag{8}$$

that is basically determined by initial energy, as expected.

This estimate was supported by computer experiments. Below, the results for the cases of cubic nonlinearity (s=2) and dimensions d=3 and d=4 are presented. Parameters of Gaussian initial condition, $\psi(r,0)=\exp(-r^2/16)$, correspond to the values of initial wave energy $N_0 \approx 10$ (d=3) and N_0=32 (d=4). One can see in Fig. 1 that, starting at some point $\xi_0(\tau)$, sharp failure of self-similarity takes place. With high accuracy, the motion of self-similarity boundary is uniform, that corresponds, taking account of Eq. (5), to stationary boundary point r_0 in physical variables. The immobile boundary points are clearly seen, for both cases, in Fig. 2. The values $r_0 \sim 5$ (d=3) and

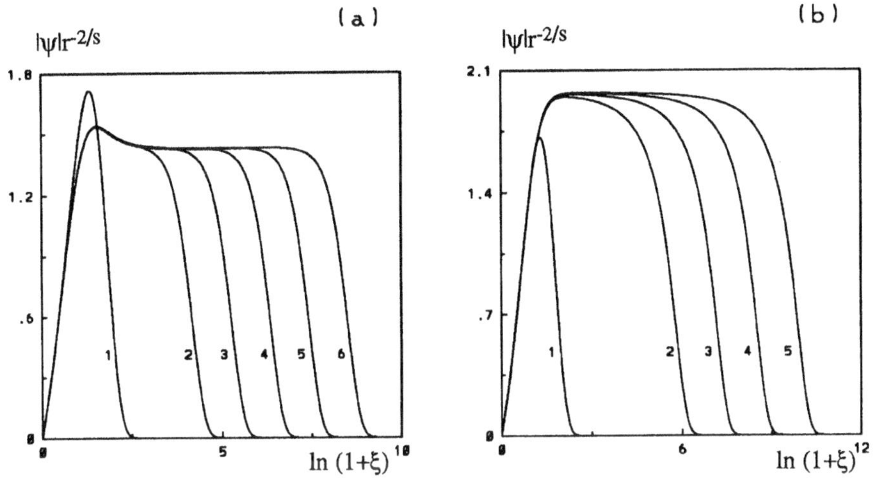

Figure 1. Structure of solution in adaptive variables at different moments of time. (a) d=3. 1, τ=0; 2, τ=8; 3, τ=12; 4, τ=16; 5, τ=20; 6, τ=24. (b) d=4. 1, τ=0; 2, τ=12; 3, τ=16; 4, τ=20; 5, τ=24.

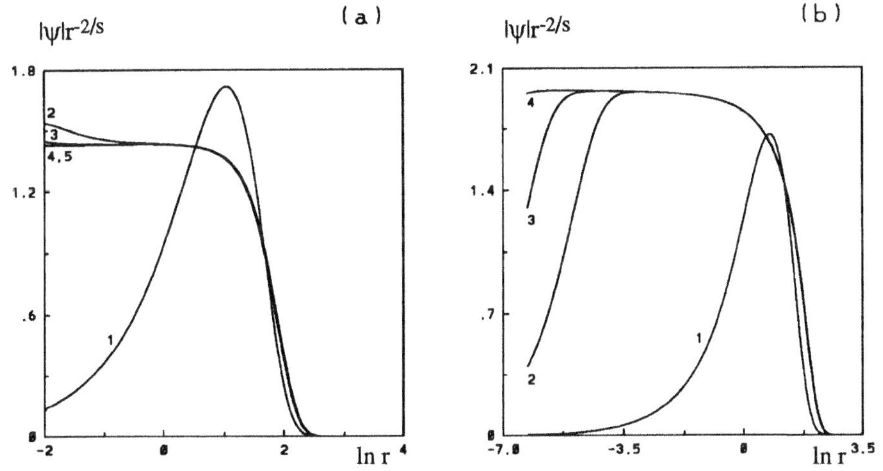

Figure 2. Structure of solution in physical space at different moments of time. (a) d=3. 1, τ=0; 2, τ=12; 3, τ=16; 4, τ=20; 5, τ=24. (b) d=4. 1, τ=0; 2, τ=16; 3, τ=20; 4, τ=24.

$r_0 \sim 4$ (d=4), given by the estimate (8), correspond approximately to those observed in Fig. 2 (a) and Fig. 2 (b), respectively. Thus, computational results confirm analytical estimate (8) and underlying qualitative conjecture.

3. Long Distant Behavior of the Solution

Beyond the self-similarity region, solutions (2) and (3) are not valid anymore. At large distances, nonlinearity is negligible, and one can consider linear equation

$$i\psi_t + \Delta_r\psi = 0, \quad \psi(r,t) \underset{r \to \infty}{\to} 0. \tag{9}$$

The solution of this equation is given by

$$\psi(r,t) = f(r)\exp(i\lambda^2 t), \tag{10}$$

where the function $f(r)$ is expressed in terms of Bessel function of imaginary argument and has asymptotical behavior

$$f(r) \sim r^{(1-d)/2}\exp(-\lambda r) \quad \text{as} \quad r \to \infty. \tag{11}$$

Correspondingly, wave energy density falls off exponentially

$$n(r,t) = |\psi|^2 r^{d-1} \sim \exp(-2\lambda r) \quad \text{as} \quad r \to \infty, \tag{12}$$

providing the convergency of the integral of wave energy.

The results of computations confirm the exponential decay of wave energy density as $r \to \infty$ (see Fig. 3). It is also seen that λ depends adiabatically on time and approaches $\lambda_0 \approx 0.7$ as $t \to t_0$ for both d=3 and d=4. Although λ seems to be parameter independent, we believe that additional studies are needed for ultimate conclusions about the rate of decay as $r \to \infty$.

4. Conclusions

We have demonstrated, both qualitatively and numerically, that self-similarity of a weak collapse fails in the vicinity of some boundary point that is determined by initial intensity and can be estimated from Eq. (8). Beyond this point, solution drops exponentially, and wave energy integral converges. The rate of exponential decay, that was found numerically, turned out to be approximately the same (about 0.7) at different dimensions.

Acknowledgment

One of the authors (V.F.S.) would like to thank E. A. Kuznetsov for valuable discussion.

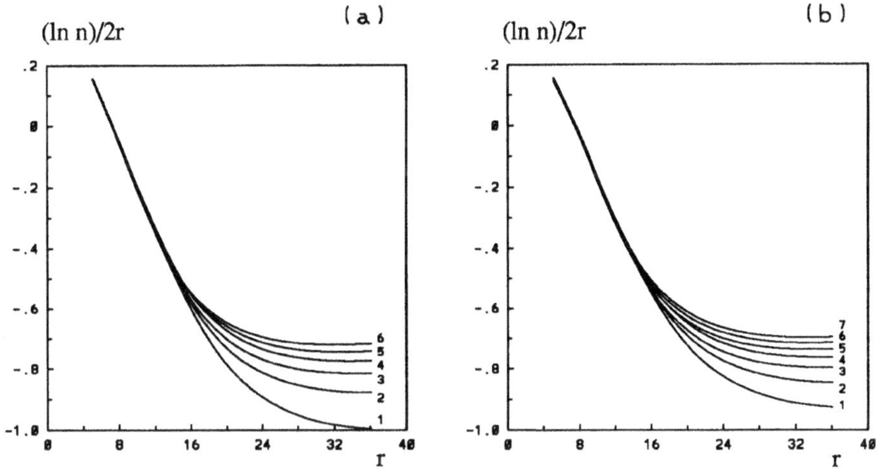

Figure 3. Asymptotical behavior of solution at large distances. (a) d=4. 1, τ=4; 2, τ=8; 3, τ=12; 4, τ=16; 5, τ=20; 6, τ=24. (b) d=3. 1, τ=6; 2, τ=10; 3, τ=14; 4, τ=18; 5, τ=22; 6, τ=26; 7, τ=30.

References

[1] Rypdal, K. and Rasmussen, J. Juul (1986) 'Blow-up in Nonlinear Schroedinger Equation-II. Similarity structure of the blow-up singularity', Physica Scr. 33, 498-504.

[2] Zakharov, V. E. and Kuznetsov, E. A. (1986) 'Quasiclassical theory of three-dimensional wave collapse', Sov. Phys. JETP 64, 773-780.

[3] McLaughlin, D. W., Papanicolaou, G. C., Sulem, C., and Sulem, P. L. (1986) 'Focusing singularity of the cubic Schroedinger equation', Phys. Rev. A 34, 1200-1210.

[4] Kosmatov, N. E., Petrov, I. V., Shvets, V. F., and Zakharov, V. E. (1988) 'Large - amplitude simulation of wave collapses in nonlinear Schroedinger equation', Space Research Institute, Moscow, Preprint No. 1365.

[5] Zakharov, V. E., Litvak, A. G., Rakova, E. I., Sergeev, A. M. and Shvets, V. F. (1988) 'Structural stability of wave collapse in media with a local nonlinearity', Sov. Phys. JETP 67, 925-927.

[6] LeMesurier, B. J., Papanicolaou, G. C., Sulem, C., and Sulem, P. L. (1988) 'Local structure of the self-focusing singularity of the nonlinear Schroedinger equation', Physica D 32, 210-226.

[7] Kosmatov, N. E., Shvets, V. F., and Zakharov, V. E. (1991) 'Computer simulation of wave collapses in the nonlinear Schroedinger equation', Physica D 52, 16-35.

Sharp Criteria for Wave Collapse and Wave Collapse Ideology in Application to Optical Pulses Compression

Sergei K. Turitsyn (*)
Institüt für Theoretische Physik I
Heinrich-Heine-Universität Düsseldorf
D-4000 Düsseldorf 1
Germany

ABSTRACT. Sharp sufficient critria for collapse is found for several well-known nonlinear equations. It is suggested to use wave collapse mechanism for optical pulses compression. An example of the nonlinear optical medium, which is described by the Ginzburg-Landau equation for a subcritical bifurcation, is ivestigated and the exact sufficient criterion of the blow-up is obtained.

1. Introduction

This report consists of three fairly independent, but at the same time internally connected parts.

The first one is a brief introduction and description of mathematical tool which will be used later. In this part I will demonstrate how some known results concerning a sufficient conditions for a wave collapse may be improved.

The second part is a presentation of the main idea: a possibility to use a wave collapse phenomenon to optical pulses compression.

And the third part is devoted to the investigation of optical pulses propagation and blow-up in the nonlinear medium, modeling by the Ginzburg-Landau equation.

Let us remind several examples of the analytical proofs for the blow-up. More precisely, the examples of the nonlinear equations for which sufficient criteria of the wave collapse were obtained. There are no rigorous mathematical methods for solving such type problems. One may notice that the most often used approach is the so-called majoring equation method [1] . This approach can be used for the integrable equations as well as for the

R. E. Caflisch and G. C. Papanicolaou (eds.), Singularities in Fluids, Plasmas and Optics, 323–332.
© 1993 *Kluwer Academic Publishers.*

nonintegrable ones. On this way one may obtain a sufficient criterion of the blow-up for the following equations:

1. Nonlinear Schrödinger equation:

$$i\Psi_t + \Delta\Psi + (\sigma+1)|\Psi|^{2\sigma+2}\Psi = 0. \tag{1}$$

Majoring equation is a linear, second-order :

$$\frac{\partial^2}{\partial t^2}\int \mathbf{r}^2|\Psi|^2 d\mathbf{r} = 8H - 4(\sigma d - 2)\int |\Psi|^{2\sigma+2} d\mathbf{r} \tag{2}$$

2. Nonlinear heating equation:

$$U_t = \Delta U + U^{1+\beta}, \quad \beta > 0. \tag{3}$$

Majoring equation is a nonlinear, second order:

$$\Phi_{tt}\Phi - (1+\alpha)\Phi_t^2 \geq 0, \quad \Phi_t = \int U^2 d\mathbf{r}. \tag{4}$$

3. Boussinesq (nonlinear string) equation:

$$U_{tt} - U_{xx} + U_{xxxx} + \frac{\partial^2}{\partial x^2}U^2 = 0 \tag{5}$$

Majoring equation is the same as for the nonlinear heating equation, role of the function Φ is played by $\int W^2 dx$, where $W_x = U$.

4. Modified Kadomtsev-Petviashvili equation, describing, in paricular, an evolution of magnetoelastic waves in antiferromagnets:

$$\frac{\partial}{\partial x}(U_t + (n+2)(n+1)U^n U_x + U_{xxx}) = \Delta_\perp U \tag{6}$$

Majoring equation was found in the [2]:

$$\frac{\partial^2}{\partial t^2}\int r_\perp^2 U^2 d\mathbf{r} \leq 48H \tag{7}$$

Depending on concrete problem, one may use a decreasing majoring functions (for example, parabola for the NLS equation), as well as, increasing ones, which tend to infinity as t tends to some certain finite point t_0.

For example, from the mentioned above majoring equation:

$$\Phi_{tt}\Phi - (1+\alpha)\Phi_t^2 \geq 0 \tag{8}$$

it follows that:

$$\Phi(t) \geq \Phi_0 \left(\frac{t_0}{t_0 - t}\right)^{\frac{1}{\alpha}}, \tag{9}$$

so that, function $\Phi(t)$ becomes singular for a finite time.

Equations 1, 3 and 4 are examples of Hamiltonian systems and the sufficient criterion for the blow-up has been found as a negativeness of corresponding Hamiltonian.

1. Nonlinear Schrödinger equation.

$$i\Psi_t = \frac{\delta H}{\delta \Psi^*}, \quad H = \int |\nabla \Psi|^2 d\mathbf{r} - \int |\Psi|^{2\sigma+2} d\mathbf{r}. \tag{10}$$

Sufficient criterion for the collapse: $H < 0$, provided that $\sigma d \geq 2$

2. Nonlinear heating equation.

$$U_t = \frac{\delta H}{\delta U}, \quad H = \frac{1}{2}\int (\nabla U)^2 d\mathbf{r} - \frac{1}{2+\beta}\int U^{2+\beta} d\mathbf{r}. \tag{11}$$

Sufficient criterion for the collapse : $H < 0$.

3. Boussinesq equation.

$$W_t = -\frac{\delta H}{\delta U}, \tag{12}$$

$$U_t = \frac{\delta H}{\delta W}, \tag{13}$$

here Hamiltonian $H = \int (\frac{1}{2}W_x^2 + \frac{1}{2}U^2 + \frac{1}{2}U_x^2 - \frac{1}{3}U^3)dx$.

Sufficient criterion for the collapse : $H < 0$.

4. Modified Kadomtsev-Petviashvili equation.

$$U_t = \frac{\partial}{\partial x}\frac{\delta H}{\delta U}, \quad H = \int (\frac{1}{2}U_x^2 + \frac{1}{2}(\nabla_\perp W)^2 - U^{n+2})dxdydz, \tag{14}$$

here $W_x = U$.

Sufficient criterion for the collapse : $H < 0$.

Equation 2 is not Hamiltonian, but there exists Lyapunov functional monotonically decreasing as t increas, and sufficient criterion for the blow-up is the negativeness of this functional H at the initial moment of time.

I would like to show how these criteria may be improved. The idea how it can be done is connected very closely with the answer on the following question. Everybody knows a role, playing by the stable solitons in nonlinear dynamics. Often, however, nonlinear systems admit soliton solutions, but these solutions are unstable. Practically always, that corresponds to the situation when a nonlinear system supports collapse existence. What is a role playing by unstable soliton solutions in dynamics of nonlinear equations admitting collapse? At first glance, because unstable stationary solutions can not be attractors for other solutions it seems they can not play a role of any importance.

A simplified formulation of the answer on the mentioned above question is that nonstable soliton solutions, under some additional assumptions, devide the phase space of a nonlinear system admitting collapse into two parts: the first part consists of solutions remaining bounded for any moment of time, and the second one consists of solutions for which collapse occurs.

I would like to declare this statement as a gypothesis for any nonlinear system which demonstrates the blow-up-type dynamics and possesses unstable soliton solutions.

For several concrete physical models one can prove this statement as a theorem.

Let consider, as an example, NLS with power nonlinearities for the case then this equation has soliton solutions, but unstable; it correspons to the so-called supercritical case $2/D < \sigma < 2/(D-2)$.

For the critical case $\sigma = 2/D$ the sharp sufficient condition for the blow-up was obtained by Weinstein. For simplicity (but without loss of generality) I limit myself by considering the most interesting (for physical application) case $\sigma = 1$, and $D = 3$. More detailed analysis will be presented in [3].

From the well-known relation:

$$\frac{\partial^2}{\partial t^2} \int r^2 |\psi|^2 dr = 8H - 4I_2, \tag{15}$$

(where Hamiltonian $H = I_1 - I_2$) it is fairly obvious, that the negativeness of the Hamiltonian is a sufficient condition for collapse, but this condition is not sharp, because in the right-hand side of this relation we have some evident spare. Now our intermediate goal is to estimate from below the Hamiltonian by some function of I_2. To do this, one may use the following inequality:

$I_1 \geq C_0 N^{-3} I_2^2$,

here the best interpolation constant $C_0 = \frac{3}{4} N_{cr}^2$ was calculated by M.Weinsein (N_{cr} is a value of integral N on a solution of the following equation: $\Delta f - f + f^3 = 0$).

After substitution of this estimation into Hamiltonian we get interesting inequality:

$$H \geq F(I_2) = C_0 N^{-3} I_2^2 - I_2 \tag{16}$$

The remarkable fact is that maximum of the function $F(I_2)$ is equal to $max F(I_2) = H_s$ exactly, where H_s is a value of Hamiltonian on the soliton solution and this maximum is reached at the point $I_2 = I_{2s} = 2H_s$ I_{2s} being a value of the integral I_2 on a soliton solution.

Thus for $H < H_s$ the equation $H = F(I_2)$ has two solutions $I_2^{(-)}$ and $I_2^{(+)}$. Inequality (16) is satisfied for two intervals $0 \leq I_2 \leq I_2^{(-)}$ and $I_2 \geq$

$I_2^{(+)}$. Our basic equation implies that any initially smooth solution evolves continuously in the time interval of existence, hence the integral $I_2(t)$ is continuous for any t before the collapse point. Since the intervals $(0, I_2^{(-)})$ and $(I_2^{(+)}, \infty)$ are disjoint, $I_2(t)$ cannot evolve from one interval to the other. Let at the initial moment $I_2|_{t=0} > I_{2s}$, then $I_2(t) > I_{2s}$ for all t. Now, substituting the estimate for $I_2(t)$ into the equation (15) we obtain the final formula:

$$\partial^2/\partial t^2 \int r^2|\psi|^2 dr \leq 8(H - H_s) \tag{17}$$

Thus, the sharp sufficient criterion for the blow-up dynamics is the condition $H < H_s$.

We may conclude, that if there is no some special unphysical focusing at the initial moment (In this report I limit myself by consideration tha case (without loss of generality) then $\partial/\partial t \int r^2|\psi|^2 dr \leq 0$ is not positive, or positive , but amplitude is small) then solutions for which Hamiltonian is less than the value of H on soliton (nonstable) solution collapsing for the finite time. So that, nonstable soliton solutions in some sence play a role of boundary between collapsing and noncollapsing regimes.

The same result may be proved for other models [4]. One of them I will discuss in the last part of the report.

2. Wave collapse and optical pulses compression

The main goal of this part is to attract attention to the possibility of using collapse for compression of optical pulses.

The last two decades of development of nonlinear optics clearly showed an importance of universal nonlinear models in this field of physics. Simply formulated, but, at the same time, rich in content, the models such as the nonlinear Schrödinger equation, the sine-Gordon equation, the system describing three-waves resonant interaction and others are used in a wide array of different physical contexts. In recent years a number of investigations performed was devoted to the study of nonlinear dynamics of soliton pulses in the frameworks of these basic models and their numerous generalizations. From a general point of view, the most of applications of the above mentioned and related to them models are based on using of nonlinear systems admitting *stable soliton dynamics.*

The most well-known (and one of important for the technical applications) example is propagation of short optical pulses through a single-mode optical fiber. In 1973 Hasegawa and Tappert in their pioneer work [5] proposed to compensate the dispersive broadening of the optical pulses by the nonlinear change of the refractive index of the fiber material (the so-called Kerr-effect). When the frequency shift due to the Kerr effect is balanced with that due to the dispersion effect, a stable propagation of optical soliton is possible. This idea drew excited attention after the experimental observation of optical solitons by Mollenauer, Stolen and Gordon [6]. The time period between putting forward idea by Hasegawa and Tappert and experimental observation of the effect was about seven years.

The using of the nonlinear effects in optical devices has allowed to demonstrate such applications of the solitons as the pulse compression [7], the soliton laser [8], long-distance soliton transmission through high-bit-rate communication systems [9-11], soliton amplification in erbium doped fibers [12] etc.. Thus, one may conclude that wide using of ideology of *stable soliton dynamics* in nonlinear optics is nowdays reality.

The primary objective of my report is to attract attention to the potential possibility of using in optical applications an another interesting phenomenon habitual to nonlinear systems - the so-called *wave collapse.*

There exist to date many examples of importance for the collapse-type dynamics in nonlinear medium: creation of shock waves from an initially smooth distribution in hydrodynamics and magnetohydrodynamics; self-focusing of light beams; collapse of the plasma waves; the blow-up processes in chemical reactions; finite time singularity formation in the boundary layer equations of fluid flow and others.

From the mathematical point of view, one may define wave collapse as appearence of a singularity for the finite time in initially smooth solutions of the nonlinear equations describing wave dynamics in a linearly stable medium. Of course, appearence of the singularities means, that one must take into acount other physical effects, which may be droped before, but they do play an important role in a vicinity of the collapse point. It should be noticed that important point is not only the singularities themselves, but a very possibility for a weakly nonlinear system to create an explosive-type growing patterns.

Collapse may be viewed as natural and a very nonlinear mechanism of amplification of an initially weak nonlinear pulse up to the limit of aplicability of a weakly nonlinear model.

Let us illustrate this idea on the following example. Putting a localized initial optical pulse in the linear dispersive medium (considering the last as a "black box"), one obtains (in general) dispersively broadened pulse on the output.

Realizing the same procedure for a medium with soliton properties (e.g. nonlinear optical fibers) one may obtain an efficient and elegant way to compress initial pulse consistented from several fundamental solitons (Mollenauer et al [7]). In another words one may use weakly nonlinear medium possesing stable solitons for the optical pulse compression.

Let us imagine now, that we have a nonlinear medium which possesses the collapse phenomenon in the limit of a weak nonlinearity for the some class of initial field distributions. Putting in such medium a pulse from this class, one obtains regime of the pulse amplitude increasing for a finite time (or distance) up to a limit of applicability of a weakly nonlinear model. The most of the collapses studied to date have the self-similar (quasi-self-similar) asimptotics near the collapse point. It means that increasing of a peak amplitude of the pulse is accompanied by the collapsing pulse narrowing. Thus, in principle, it is possible to use a collapse mechanism to compress pulses.

3. Blow-up of the optical pulses

In order to demonstrate the idea performed above it is necessary to give a concrete example of a weakly nonlinear medium exhibitng the collapse dynamics. We will discuss here the distributed model which recently has been described by Belanger [14] in context of generation of short optical pulses:

$$\frac{\partial S}{\partial x} = -aS + b\frac{\partial^2 S}{\partial t^2} + 2c|S|^2 S, \qquad (18)$$

with the boundary condition $S \to 0$ as $t \to \infty$.

Equation (18) describes a distributed system modeling dynamics of a pulse envelope propagating through a coupled-cavity mode-locked laser, for a phase-mismatch angle of $\pi/2$ between the two cavities. In the Eq. (18) the coefficients a, b, c are defined by the following relations: $a = -g + \gamma > 0$, $b = B - \Delta A > 0$, $c = n_2 k_0 \Delta^2 > 0$, where g is a small saturated signal gain, γ is the loss parameter, $B = gT_0^2$, T_0 is a time parameter related to a gain bandwidth, Δ is a parameter describing coupling between the main cavity and the external cavity, A is the group-velocity dispersion of the medium, n_2 is nonlinearity parameter, and k_0 is a wavenumber of the pulse. Equation (18) describes the evolution of the weak- nonlinear pulse envelope during propagation into equivalent medium that is composed of a gain medium and a nonlinear dispersive medium [14]. Equation (18) is the well-known Ginzburg-Landau equation for the subcritical bifurcation, which appears in different physical problems [15-16].

A homogenious solution $S = 0$ is stable against small perturbations , provided the coefficient a is positive. Thus, under sufficiently weak initial disturbances the bound state ($S = 0$) remains stable and the disturbances die out. But for the pulses for which nonlinearity is sufficient enough the situation changes. A perturbation will experience explosive growth when it becomes sufficiently large for the nonlinear term to come into play. Then the cubic term in Eq.(18) dominates , causing an amplification of the rate of increase of the maximum value of $|S|$, a focusing phenomenon occurs and S become singular at a finite value of x.

There exist an exact soliton solution of the Eq.(18) in the following form:

$$S(t,x) = S_0 \ \mathrm{sech}(t/T), \qquad (19)$$

where the pulse width is given by $T = \sqrt{b/a}$ and the soliton amplitude is $S_0 = \sqrt{a/c}$. However, the soliton pulse is unstable in the framework of Eq.(18), i.e. it means that although an interplay between dispersion and nonlinearity admits to construct a steady-state solution, there exist perturbations of the solitons for which the nonlinear effects are predominant and a collapse may appear for a some class of the initial distributions.

As was shown in [17], there exist the self-similar solutions of Eq.(18) which describe the increasing of initial pulse accompanying self-similar compression:

$$S_s = \frac{1}{(x_0 - x)^{1/2}} R(\xi), \ \xi = \frac{t}{f(x_0 - x)}, \ f(z) = (z \ln z^{-1})^{1/2},$$

where R is a monotonic function of ξ, and $R \to 0$, as $\xi \to \infty$. One may see from the last expression that near the collapse point the maximum pulse amplitude is increasing but its characteristic width ($\tau = f(x_0 - x) \to 0$) is narrowing. In order to find a sufficient criterion of the wave collapse (a class of initial distributions leading to the collapse dynamics), one may use the method of a majoring equation [1, 18]. Let us define two functionals H and P by the formulas:

$$H = \int (b|S_t|^2 + a|S|^2 - c|S|^4) dt \tag{20}$$

and

$$P = \int |S|^2 dt. \tag{21}$$

An important aspect of Eq.(18) is an existence of a Lypunov-type majoring functional, because H obeys the following equation:

$$\frac{dH}{dx} = -2 \int |S_x|^2 dt. \tag{22}$$

It means that one possible limiting asymptotic for the Eq.(18) is : $S \to 0$ as $x \to \infty$. Provided S is sufficiently small at the point $x = 0$, this limit will be achieved. The change of P vs x is governed by:

$$\frac{d}{dx} P = -2b \int |S_t|^2 dt + 2a \int |S|^2 dt + 4c \int |S|^4 dt = -2H + 2c \int |S|^4 dt \tag{23}$$

One may notice from Eq.(23), that $\frac{d}{dx} P > 0$, at least for the $H < 0$. This observation plays a significant role in determination of a sufficient criterion for the origin of collapse dynamics. Now our goal is to prove that some positive growing function of x (the majoring function) tends to infinity as x

tends to some fixed x_0. Let us evaluate the changing on x of the following quantity:

$$\frac{d}{dx}(-\frac{H}{P}) = 2P^{-1}\int |S_x|^2 dt + \frac{HP_x}{P^2} \tag{24}$$

The first term in Eq.(24) can be estimated by using Hölder inequality $\frac{d}{dx}P = 2\int |S||S|_x dt \leq 2(\int |S|^2 dt)^{1/2}(\int |S_x|^2 dt)^{1/2}$ in combination with the formula (23). Inserting the estimation for $\int |S_x|^2 dt$ into the Eq. (24), we get an inequality:

$$\frac{d}{dx}(-\frac{H}{P}) \geq \frac{1}{2}\frac{P_x}{P^2} + \frac{HP_x}{P^2} = \frac{1}{2}\frac{P_x}{P^2}(P_x + 2H). \tag{25}$$

Supposing function H to be negative at the moment $x = 0$, i.e. $H(x)|_{x=0} = H(0) < 0$ we may obtain the resulting estimation:

$$\frac{d}{dx}\left(-\frac{H}{P}\right) \geq 4(-\frac{H}{P})^2. \tag{26}$$

Integrating now the inequality (26) and defining $R = -\frac{H}{P}$, we find:

$$\int_{R_0}^{R}\frac{dR}{R^2} = \frac{1}{R_0} - \frac{1}{R} \geq 4x,$$

and rewriting the last formula, we obtain:

$$4R \geq \frac{1}{(x_1 - x)},$$

i.e. it means, that at the point $x_0 \leq x_1 = \frac{1}{4R_0}$ the function $R(x)$ tends to infinity. We conclude from this, that, if the initial pulse on the input is sufficiently large, the disturbances grow indefinitely [in the framework of Eq. (18)] on a finite distance x_0.

Thus, an initial pulse obeying the requirement $H = \int (b|S_t|^2 + a|S|^2 - c|S|^4)dt < 0$ will grow and compress up to the moment then increasing will be stopped. A pulse which will be formed obviously will have a shape differs from the sech-type shape of the fundamental soliton.

It can be proved , that a more precise estimation of a collapse criterion for Eq.(18) has the following form:

$$H = \int (b|S_t|^2 + a|S|^2 - c|S|^4)dt \leq H[S_0] = \frac{4}{3}a^{3/2}b^{1/2}c^{-1},$$

where right-hand side is a value of H on the soliton solution S_0. Thus, an unstable soliton solution in this model also plays a role of a boundary between the collapsing and noncollapsing regimes.

In conclusion , it was suggested to use the collapse mechanism for the compression of optical pulses. Forming of a self-similar peak leads to a compression of a pulse up to the moment when one has to take into consideration effects which are not taking into account by weak-nonlinearity limit. Formation of a singularity over a finite time means of course that we are going beyond the range of applicability of Eq.(18), so that this result cannot have a literal physical meaning. The important point here is very possibility that the amplitudes can grow in a finite time to this extent , with nonlinearity no longer remaining small. Of particular interest is determination of physical mechanisms responsible for a stopping of the collapse and therefore giving an estimation of the parameters of the amplified pulses.

It should be pointed out that the example of a medium possesing a collapse which has been considered in this report is only one of the possible variants, but the main idea suggested may be used in a variety of physical contexts and not in optics only.

References

On leave from Institute of Automation and Electrometry, Russian Academy of Sciences, Novosibirsk 630090, Russia.

1.H. A. Levine, SIAM Review, **32**, 283 (1990).
2.S. K. Turitsyn, G. E. Fal'kovich, Sov. Phys. JETP, **62**, 146 (1985).
3.E. A. Kuznetsov, J. J. Rasmussen, K. Rypdal and S. K. Turitsyn (to be published).
4.S. K. Turitsyn (to be published).
5.A. Hasegawa, F. Tappert, Appl. Phys. Lett. **23**, 142 (1973).
6.L. F. Mollenauer, R. H. Stolen and J. P. Gordon, Phys. Rev. Lett. **45**, 1095 (1980).
7.L. F. Mollenauer, R. H. Stolen, J. P. Gordon and W. J. Tomlinson, Optics Lett. **8**, 289 (1983).
8.L. F. Mollenauer and R. H. Stolen, Optics Lett. **9**, 13 (1984).
9.L. F. Mollenauer and K. Smith, Optics Lett. **13**, 675 (1988).
10.K. Smith, J. R. Armitage, R. Wyatt, N. J. Doran, S. M. J. Kelly, Electron. Lett. **26**, 1149 (1990).
11.L. F. Molenauer, S. G. Evangelides, H. A. Haus, J. Lightwave Technol. **9**, 194 (1991); N. A. Olsson, P. A. Andrekson, P. C. Becker, J. R. Simpson, T. Tanbun-Ek, R. A. Logan, H. Presby and K. Wecht, IEEE Photonics Technol. Lett. **2**, 358 (1990).
12.M. Nakazawa, Y. Kimura, K. Suzuki and H. Kubota , J. Appl. Phys. **66**, 2803 (1989).
13.V. E. Zakharov, Sov. Phys. JETP **32**, 1 (1990).
14.P. A. Belanger J. Opt. Soc. Am. **8B**, 2077 (1991).
15.A. C. Newell, in *Propagation in Systems Far from Equilibrium*, edited by J. E. Wesfreid, H. R. Brand, P. Manneville, G. Albinet, and N. Boccara (Springer, Berlin, 1988), p. 122, and references therein.
16.A. C. Newell, *Pattern Formation and Pattern Recognition* (Springer-Verlag, Berlin, 1979); K. Stewartson and J. P. Stuart, J. Fluid Mech. **48** 529 (1971).
17.L. M. Hocking, K. Stewartson and J. P. Stuart, J. Fluid Mech. **51**, 705 (1972).
18.L. E. Payne and D. H. Sattinger, Israel Journal of Mathematics **22**, 273 (1975).
19.M. I. Weinstein, Commun. Math. Phys. **87**, 567 (1983).

FLUID ANALOGIES IN OPTICS

C.O. Weiss*, K. Staliunas**, C.P. Smith***, G. Slekys**, M.F.H. Tarroja****

**Physikalisch-Technische Bundesanstalt*
Bundesallee 100
D-3300 Braunschweig,
Germany

ABSTRACT. Recent experimental as well as analytical/numerical investigations of lasers have shown surprising analogies with the fields of fluid dynamics, superfluids, and superconductors. The phenomena observed include quantized vortices as in superfluids, vortex lattices as in superconductors, vortex annihilation, and creation, and obstacle flows as in fluids. We show that our observations can be explained by the possibility to recast the class-A-laser equations in forms of a complex Ginzburg-Landau equation. The convective instability of these equations leads to formation of radiation patterns in the form of "modes". Further instabilities like Benjamin-Feir can lead to "defect-mediated turbulence". Moving vortices are found in class-B-lasers, relating optics to the dynamics of excitable media with 2- and 3- dimensional rotors.Optics is thus found at this point in time to integrate itself into the wider general field of non-linear physics, whose various subfields such as chemistry, fluid dynamics, plasmas are related to one another by common equations, common instabilities and common phenomenology.

I. INTRODUCTION

A relation of optics and fluid physics is not entirely new. 1969 Graham and Haken [1] showed that the Maxwell-Bloch equations describing a laser, can by straight adiabatic elimination of the material variables, be reduced to a complex Ginzburg-Landau equation and pointed out the correspondence of variables e.g. for lasers and superfluids. This early work was not followed by an investigation into the question whether typical phenomena e.g. vortices would have their counterpart in lasers - presumably due to the preoccupation of the time with the single-mode laser.

Equally, Haken [2] showed 1975 the isomorphy of the Maxwell-Bloch equations in the plane wave case with the Lorenz equations [3] which were initially derived to describe Bénard convection.

This latter equivalence between lasers and fluids remained an academic one until in 1984 laser systems were identified which permitted to realize the parameters required for the occurrence of the dynamics and the chaos of the Lorenz model [4] which has then been shown in detail in a number of investigations of the chaotic dynamics of these laser systems [5]. The investigation of the former equivalence necessitated abandoning the 30 year lasting restriction of research interest to single-mode lasers.

First suggestions of spatial instabilities of lasers were given in 1987 [6]. They resulted in the first experimental demonstration of helical waves (or vortices) in lasers [7]; based on the understanding that a process vital to structure formation in lasers is phase-locking among transverse modes.

R. E. Caflisch and G. C. Papanicolaou (eds.), Singularities in Fluids, Plasmas and Optics, 333–348.
© 1993 *Kluwer Academic Publishers.*

Numerically, the evidence of "screw-defects" (vortices) in lasers was given in 1989 [8]. Lattices of vortices in lasers reminiscent of Abrikosov lattices in type-II superconductors were found experimentally in 1989 [9]. A simple procedure showed the quantization of vortices in lasers, reminiscent of superfluids [10]. Defect-mediated turbulence was reported 1991 in a laser-like system [11]. Presently the efforts are directed to explore mathematically and experimentally the correspondences between optics and other fields of physics and to identify (hopefully) the unique contributions optics can make to the general field of non-linear physics.

We organize this article as follows: In part II the relation of optics and hydrodynamics is illustrated in the framework of ordinary differential equations, i.e. showing Lorenz model dynamics in laser experiments. Part III gives the simplest example of "transverse" structure formation (corresponding to 2- and 1- dimensional partial differential equations). Among the prominent features here are phase singularities in the optical field . Part IV shows how these phase singularities can arrange into regular arrays, their quantization , the correspondence to vortices and some dynamical behaviour of vortex motion. Part V gives an explanation of the similarities between lasers and superconductors/superfluids/fluids by deriving a set of Ginzburg-Landau-type equations for class-A-lasers and a fluid analogy. The fluid analogy is discussed in detail in particular to explain motions of the phase-singularities in lasers and to show a type of obstacle flow with vortex generation as is suggested by the laser fluid analogy.

II.

A first relation between fluids and lasers was found 1975 by H. Haken [2] who recognized the isomorphy of the laser equations with the Lorenz equations [3] initially derived to describe Bénard convection flow. The dynamics of the Lorenz equations has been studied in much detail and the model became the prototype for systems showing chaotic dynamics. Experiments on lasers to show this dynamics, however, were considered unrealistic because the conditions for dynamics in the Lorenz model to occur correspond to high losses of a laser resonator and as a consequence an extremely high pump parameter which is not in reach - for the better known types of lasers. After our first observation of chaotic dynamics in lasers (with a multimode laser system) [12], the necessity to have a more tractable system led us to reconsider the Lorenz model. We found that in certain types of optically excited gas lasers the parameters necessary for the dynamical laser emission are in reach. The principal reason for this lies in the narrow homogeneous gain line widths of these laser, which permits to keep the resonator losses relatively small, while still maintaining the conditions to exceed the instability threshold. As a consequence no excessive pump parameter is required for dynamics [13]. In a series of experiments [5] we have shown that dynamics does occur, and under which conditions it corresponds exactly to the dynamics of the Lorenz model [14] and under which (other) conditions it shows subtle differences from the solutions of the Lorenz model [15].

Fig. 1a shows the attractor of the Lorenz equations in the regime of chaotic dynamics, The x-, y-, z- variables of the model correspond to the population inversion, the material polarization and the optical field resp. Fig. 1b then shows the time dependence of the optical field (x) and the laser intensity (x^2) corresponding to the motion on the Lorenz attractor Fig. 1a. For comparison Fig. 2 shows the time-varying intensity measured on a NH_3-gas laser which emits in the far infrared and which is optically pumped by another gas laser (an N_2O-laser). The good agreement between model and laser dynamics is evident. As the intensity is the square of the field, it does not permit to judge whether the motion is really due to the two unstable saddle foci or only due to one. This can only be judged by the optical field which shows the symmetry about the zero axis (Fig. 1b). Although it is not common practice to measure optical fields instead of intensities, the former is technically feasible by superimposing another (steady) reference laser field (from a second laser-pumped NH_3-

laser system) with the dynamical laser field to be measured and then to infer the (in general complex) optical field from two quadrature interference signals [16].

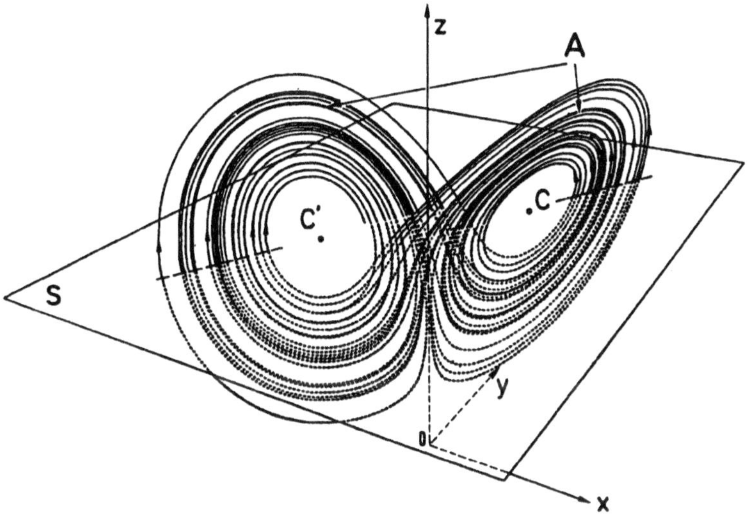

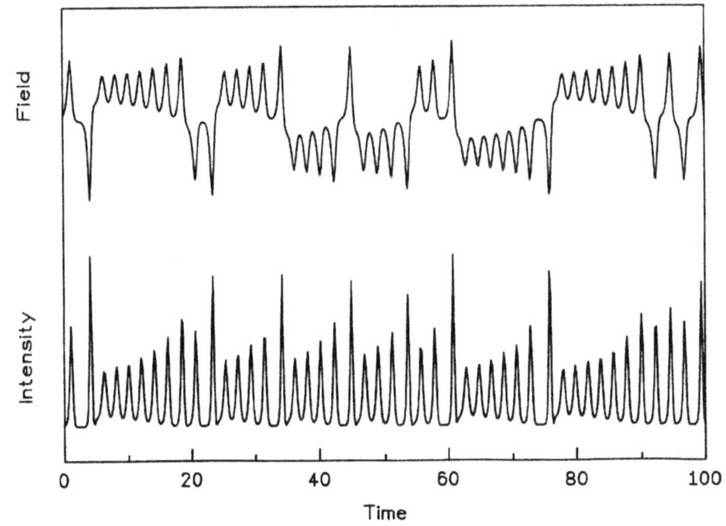

Figure 1a). The Lorenz attractor (A), and C' are the unstable steady states corresponding to steady laser emission. x,y,z corresponds to laser field, -material polarization and -population inversion resp.

Figure 1b. The laser field and the laser intensity as a function of time, corresponding to the motion on the attractor Fig. 1a), calculated from the Lorenz equations.

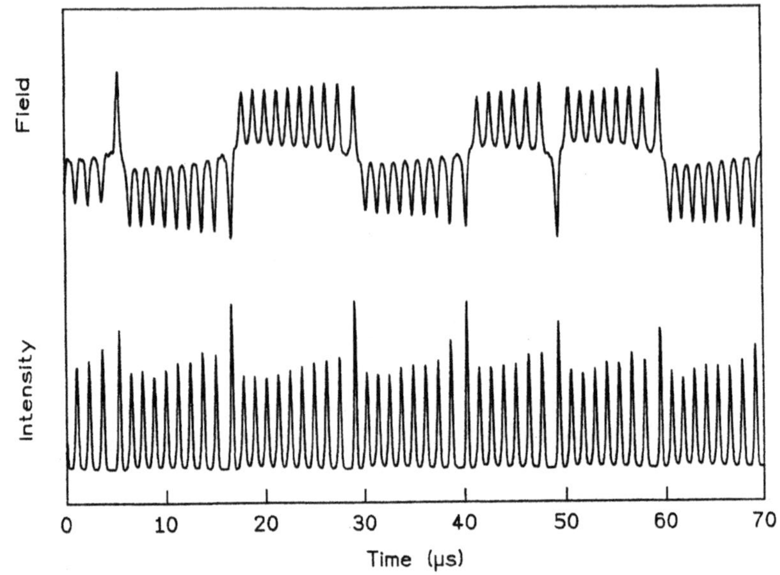

Figure 2. The measured laser intensity of the NH_3-laser corresponding to Figure 2 (lower trace).
Figure 3. The optical field as a function of time measured on a NH_3-laser (upper trace).

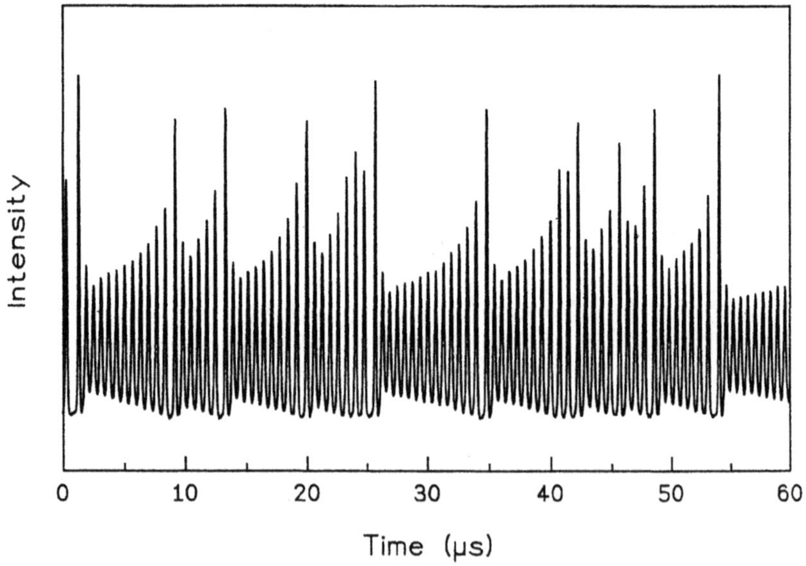

Figure 4. NH_3 laser pulsing showing "intermediate" height pulses between spirals.

Fig. 3 shows a time dependent optical field of the NH_3-laser thus measured, the comparison of which with the field of the Lorenz model shows that equally the laser field is symmetric with respect to zero and thus represents the motion corresponding to the two saddle foci as found in the Lorenz model. This good agreement between laser- and Lorenz model- dynamics has been tested quantitatively in a series of experiments and found to occur not for all laser parameter ranges. There can be pronounced differences between the laser dynamics and the Lorenz model which are attributable to experimentally controllable differences between the conditions of the laser operation and the assumptions in the model; like 2-mode emission [15], or due to subtleties in the physics of interaction of radiation with the active medium, such as coherences between the atomic wave function and the optical pump field. In this latter case, the time evolution (of intensity) is "almost " the one of the Lorenz model (Fig. 4), however, we find characteristic the appearance of an intensity pulse of intermediate height at the end of the "spirals" which does not seem to influence the general dynamics strongly but corresponds to qualitative changes of the dynamics which can e.g. be seen in the intensity maxima return map (the "Lorenz map" [4]) Fig. 5. The pure Lorenz case (5a) is characterized by the "tent"-shaped map [15], which appears doubled in the presence of quantum coherence effects in the active NH_3-medium, Fig. 5b [15]. Generally, it can thus be said that the NH_3-laser experiments have proven that a relation between optics and fluids on the level of ordinary differential equations exists.

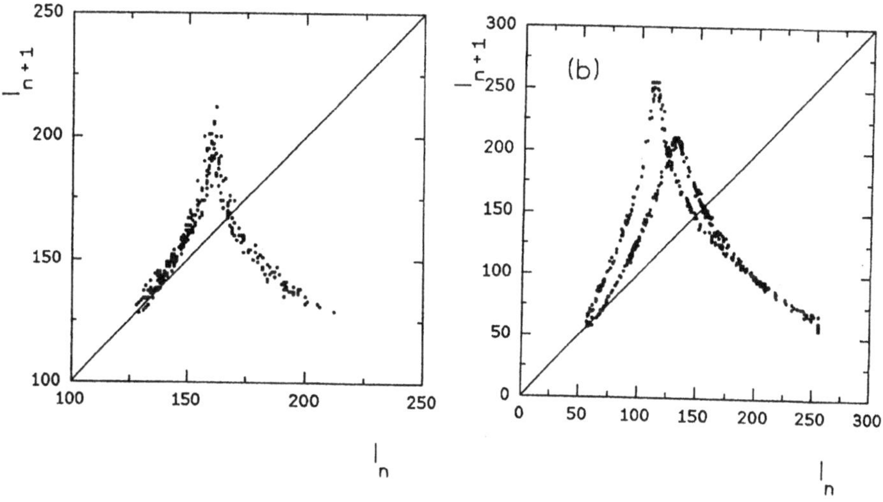

Figure 5. a) The intensity maxima return map for Lorenz-like NH_3-laser emission; b) for NH_3-laser emission with the "intermediate" pulses of Figure 4.

III.

Throughout the 80s in which the experiments on chaotic laser dynamics were carried out, there was, however, an uneasy feeling about plane wave- (ODE-) models to describe the dynamics of lasers, since even in the most conceptually simple case - the single fundamental mode laser - the field and the other variables have a dependence on the radius in a cylindrically symmetric laser and then obviously the assumption of plane waves and homogeneous parameter distributions (as in the Lorenz model) are not valid.

In fact the reasons for the good agreement which has been found between the dynamics of lasers and the one of plane wave models to this day are not completely understood.

Lugiato and collaborators finally approached the problem of the transverse structure of laser fields, with a particular case - the Turing instability - in mind; by using expansions in finite families of frequency - degenerate modes [6]. They showed - in laser parlance - that excited modes of different transverse field structures can synchronize in phase thus forming a steady state pattern. With this mechanism, in the plane transverse to the laser light propagation direction a structure formation related to the Turing instability occurs when a laser parameter (e.g. the pump) is changed. A structure emerges with a breaking of the rotational symmetry out of a homogeneous background (Fig. 6), [17] or in a rectangular geometry the transmission for laser light of a passive non-linear resonator changes from uniform to periodically spatially modulated ("striped") [6].

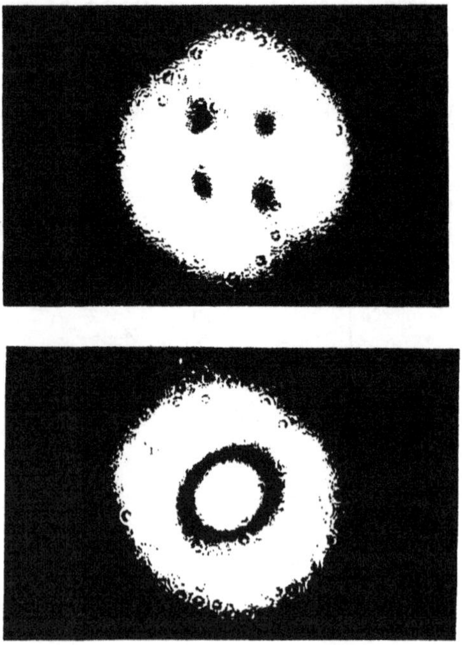

Figure 6. Structure formation with spatial symmetry breaking. The dark ring breaks up into 4 vortices when the pump of an Na_2-laser is increased [17].

a)

b)

c)

Figure 7. Two constituent modes TEM (0,1) (a) and TEM (1,0) (b) can phase-lock to form a mode with a central phase singularity (c).

As the fundamental mechanism at the root of these Turing instabilities is the phase synchronisation of different transverse excited laser modes ("transverse mode locking") it was decided to test this phenomenon experimentally. The lowest order family of transverse laser modes was used, which consists of more than one member (q=1). This family is comprised of the TEM(0,1) and the TEM(1,0) Gauss-Hermite modes (Fig. 7). In an astigmatism-free resonator their frequencies are equal, in practical resonators not usually. It was possible to decrease the astigmatism of a laser resonator sufficently to permit the nonlinearity of the laser medium to lock the phases of the modes. Fig. 8 shows the dynamics on the way to locking. When the natural frequencies of the constituent modes are made to approach each other,a complete period-doubling sequence, which terminates in chaos, is observed. The existence of the chaotic attractor terminates finally [18] - at the phase-locking point - in a crisis [19].

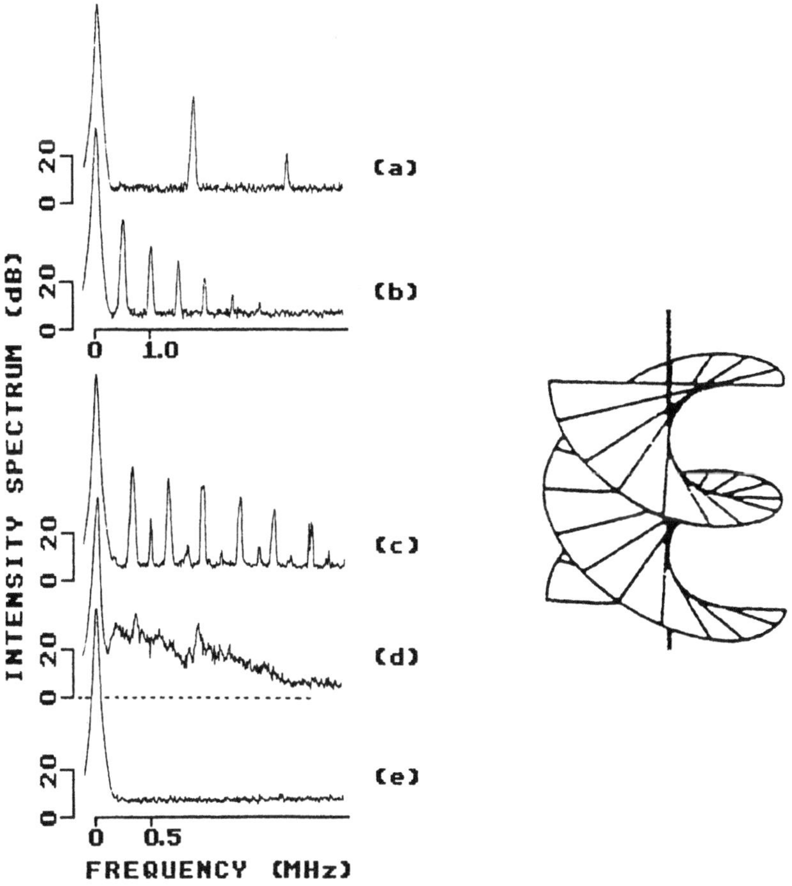

Figure 8. Spectra of the intensity fluctuations of the laser near to the phase-locking point, when the laser astigmatism is progressively reduced. Period-doubling: trace (c) and chaos: trace (d) occur before the locked state: trace (e). The locked state has a helical wave surface (inset).

The locked wave has helical wave fronts (inset Fig. 8) and a phase singularity at the center (where the light intensity due to all the destructive interference is zero) [18]. Plausibly enough, the righthand- and lefthand helical waves can coexist in a laser; with hysteresis between the two states [7]. This helical wavefront with intensity zero at the phase singularity appeared related to the "optical vortices" numerically discovered in reduced laser equations of the general form of a complex Ginzburg-Landau equation (CGLE) [8] where they appear as "defects" in the homogeneous laser field in the sense that they develop out of zeros in the field existing before the laser is switched on, (subsequently they can attract one another and annihilate) (Fig. 9).

Figure 9. Defects (vortices) calculated numerically for a large area laser.

IV.

In order to prove the existence of vortices in lasers, a photograph of the beam cross section of an optically pumped Na_2-laser which we took at the occasion of trying to prove the existence of spatio-temporal chaos in lasers [20] was examined more closely (Fig. 10). Since the picture was taken in the far field there is no possibility that the dark spots represent unilluminated areas (e.g. in the Fresnel diffraction range). An attempt to properly arrange the equiphase lines around the presumed phase singularities showed that this is possible. In a simple linear picture the phase singularities can be found as interference nodes generated e.g. by superposition of Hermite laser modes.
Fig. 11 shows such a superposition using TEM (2,3) Hermite modes.
Fig. 12 shows the equiphase lines.

Fig. 13 shows the equiphase lines of a superposition of TEM (4,5) Hermite modes.The adjacent phase singularities (vortices) have opposite topological charge so that the rotation of the vortices resembles toothwheels. An analogy of these vortex arrays with ion crystals suggests itself.
In Fig. 14 a superpositon of Laguerre modes is shown which generates a vortex of topological charge 3. By generalization it is clear that vortices in lasers will be quantized like in superfluids.
Fig. 15 shows intensity patterns observed for a Na_2-laser tuned to the transverse frequency-degenerate mode family q=2. When the pump strength of the laser is increased, 4 different steady states are encountered [21]:

a) a gaussian intensity distribution with a dark circular line,

b) this line breaks up and 4 vortices appear (in a Turing instability) with angular symmetry breaking,

c) two of the vortices of equal charge move off to infinity,

d) the two remaining vortices combine to form a vortex of topological charge 2 (which is stable!).

These 4 steady states observed correspond precisely to the ones calculated by minimizing the laser free energy [21].

Tuning the laser to the mode family q=3 results in the same pictures with the addition of another central phase singularity. Thus the doubly charged vortex in this case becomes a triply charged vortex (which is again stable!) [21].

A simple case of vortex dynamics in a laser was also investigated using the same Na_2-laser. In this case one vortex exists within the laser field at a certain distance from the laser axis (the vortices in lasers are always vortex lines which are parallel to the laser axis) and spins about the laser axis at a frequency of $\sim$100 MHz. This moving vortex was observed in the Na_2-laser [22]. Its motion can be described by a simple expansion in non-degenerate modes or by a hydrodynamic analogy (see section V). More complex motion of vortices in lasers including creation and annihilation of vortex pairs and turbulent motion have been calculated [23]. In the next section a hydrodynamic analog is shown in which a numerical calculation gives a periodic creation of vortex pairs behind an optical "obstacle" in an optical "flow".

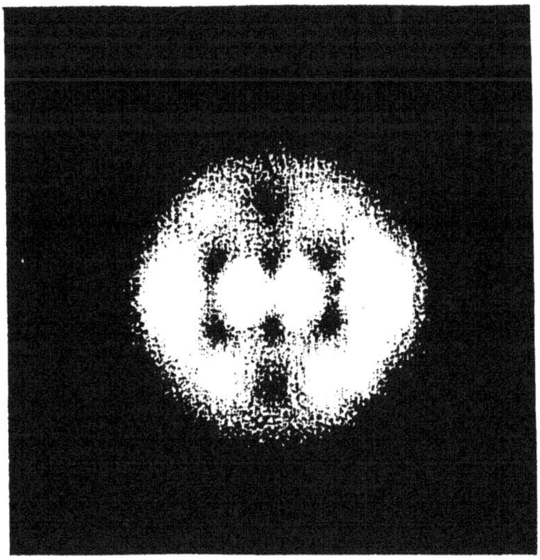

Figure 10. Photograph of a laser beam emitted from a Na_2-laser. 8 dark spots corresponding to phase-singularities are visible.

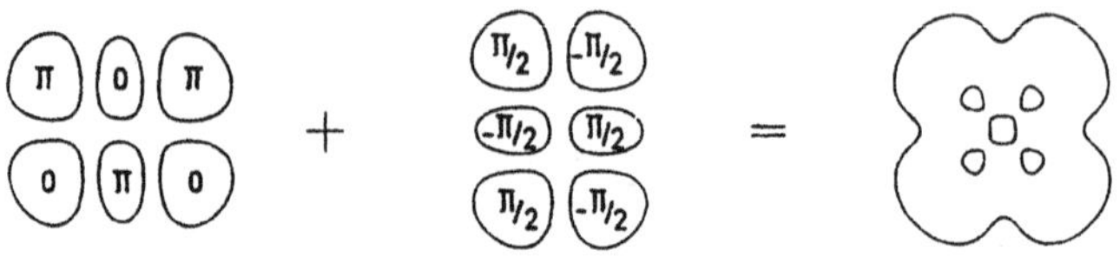

Figure 11. The superposition of rotated and phase-shifted TEM(2,1) modes produces 5 phase singularities.

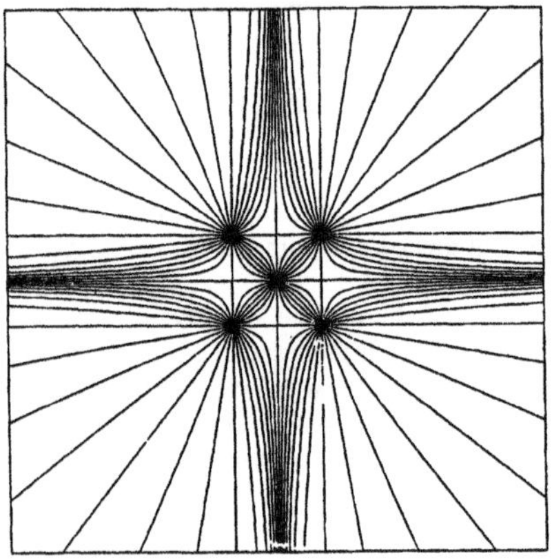

Figure 12. The equiphase lines corresponding to Figure 11.

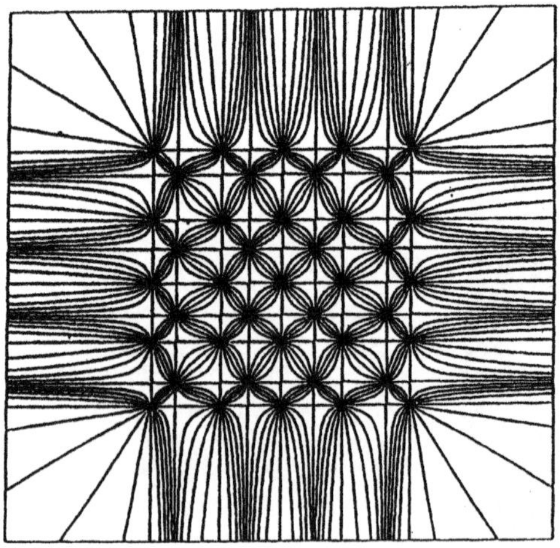

Figure 13. The equpiphase lines of a TEM(5,4) mode superposition.

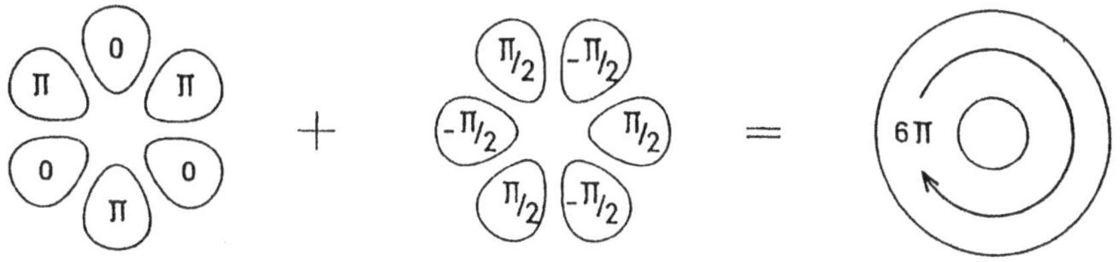

Figure 14. The superposition of two Laguerre modes produces a triply charged phase singularity.

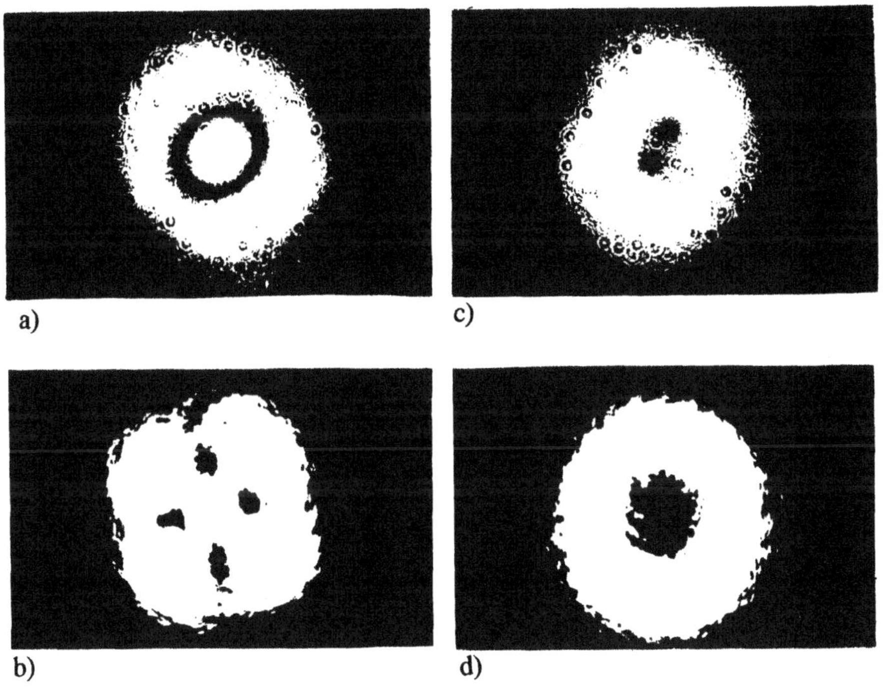

Figure 15. Steady state patterns of the mode family q = 2 in a Na_2-laser for different pump strengths.

V.

The existence of vortices in lasers suggests a fluid analogy. The "crystal-like" arrangements of vorties in lasers is reminiscent of type-II superconductor flux lattices (Abrikosov-lattices), and the quantization of the laser vortices suggest a relation of lasers and superfluids.

As noted above, it was shown early already that the laser equations can be reduced to a complex Ginzburg-Landau equation when the polarization and the inversion of the active medium are adiabatically eliminated [1], which means that the optical field has the smallest relaxation rate of all laser variables (class-A-laser). Unfortunately this Ginzburg-Landau equation has no finite solution and is thus not suited to investigate laser properties.

To obtain an equation of the Ginzburg-Landau type (basically to retain to quantization of the vortices) which is useful as a laser model, the polarization was eliminated in higher perturbation order [24].

One obtains the following equations:

$$\frac{\partial A}{\partial \tau} = (D_0 - D_{th}) \cdot (1+i\beta) \cdot A + id \cdot \vec{\nabla}^2 A - (1+i\beta)|A|^2 A; \tag{1}$$

CGLE with straight elimination. This equation was first derived in [1]. Its solutions are unstable for all detunings.

$$\frac{\partial A}{\partial \tau} = (D_0 - D_{th}) \cdot (1+i\beta) \cdot A + (id+d_2) \cdot \vec{\nabla}^2 A - (1+i\beta)|A|^2 A; \tag{2}$$

This equation is obtained by using adiabatic elimination of the polarization in first order [24]. It was first derived in [8]. It has stable solutions for positive detuning.

$$\frac{\partial A}{\partial \tau} = (D_0 - D_{th}) \cdot (1+i\beta) \cdot A + (id+d_2) \cdot \vec{\nabla}^2 A + d_4 \cdot \vec{\nabla}^4 A - (1+i\beta)|A|^2 A; \tag{3}$$

This equation ("CGLE IV") contains a 4th order diffusion term which is obtained by eliminating the polarization in second order [24]. It has stable solutions for all detunings.

$$\frac{\partial A}{\partial \tau} = (D_0 - D_{th}) \cdot (1+i\beta) \cdot A + id \cdot \vec{\nabla}^2 A - (1+i\beta)|A|^2 A - D_0 \cdot \left[\frac{(1+i\beta) \cdot \sigma \cdot \vec{\nabla}^2}{1+\sigma \cdot \vec{\nabla}^2} \right] \cdot A; \tag{4}$$

This equation ("Laser CGLE") is derived with polarization eliminated to ∞^{th} order [24]. It is the most precise description of a class-A-laser. Its spectrum of growth exponents for waves with finite transverse wave number corresponds with high precision to the full 3 Laser-Maxwell-Bloch-equations.

It is found that structure formation occurs in (3) and (4) at low pumps due a convective instability as in heated fluids. The stability analysis of the zero solution of (3) and (4) yields the "tilted waves" found for the full set of the 3 laser equations [25] which arrange to form the "mode-patterns" excited in lasers. The growth coefficients for spatial waves which result from the stability analysis of (4) agree completely with those obtained from the full 3 laser equations [25]. Eq. (3), in particular, permits a direct analogy with fluids. Separating the phase and the intensity yields a continuity equation (5) and an Euler equation (6) for a compressible viscous fluid [24].
This analogy permits to interpret the laser variables as fluid variables.
In particular corresponds:
Phase gradient ($\nabla\Phi$): flow velocity
Intensity : fluid density
Order of highest excited transverse mode: Reynolds number
This analogy permits to easily interpret motion and interaction of vortices (and other structures) in lasers. They drift with the flow ($\nabla\Phi$), vortices exhibit buoyancy (because they are "empty") in intensity gradients. Vortices experience a Magnus force (because they are rotating).

In this way e.g. the spinning vortex observation (section IV) is easily interpretable as a Magnus drift in the intensity gradient of the (gaussian shaped) laser beam. Vortices and turbulence appear at the same Reynolds numbers as in fluids [24].

As a numerical example Fig. 16 shows the "v. Karman vortex street" in a laser as calculated from the laser CGLE (4). A flow (phase gradient) is created by a small tilt of the laser resonator mirrors. An obstacle is created at the center of the laser by a region of high loss.

In Fig. 16 the obstacle is indicated by the circle, the arc indictes the half power contour of the laser beam. At the location of the crosses vortex pairs are generated which are annihilating later - directly behind the obstacle and further behind resp. The phase gradient (flow velocity) is from left to right [24].

This numerical example illustrates how a realistic laser model the laser-CGLE (4) shows phenomena well-known from fluids.

We remind that the fluid phenomena described result in the frame of a laser being describable by an equation of the Ginzburg-Landau type. This means class A-lasers. In class B-lasers where two of the variables have comparable relaxation rates and consequently only one variable can be adiabatically eliminated additional phenomena can be expected to occur.

We have found recently [26] that these class-B lasers make contact with the dynamics of "excitable media" [27]. Vortices e.g. in these lasers can never be stationary but move continuously.

The physical reason for this motion is (in laser parlance) "spatial hole burning" (or more correctly spatial "anti hole burning").

The vortex motion comes about since at the vortex core a peak of population inversion builds up. As there is no field at the vortex core, there can be no stimulated emission, consequently the inversion which does not decay naturally in a class-B-laser develops a local maximum. This inversion does not correspond to a state of minimum free energy. To remove the excess inversion the laser can only move the vortex aside. (The laser cannot destroy it since it is topologically stable.) At the new position the vortex builds up another peak of excess inversion such that the motion has to continue infinitely. The symmetry of the vortex requires that its motion starts with a symmetry breaking of the directions of motion.

Numerically it is found that the motion of the vortex for typical class-B-lasers starts with a toroidal transient (Fig. 17). Asymtotically a motion on a circle is reached. It can be imagined that the location of that circle will be neutrally stable, thus, under the influence of noise it is to be expected that the circle location will change. In real laser systems, as has been found by mode expansion, the motion is primilarily about the laser axis with flower-like trajectories. Thus we believe that this class-B-laser phenomenon is related to Winfree's "meandering rotors" [28] as encountered in chemical and biological excitable systems.

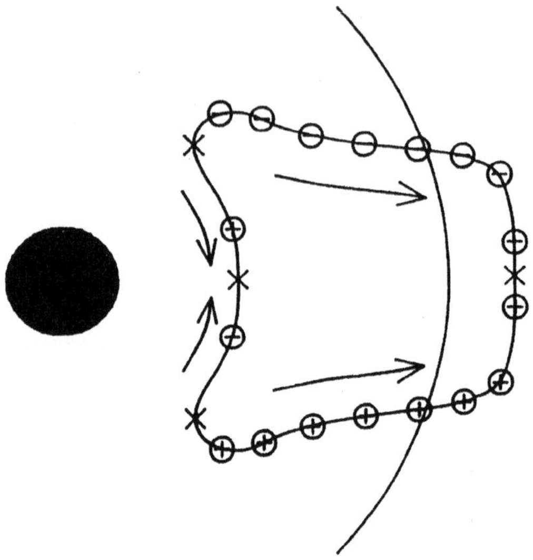

Figure 16. A vortex "flow" behind an "obstacle" in a laser. The dark spot is a region of high loss at the center of the laser. A "flow" is set up as a phase gradient by tilting the laser mirrors. At the crosses, vortices are generated and annihilated.

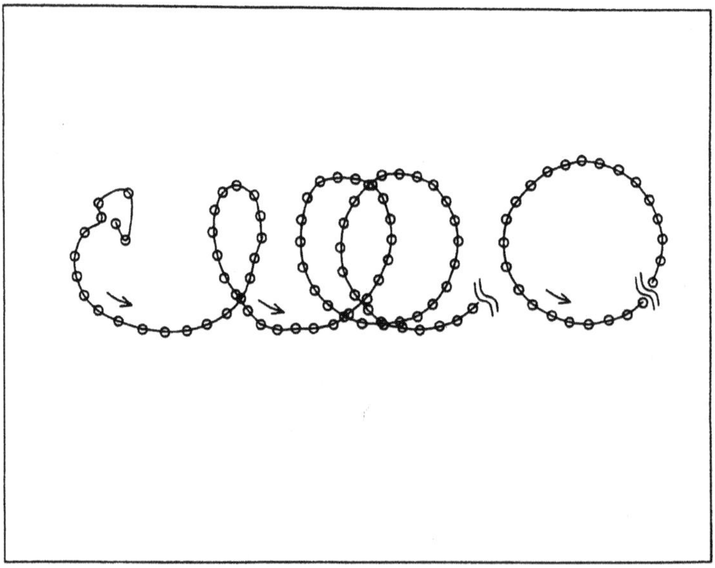

Figure 17. The motion of a vortex in a class-B-laser. After a toroidal transient, it ends on a circle.

Acknowledgements

This work was supported by Deutsche Forschungsgemeinschaft and ESPRIT Project 7118 (TONICS). K. Staliunas and M.F.H. Tarroja are recipients of Alexander v. Humboldt fellowships.

Permanent addresses

** Deptartment Quantum Electronics, Vilnius University, Lithuania.
*** Physics Department, University of Queensland, St. Lucia, Brisbane, Australia.
**** National Institute of Physics, University of the Philippines, Quezon City, Philippines.

References

[1] Graham, R.; Haken, H. (1970) Z. Phys. **237**, 31.
[2] Haken, H. (1975) Phys. Lett. **53A**, 77.
[3] Lorenz, E. (1963) J. Atm. Sci. **20**, 130.
[4] Sparrow, C. (1982) The Lorenz-Equations, Springer Verlag New York-Heidelberg-Berlin.
[5] Weiss, C.O.; Brock, J. (1986) Phys. Rev. Lett. **57**, 2804.
Weiss, C.O.; Abraham, N.B.; Hübner, U. (1988) Phys. Rev. Lett. **61**, 1587.
Tang, D.Y.; Li, M.Y.; Weiss, C.O. (1991) Phys. Rev. **A44**, 7597.
Li, M.Y., Weiss, C.O.; Heckenberg, N.R. (1992) Opt. Comm. **87**, 254.
Hübner, U.; Klische, W.; Weiss, C.O. (1992) Phys. Rev. **A45**, 2128.
[6] Lugiato, L.A.; Lefever, R (1987) Phys. Rev. Lett. **50**, 220.
[7] Tamm, Chr.; Weiss, C.O. (1990) JOSA **B7**, 1034.
[8] Coullet, P., Gil, L., Rocca, F. (1989) Opt. Comm. **73**, 403.
[9] Weiss, C.O. Tamm, Chr.; Coullet, P. (1990) Journ. Mod. Opt. **37**, 1825.
[10] Tamm, Chr.; Lugiato, L.A.; Brambilla, M.; Coates, A.B.; Weiss, C.O.; McDuff, R. (1991) in R. Vilaseca, R. Corbalan (eds.), 'Non-linear Dynamics and Quantum Phenomena in Optical Systems', Springer Proceedings in Physics **55**, Springer Verlag, pp. 225.
[11] Arecchi, F.T.; Giacomelli, G.; Ramazza, P.L.; Residori, S. (1991) Phys. Rev. Lett. **67**, 3749.
[12] Weiss, C.O.; Godone, A.; Olafsson, A. (1983) Phys. Rev. **A28**, 892.
[13] Weiss, C.O.; Klische, W. (1984) Opt. Comm. **51**, 47.
[14] Weiss, C.O.; Brock, J. (1986) Phys. Rev. Lett. **57**, 2804.
[15] Tang, D.Y.; Weiss, C.O.; Roldan, E.; de Valcarcel, G.J. (1992) Opt. Comm. **89**, 47.
[16] Tang, D.Y.; Li, M.Y.; Weiss, C.O. (1991) Phys. Rev. **A44**, 7597.
[17] Tamm, Chr., Weiss, C.O. (1990) Opt. Comm. **78**, 253.
[18] Tamm, Chr. (1988) Phys. Rev. **A38**, 5960.
[19] Ott, E. (1981) Rev. Mod. Phys. **53**, 655.
[20] Klische, W.; Weiss, C.O.; Wellegehausen, B. (1989) Phys. Rev. **A39**, 19.
[21] Brambilla, M.; Battipede, F.; Lugiato, L.A.; Penna, V.; Prati, F.; Tamm, Chr.; Weiss, C.O (1991) Phys. Rev. **A43**, 5090.
Brambilla, M.; Lugiato, L.A.; Penna, V.; Prati, F.; Tamm, Chr.; Weiss, C.O. (1991), Phys. Rev. **A43**, 5114.

[22] Brambilla, M.; Cattaneo, M.; Lugiato, L.A.; Pirovani, R.; Prati, F.; Kent, A.J.; Oppo, G.L.; Coates, A.B.; Weiss, C.O., Green, C.; D'Angelo, E.J.; Tredicce, J.R. (1993) 'Dynamical transvere laser patterns II: Experiments', to be published in Phys. Rev. A.
[23] Numerical calculations of M. Brambilla and L.A. Lugiato.
[24] Staliunas, K. (1992) 'The Laser-Ginzburg-Landau Equation and Laser Hydrodynamics', Phys. Rev. A submitted.
[25] Jacobsen, P.K.; Moloney, J.V.; Newell, A.C.; Judik, R. (1992) 'Space Time Dynamics of wide gain section lasers', preprint.
[26] Weiss, C.O.; Telle, H.R.; Staliunas, K.; Brambilla, M.(1992) 'The restless vortex', Phys. Rev. A, submitted.
[27] Winfree, A.T. (1991), Chaos **1**, 3.
[28] Jahnke, W.; Winfree, A.T. (1991) Internat. Journ. of Bifurcat. and Chaos.

9 780792 323334

GPSR Compliance

The European Union's (EU) General Product Safety Regulation (GPSR) is a set of rules that requires consumer products to be safe and our obligations to ensure this.

If you have any concerns about our products, you can contact us on ProductSafety@springernature.com

In case Publisher is established outside the EU, the EU authorized representative is:

Springer Nature Customer Service Center GmbH
Europaplatz 3
69115 Heidelberg, Germany

Batch number: 10816188

Printed by Printforce, the Netherlands